Innovative Economic Symposium (IES2018)

Milestones and Trends of World Economy

SHS Web of Conferences Volume 61 (2019)

Beijing, China
8-9 November 2018

Editor:

Jakub Horak

ISBN: 978-1-5108-8076-4

Printed from e-media with permission by:

Curran Associates, Inc.
57 Morehouse Lane
Red Hook, NY 12571

Some format issues inherent in the e-media version may also appear in this print version.

No changes have been made to the content of these proceedings. There may be changes to pagination and minor adjustments for aesthetics.

Printed by Curran Associates, Inc. (2019)

For additional information, please contact EDP Sciences – Web of Conferences
at the address below.

EDP Sciences – Web of Conferences
17, Avenue du Hoggar
Parc d'Activité de Courtabœuf
BP 112
F-91944 Les Ulis Cedex A
France

Phone: +33 (0) 1 69 18 75 75
Fax: +33 (0) 1 69 28 84 91

contact-edps@webofconferences.org

Additional copies of this publication are available from:

Curran Associates, Inc.
57 Morehouse Lane
Red Hook, NY 12571 USA
Phone: 845-758-0400
Fax: 845-758-2633
Email: curran@proceedings.com
Web: www.proceedings.com

TABLE OF CONTENTS

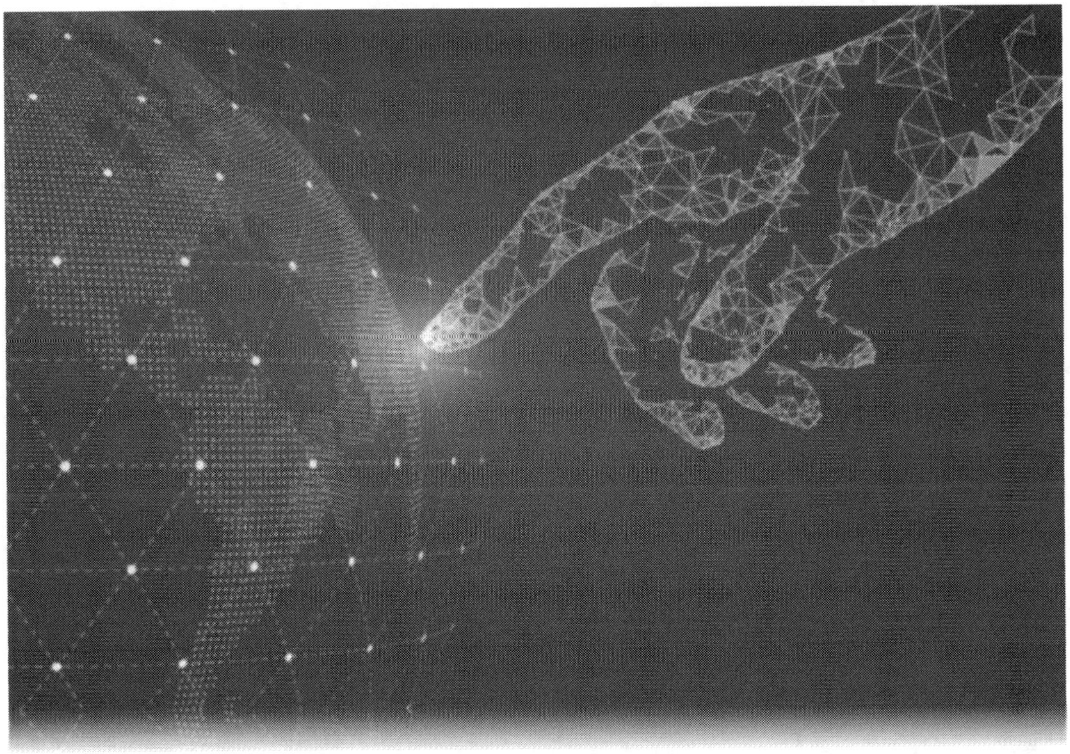

Innovative Economic Symposium 2018

Milestones and Trends of the World Economy

Innovative Economic Symposium 2018

Milestones and Trends of the World Economy

http://ies.vstecb.cz/

Editor: Jakub Horák

About the conference IES 2018

Date of the conference: 2018-11-08 (North China University of Technology, No. 5 Jinyuanzhuang Road, Shijingshan District, Beijing)

The conference will be held under the auspices of President of the Czech Republic Ing. Miloš Zeman, CSc.

Motto

"The only preparation for prospering in the global economy is investing in ourselves.,, William J. Clinton

Target

IES 2018 brings together scientists from general economics, econometrics, knowledge economy, economic policy, corporate finance, and marketing in order to exchange knowledge and discuss issues related to major events and world politics at the macro and micro levels. Globalization is an indisputable trend. This is related to increasing cross-border links such as international cooperation, international trade, international investment, multinational corporations and global institutions. This long-term trend seeks to address the topic of this international scientific conference.

Thematic Areas

Primary Issues

- The strengths and weaknesses of the 21st century global economy.
- Economic and social aspects of the Fourth Industrial Revolution.
- Current world economy and economic inequality.
- Can the next financial crisis be predicted?
- Cryptocurrencies as a modern tool to expand the investment portfolio.
- Investment decision making on a global basis.

Alternative Issues

- Decision models under uncertainty and risk conditions.
- Fundamental versus technical approaches to business investment management.
- Monetary policy, euro area and currency zone theory.
- Current and predicted policy and economy of renewable energy sources.
- Challenges and challenges of innovation management.
- Behavioral Finance.
- Financial planning of international investment.
- Modern approaches to international investment calculations.

About organizers

North China University of Technology (NCUT), located on the western side of Beijing, is a multi-disciplinary university that combines the natural sciences and engineering with liberal arts, economics, management and law. NCUT targets the development of qualified, high-level, application-oriented professionals with solid basic theoretical knowledge and strong practical skills and creative abilities. A quality education is the university's lifeline. Teaching is a primary focus at NCUT and the center of its activities, while the university also emphasizes the importance of research and the contribution research can make both to the educational environment and to address local and national needs. The under graduate program offers a full range of undergraduate degrees; there are also numerous graduate degree programs, as well as continuing education programs. The academic education is complemented with moral, intellectual, physical and aesthetic training in order to build well-rounded students prepared to meet the challenges of modern society. More information about NCUT at www.ncut.ciss.org.cn.

The primary focus of the Institute of Technology and Business in České Budějovice (ITB) activities is the implementation of full-time and part-time studies for professional bachelor's and master's degree programmes. The professional study programmes are supported by cooperation with regional entrepreneurs. ITB´s other priorities include applied research, the implementation of innovative technologies and knowledge transfer into areas that are current and crucial with regards to the regions needs. More information about ITB at www.vstecb.cz.

Conference report

Conference Innovative Economic Symposium 2018 took place on 8th November 2018 in the North China University of Technology (NCUT), Beijing, China. The thematic focus of this year was Milestones and Trends of the World Economy. The whole conference started with opening speeches of prof. Shen Zhilli, Ph.D., vice president of NCUT and Pavel Rousek, Ph.D., promoter of Innovative Economic Symposium 2017. The participants of the conference were academics, officials and entrepreneurs from China, Russia, Poland and the Czech Republic. The latest trends in the world economy were introduced. The topic of modern information technologies has been repeatedly mentioned as one of the promising opportunities for world economy trends.

Expert guarantors

prof. Ding Hui, PhD.
President of the North China University of Technology (CHN)

doc. Marek Vochozka
Rector of the Institute of Technology and Business (CZE)

The Organizing Committee

doc. Marek Vochozka, Rector of ITB, School of Expertness and Valuation, the Institute of Technology and Business in Ceske Budejovice (CZE)

prof. Zhao Jixin, Dean of the School of Economics and Management, North China University of Technology (CHN)

Jakub Horák, MSc., ITB in Ceske Budejovice (CZE)

Song Ping, NCUT in Beijing (CHN)

Sarah Wu, NCUT in Beijing (CHN)

Pavel Rousek, Ph.D., ITB in Ceske Budejovice (CZE)

Simona Hašková, Ph.D., ITB in Ceske Budejovice (CZE)

Veronika Machová, MSc., ITB in Ceske Budejovice (CZE)

Jaromír Vrbka, MSc., ITB in Ceske Budejovice (CZE)

The Scientific Committee

Innovative Economic symposium 2018

8th Nov., Beijing

Location: North China University of Technology, No.5 Jinyuanzhuang Road, Shijingshan District, Beijing

Thursday 8th November		
8:00-9:00	Participant registration	Cultural Hall, at the west side of Lixue Building of NCUT
9:00-9:20	Opening Ceremony	Welcome Speech: prof. Shen Zhili, Ph.D., vice president of NCUT prof. Pavel Rousek, promoter of IES 2017 Host: prof.Zhao Jixin, Dean of the School of Economics and Management, NCUT
9:20–10:00	Plenary Section (1)	Keynote Speaker: Ing. Tsolmon Jambal, Ph.D.,Institute of Technology and Business in České Budějovice Wang Xuecheng, Ph.D. , School of Economics and Management, NCUT Host: Huang Xiaobei, Ph.D. , School of Economics and Management, NCUT
10:00-10:20	Tea Break	
10:20-11:30	Plenary Section (2)	Keynote Speaker: prof.Mateusz Grzesiak, Faculty of Management, WSB University Deng Xin, Ph.D. , School of Economics and Management, NCUT Host: Luo Wenhao, Ph.D. , School of Economics and Management, NCUT
11:30-12:30	Lunch	Buffet, Staff Canteen
13:00-14:40	Special Topic (Section 1)	Keynote Speaker: Bogdan Wierzbinski, PhD, Economics, University of Rzeszow Marek Cierpiał - Wolan, PhD, director of statistics bureau in Rzeszow Grzegorz Ślusarz, PhD, Vice-Rector of University of Rzeszow Ing. Tomas Krulicky, Institute of Technology and Business in České Budějovice Wang Xinyu, Ph.D. , School of Economics and Management, NCUT prof.Katarzyna Bilinska-Reformat, PhD, University of Economics in Katowice Host: Pavel Rousek, Ph.D., Institute of Technology and Business inČeské Budějovice
14:40-15:00	Tea Break	
15:00-16:20	Special Topic (Section 2)	Keynote Speaker: Ing. Pavel Rousek, Ph.D., Institute of Technology and Business in České Budějovice Han Jie, Ph.D. , School of Economics and Management, NCUT Mgr.Barbora Wernerova, Institute of Technology and Business in České Budějovice Host: prof.Katarzyna Bilinska-Reformat,PhD,University of Economics in Katowice Izabela Sztangret, PhD, University of Economics in Katowice
16:20-16:40	Closing ceremony for the main forum	Closing Speech: Prof. Grzegorz Ślusarz, PhD, Vice-Rector of University of Rzeszow prof. Ding Hui, PhD., President of the NCUT Host: Xiong Jiaquan, executive vice dean of the International School of NCUT

Innovative Economic Symposium 2018
Milestones and Trends of the World Economy

Under the auspices of Ing. Miloš Zeman, President of the Czech Republic

http://ies.vstecb.cz/

8th November 2018
North China University of Technology
Beijing, China

Motto

"The only preparation for prospering in the global economy is investing in ourselves.,, (William J. Clinton)

Target

IES 2018 brings together scientists from general economics, econometrics, knowledge economy, economic policy, corporate finance, and marketing in order to exchange knowledge and discuss issues related to major events and world politics at the macro and micro levels. Globalization is an indisputable trend. This is related to increasing cross-border links such as international cooperation, international trade, international investment, multinational corporations and global institutions. This long-term trend seeks to address the topic of this international scientific conference.

Thematic Areas

Primary Issues

- The strengths and weaknesses of the 21st century global economy.
- Economic and social aspects of the Fourth Industrial Revolution.
- Current world economy and economic inequality.
- Can the next financial crisis be predicted?
- Cryptocurrencies as a modern tool to expand the investment portfolio.
- Investment decision making on a global basis.

Alternative Issues

- Decision models under uncertainty and risk conditions.
- Fundamental versus technical approaches to business investment management.
- Monetary policy, euro area and currency zone theory.
- Current and predicted policy and economy of renewable energy sources.
- Challenges and challenges of innovation management.
- Behavioral Finance.
- Financial planning of international investment.
- Modern approaches to international investment calculations.

Conference program

8:00-9:00	Participant registration
9:00-9:20	Opening Ceremony
9:20-10:00	Plenary Section (1)
10:00-10:20	Tea Break
10:20-11:30	Plenary Section (2)
11:30-12:30	Lunch
13:00-14:40	Special Topic (Section 1)
14:40-15:00	Tea Break
15:00-16:20	Special Topic (Section 2)
16:20-16:40	Closing ceremony for the main forum

Expert Guarantors

Prof. Ding Hui,
President of the North China University of Technology (CHN)
doc. Marek Vochozka,
Rector of the Institute of Technology and Business (CZE)

Scientific Committee

prof. Svetlana I. Ashmarina,
Samara State University of Economics (RUS)
prof. Stefan Forlicz,
Wrocław School of Banking (POL)
dr. Simona Hašková,
Institute of Technology and Business (CZE)
prof. Leonid D. Gitelman,
Ural Federal University (RUS)
prof. Fariborz Haghighat,
Concordia University (CAN)
prof. Alois Heißenhuber,
Technical University of Munich (GER)
prof. Branislav Mičieta,
University of Žilina (SVK)
doc. Maxim A. Ponomarev,
Plechanov University (RUS)
prof. Amarjit Singh,
University of Hawai'i at Manoa (USA)
prof. Bingtao Song,
Henan University and the Chinese Economic Society (CHN)
prof. Oleksandr M. Trofymchuk,
Ukrainian National Academy of Sciences (UKR)
prof. Elena N. Veduta,
Lomonosov University in Moscow (RUS)
prof. Liu Yongxiang,
North China University of Technology (CHN)
prof. Shen Zhili,
North China University of Technology (CHN)
prof. Mateusz Grzesiak,
WSB University (POL)
prof. Gregorz Slusarz
University of Rzeszow (POL)
prof. Katarzyna Bilinska
University of Economics in Katowice (POL)

Organizing Committee: **doc. Marek Vochozka,** Rector of the Institute of Technology and Business in České Budějovice; **prof. Zhao Jixin,** Dean of School of Economics and Management, NCUT; Jakub Horák, MSc.; Song Ping; Sarah Wu; Pavel Rousek, Ph.D.; Simona Hašková, Ph.D.; Veronika Machová,MSc.; Jaromír Vrbka, MSc.; Zuzana Rowland, MSc.; Tomáš Krulický, MSc.

Theoretical aspects of regional sustainable development in the EU and Ukraine

Iryna Buzko[1], *Olena* Vartanova[2], *Iryna* Trunina[3,*], and *Inna* Khovrak[4,*]

[1]Doctor of Economics, professor, Volodymyr Dahl East Ukrainian National University, Ukraine
[2]Doctor of Economics, professor, Cherkasy State Technological University, Ukraine
[3]Doctor of Economics, professor, Kremenchuk Mykhailo Ostrohradskyi National University, Ukraine
[4]PhD, associate professor, Kremenchuk Mykhailo Ostrohradskyi National University, Ukraine

Abstract. Effective use of the potential of both an individual country and its regions in order to ensure sustainable development is possible only on the basis of conscious, purposeful, scientifically grounded actions on the transformation of regional systems. That is why, the aim of the paper is to provide an analysis of the theoretical aspects of regional sustainable development in the EU and Ukraine, which will create a methodological basis for the further development of practical mechanisms for ensuring balanced regional development. As a main method of research, we use the critical analysis and comparison of normative documents, analytical reports, regional development strategies and scientific publications on the issues of sustainable development. As a result, the authors substantiate the essence of sustainable development and clarified the concept of regional sustainable development. This made it possible to determine its features. The paper also summarizes the institutional framework for ensuring regional sustainable development, including the justification of informal constraints, formal rules and coercive mechanisms. The authors identify vectors for ensuring the regional sustainable development in Ukraine. However, for the effective implementation of certain strategic priorities, a change in the public consciousness is needed in order to understand the importance of regional sustainable development, as well as the definition of clear criteria for the achievement of sustainable development and continuous monitoring of its level. We consider education as one of the most effective tools that will allow the formation of a system of knowledge, skills and competences necessary for the formation of citizens responsible for the future of the country.

Key words: sustainable development, regional sustainable development, European Union (EU), sustainable development strategy, social responsibility

*Corresponding author: inna.khovrak@gmail.com, truninairina0@gmail.com

1 Introduction

Regional development is formed under the influence of the trends of socio-economic development of the country, the availability of human capital, the level and quality of life of the population, the ability to rational use and reproduction of natural resources. At the same time, the imbalance of development both within individual regions and the disproportions in the level of development of different regions can disrupt national security. That is why, not only sustainable development of the country, but also sustainable development of its regions is important. Since the adopted global goals of sustainable development [1] only reflect existing global problems and priorities, they can be achieved through international and multilateral cooperation. At the same time, the development of national strategies is only an outline of the priorities of the country's development. The real achievement of these priorities takes place at the local level [2], based on the individual potential of each region. It is also worth remembering that the policy of equalization of development of the regions, adopted by the EU [3], is ambiguous in countries with different rates of economic development. The concept of national selfishness is gaining popularity [4]. This complicates the achievement of the global goals of sustainable development, but gives impetus to the development of regions, which in the future will lead to sustainable development of countries and their associations.

Systemic management of sustainable development has a great theoretical and practical value, since effective use of the potential of both an individual country and its regions in order to ensure sustainable development is possible only on the basis of conscious, purposeful, scientifically grounded actions on the transformation of regional systems. That is why, the aim of the paper is to provide an analysis of the theoretical aspects of regional sustainable development in the EU and Ukraine, which will create a methodological basis for the further development of practical mechanisms for ensuring balanced regional development. Realization of the set goal requires solving the following tasks:

- Substantiation of the essence of sustainable development, clarification of the concept of regional sustainable development (this will allow the determination of its features);
- Generalization of institutional principles for ensuring regional sustainable development;
- Definition of vectors for ensuring the regional sustainable development in Ukraine.

2 Data and Methods

As a main method of research, we use the critical analysis and comparison of normative documents, analytical reports, regional development strategies and scientific publications on the issues of sustainable development.

We believe that the assessment of the level of display of the term "regional sustainable development" in comparison with the term "sustainable development" in the usual search results of Google, Google Scholar and Web of Science Core Collection suggests that the level of development of the scientific basis is not sufficient, especially in Ukraine (Table 1).

Table 1. Display of terms «sustainable development» and «regional sustainable development» " in the usual search results of Google, Google Scholar and Web of Science Core Collection

	sustainable development		regional sustainable development	
	Ukr.	Eng.	Ukr.	Eng.
Google				
Number of results, million	15.4	423	5.1	235
Time, sec.	0.44	0.55	0.49	0.53
Google Scholar				
Number of results, million	61.7	3230	28	2520
Time, sec.	0.09	0.07	0.11	0.09
Web of Science Core Collection				
Number of results, thousand	-	100.8	-	8.6

Source: Own research (Ukr.- results in Ukrainian, Eng. - in English, as of 04.10.18).

Taking into account the data in Table 1, we propose focusing on the definition of regional sustainable development, its main aspects and peculiarities of provision in the EU and Ukraine.

For the purpose of achieving the set goals, the following general scientific and specialized methods were used:

- Theoretical generalization, comparison and systematization - to study the essence and features of sustainable development of the region;
- system analysis – for determining the institutional framework for ensuring sustainable development of the region;
- abstract-logical – for theoretical summarization and conclusion;
- graphical method – for visualization of features and vectors of regional sustainable development, etc.

3 Results

3.1 The essence and characteristics of regional sustainable development

Sustainable development is a long-term process of transformation taking place in the economic, social, environmental spheres. It is aggravated by changes in the minds of citizens and increased their social responsibility. At the same time, sustainable development can lead to positive quantitative changes (growth of employment, investments, revenues, etc.) and qualitative transformations (changes in social organization, technical and technological progress).

It is important that an understanding of the need of sustainable development was shaped by the emergence of a number of imbalances and global problems. Thus, the modern world is overpopulated. According to the UN the world will be inhabited by 9.8 billion people by 2050 [5]. There is a significant interference in the natural environment and uncontrolled consumption of natural resources. According to the Food and Agricultural Organization of the United Nations, about 7.3 million hectares of forest are destroyed annually [6]. The problem of poverty and regional imbalances in development is intensifying. Thus, according to the United Nations Food and Agriculture Organization, it is about 815 million people from 7.6 billion people in the world (representing 10.7% of the total). At the same time, the vast majority of people who are suffering from hunger live in low and middle income countries [7]. These and other global problems encourage humanity to reflect on what decisions should be made and which vector of development should be chosen. The term "sustainable development" is the subject of scientific discussion in many countries (Table 2).

Table 2. Determination of the essence of sustainable development

Country	Author	Definition
China	Xuedong Liang, Dongyang Si, Xinli Zhang [8, p.2]	«Regional sustainable development emphasizes the coordinated development of the economy, society and the ecology to promote the regional economy, improve residents' lives, ensure an efficient use of resources and maintain the environment».
France	Taya Louise Owens [9, p.414]	«Sustainable development is an organising principle for global development that supports the well-being of both people and the planet».
Hungary	Mariann Szabó, Mária Szalmáné Csete, Tamás Pálvölgyi [10, p.399]	There are two levels of sustainability. «Weak sustainability allows the substitution of natural resources with another type of resource. … Strong sustainability is when we do not allow the value of natural resources to be reduced; each stock must be non-declining separately».
Korea	Yongrok Choi [11, p.1]	«Sustainable development is positively related to economic growth in China and Korea, but negatively related with energy efficiency improvements in Russia and the Democratic People's Republic of Korea, while it is relatively complicated in Japan and Mongolia due to a combination of these factors».
Poland	Zofia Wysokińska [12, p.76]	Sustainable development «is based on stable and sustainable economic growth in environmentally friendly sectors and reduces unemployment by using existing social resources in the labor market».
Romania	Rodica-Manuela Gogonea, Adrian Aurel Baltalunga, Adrian Nedelcu, Daniela Dumitrescu [13, p.3]	«Sustainable development implies that the interaction between the economy, society and the environment (ecology) should be acceptable, viable, sustainable and fair».
Slovenija	Katja Vintar Mally [14, p.33]	«The regional level is often the most important for achieving sustainable development … (from the economic, social, and environmental viewpoints)».
South Africa	Marlon E. Cerf [15, p.1]	Sustainable development is based on «public accountability and participation in priority setting for the sustainable development goals».
UK	Dawid Pearce, Giles Atkinson [16, p.103]	«The measurement of sustainable development is not without considerable difficulties, yet this should not detract from the positive advances that can be made in this direction».
USA	Robert Costanza Herman E. Daly [17, p.1]	«A minimum necessary condition for sustainability is the maintenance of the total natural capital stock at or above the current level».

Source: Own research.

In our opinion, sustainable development can be defined as a form of interaction, which maintains a balance of interests of all stakeholders and ensures the survival of future generations. So, ensuring sustainable development is not only a legal problem (the development of a regulatory framework), not only a technical problem (the invention of technical possibilities for optimal use of available resources), not only a scientific problem (the development of theories and concepts). Ensuring sustainable development primarily requires the change of consciousness on an ethical basis, which will promote cooperation and mutual understanding. The main principles of sustainable development are maintaining the integrity of ecosystems, preserving natural resources with the simultaneous development of society, achieving equality, social justice, cultural diversity and self-determination, optimal satisfaction of basic human needs.

Accordingly, regional sustainable development is a dynamic process that can save and accumulate the potential of the region, increase its long-term competitiveness, and balance development indicators in the region. This is achieved by defining and following the development priorities in the environmental, social and economic spheres (Fig. 1), efficient management of the resource base of the region, ensuring social justice and social responsibility, coherence of interests of all stakeholders, etc.

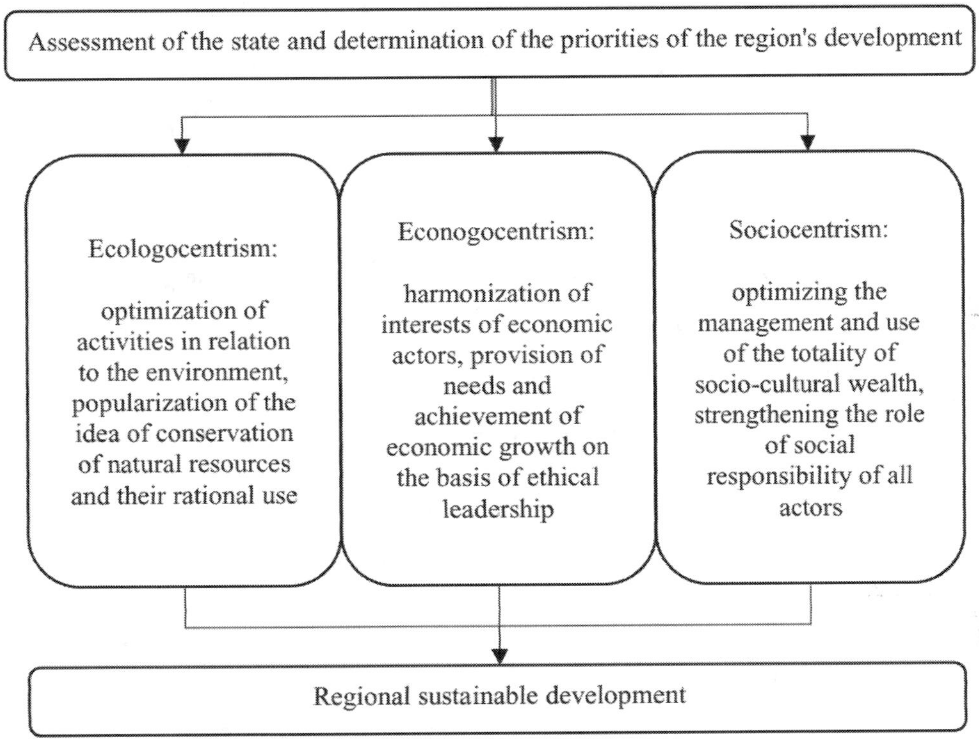

Fig. 1. Features of regional sustainable development

Source: Own research.

The sustainable development of the region takes place under the influence of numerous interrelated social, economic, spatial and other factors of internal and external character. That is why for realization of the model of regional sustainable development in Ukraine it is necessary to prepare and substantiate strategic decisions taking into account European experience.

3.2 Institutional framework for Regional Sustainable Development in the EU and Ukraine

The rules of interaction in society, according to Nobel laureate D. North [18, p. 7] define the institutions that constitute a system of restrictive conditions, namely:

- informal restrictions (combination of traditions, customs, cultural and historical features);

- formal rules (the country's legal framework, international norms and law);
- coercive mechanisms (judicial and law enforcement systems).

Thus, informal constraints are formed during long periods, fixed in the public consciousness, used in everyday life of citizens and business environment. Therefore, agents for change are needed to correct informal constraints. For example, multinational corporations introduce common standards for business in all countries of the world. Or universities can raise the future intellectual elite of the nation, which cares about sustainable development. So, in 2005, the United Nations announced the Decade for Education for Sustainable Development (DESD) (UNESCO, 2005) to integrate the principles, values and practices of sustainable development into all aspects of education and training.

Coercive mechanisms are also effective tools for shaping the consciousness of citizens and starting a new habit – to act within the norms of sustainable development. Thus, the system of restrictions imposed fines for excessive emissions into the atmosphere, stimulates the transition to environmental fuels and transport, has established product quality certificates, etc. For EU countries, it has become the norm to adhere to certain rules of the game and limitations. However, imperfection and corruption in the judicial and law-enforcement systems of Ukraine do not promote the spread of European norms in the country.

Taking into account the above-mentioned conditions, we propose to focus on the formal rules (Table 3). The emergence and development of the concept of sustainable development began in the 70s of the twentieth century.

Table 3. International norms in the field of sustainable development

Date	Event	Goal
1972	United Nations Conference on the Environment (Stockholm, Sweden)	Recognition of the relevance of environmental issues and the need to create effective international mechanisms for its solution. Bringing the relationship between economic development and the deterioration of the environment.
1980	World Strategy for the Conservation of Nature (International Union for the Conservation of Nature)	The argumentation of the thesis that the development of society should occur under the condition of preserving nature.
1982	The World Charter of Nature Defense	Proclamation of a number of principles for the conservation of nature, according to which any human activity that affects nature must be managed and evaluated.
1983	World Commission on Environment and Development WCED	Creation of the Commission, the tasks of which were to develop the basic principles, indicators of sustainable development, as well as the global ecological and economic program of action.
1987	Our Common Future (The Brundtland Report) (International Commission on Environment and Development)	A new concept of sustainable development as an alternative to development, based on unlimited economic growth, is presented.
1992	United Nations Conference on Environment and Development (Rio de Janeiro, Brazil)	Adoption of the Declaration on Environment and Development, the World Program of Action «Agenda 21» for the next 100 years, the UN Framework Convention on Climate Change, the Convention on Biological Diversity, the Declaration on the Direction for the Development, Protection and Use of Forests.
1997	Summit (special session of the UN General	Government talks were held on the documents that were adopted during the Earth Summit ("Rio + 5").

	Assembly), New York, USA	
1997	Amsterdam Treaty of the EU	Sustainable development has been identified as a key principle of all EU policies. At the same time achieving goals in one of the spheres (economic, social or environmental) should not restrain progress in others.
2000	The Millennium Development Goals (The Millennium Summit of the United Nations)	Defining eight major development goals: 1) to eradicate extreme poverty and hunger; 2) to achieve universal primary education; 3) to promote gender equality and empower women; 4) to reduce child mortality; 5) to improve maternal health; 6) to combat HIV/AIDS, malaria, and other diseases; 7) to ensure environmental sustainability; 8) to develop a global partnership for development.
2002	World Summit on Sustainable Development (Johannesburg, South Africa)	Education should stimulate change in behavior for a sustainable future. Attention was focused on the need to remove barriers to sustainable development in modern society, reduce poverty and disease, manage natural resources rationally, promote responsible consumption and production, and take advantage of globalization to achieve a balance between development and the environment.
2005	The United Nations Decade of Education for Sustainable Development (2005-2014)	The main goal is mobilization the educational resources of the world to help create a more sustainable future. This can be achieved through integration the principles, values and practices of sustainable development into all aspects of education and learning.
2015	Sustainable Development Goals (UN Summit)	Countries adopted the 2030 Agenda for Sustainable Development and its 17 Sustainable Development Goals.

Source: [19-21].

Listed in tab. international norms, as a rule, countries adapt to their own specifics. Ukraine is no exception. Thus, in order to implement the proposed Goals of Sustainable Development, a national consultation was held in Ukraine, attended by about 300 representatives of government agencies, United Nations agencies in Ukraine, the National Academy of Sciences of Ukraine, the public and expert community [22, p. 4-5]. As a result, the following goals were identified as the most priority for Ukraine: No.8 (Promote inclusive and sustainable economic growth, employment and decent work for all), No.1 (End poverty in all its forms everywhere), No.9 (Build resilient infrastructure, promote sustainable industrialization and foster innovation), No.3 (ensure healthy lives and promote well-being for everybody at all ages), No.16 (Promote just, peaceful and inclusive societies). It is important that the experts identified the most important tasks for the integration of the Sustainable Development Goals, namely: 1) the definition of national goals, targets and indicators for the medium and long term; 2) capacity building (at the national, regional and local levels) for strategic planning and prioritization (ranking) to increase the effectiveness of program / project implementation under conditions of limited resources

In recent years, many regions in Ukraine, while determining priorities for regional policy, choose regional sustainable development as a tool for effective management of the territory (Table 4). However, most often the definition of the priorities of balanced development of the region is declarative, they are not supported by scientific substantiation and methodological development of regional policy guidelines.

Table 4. Regional Strategy for Sustainable Development in Ukraine

Region	Separate strategy	Part in the overall strategy for the region	Missing information
Vinnitsa, Volyn, Dnipropetrovsk, Kharkiv and Chernihiv regions	+		
Lviv, Poltava, Sumy, Chernivtsi regions		+	
Donetsk, Zhytomyr, Zakarpatska, Zaporozhye, Ivano-Frankivsk, Kiev, Kirovograd, Lugansk, Nikolaev, Odessa, Rivne, Ternopil, Kherson, Khmelnytsky, Cherkasy regions			+

Source: Compiled by the authors on the basis of official sites of regional state administrations, as of 10.04.2018.

However, for the effective implementation of certain strategic priorities, a change in the public consciousness is needed in order to understand the importance of sustainable development of the region, as well as the definition of clear criteria for the achievement of sustainable development and constant monitoring of their level. In Ukraine, it is important to address a number of systemic issues, such as the low level of transparency and efficiency of local self-government, the lack of partnerships between government and business communities, the presence of significant asymmetry in the development of regions.

3.3 Definition of vectors for ensuring the regional sustainable development in Ukraine

Further integration into the European community requires the development and implementation of effective mechanisms of cooperation between the state, business and communities. According to experts, coopetition is rapidly becoming a key success factor for operating in the contemporary world [23]. Given the European experience of using the concept of social responsibility for regional balanced development, the main priorities are responsible production, responsible investment and responsible consumption (Fig. 2).

 https://doi.org/10.1051/shsconf/20196101001

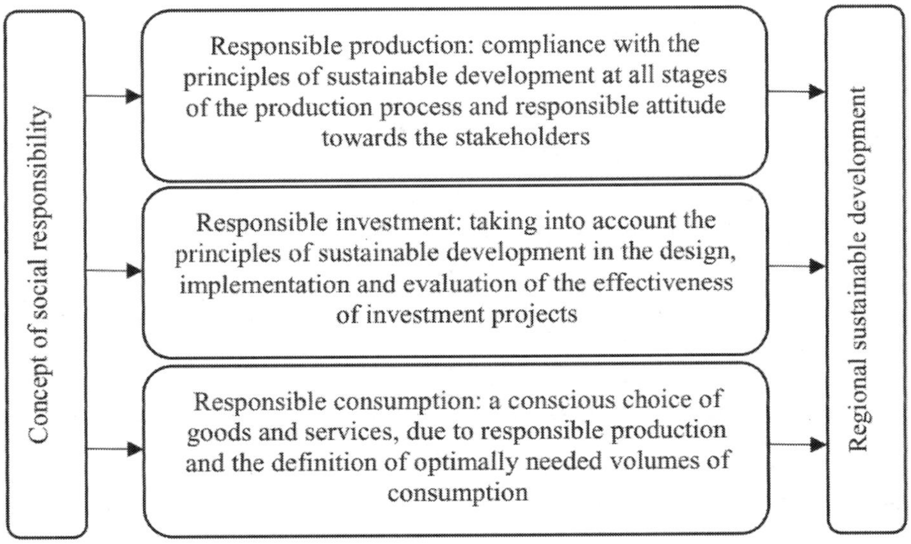

Fig. 2. Vectors of regional sustainable development of the region on the basis of social responsibility

Source: Own research.

Modern economy is characterized by intensive changes and a high risk [24, p. 530]. So one of the main prerequisites for regional sustainable development is the consideration of strategic priorities for both individuals and society. Education is one of the effective instruments for achieving regional sustainable development [25]. Thus, according to the Global Education 2030 Agenda developed by UNESCO, education helps to form the core competences of citizens needed to ensure sustainable development (Fig. 3). In our opinion, one of the most important competencies is critical thinking [27, p. 99; 28, p. 212]. This is important because in the modern conditions, it becomes more relevant to use the artificial intelligence technologies for decision making [29, p. 28].

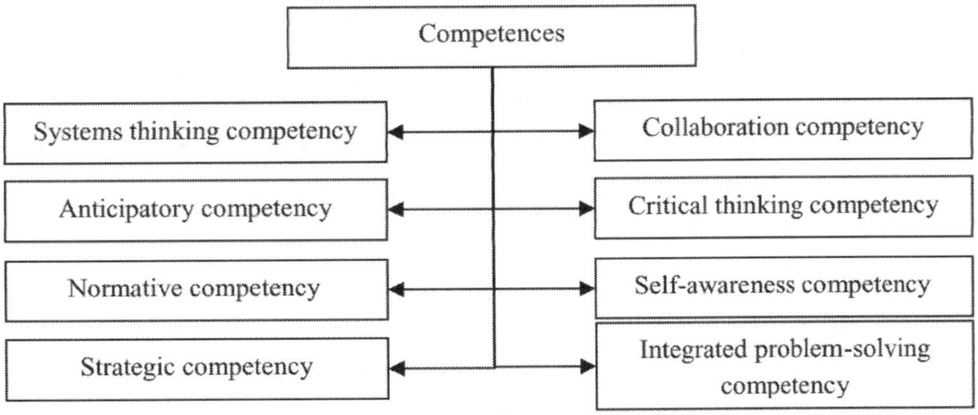

Fig. 3. Basic competences needed to ensure sustainable development

Source: Compiled by the authors on the basis of the source [26, p. 10].

Therefore, the process of changing the teaching methods and filling the training courses with information on regional sustainable development, both in higher educational establishments and in institutions of secondary and primary education, is important. This will create a system of knowledge, skills and competences necessary for the formation of citizens responsible for the future of the country.

4 Conclusion

The performed scientific research allows us to formulate the following conclusions:

1. Sustainable development is a long-term process of transformation that can lead to positive quantitative changes (growth of employment, investments, revenues, etc.) and qualitative transformations (changes in social organization, technological and technological progress). In our opinion, sustainable development can be defined as a form of interaction, which maintains a balance of interests of all stakeholders and ensures the survival of future generations. So, ensuring sustainable development is not only a legal problem (the development of a regulatory framework), not only a technical problem (the invention of technical possibilities for optimal use of available resources), not only a scientific problem (the development of theories and concepts). Ensuring sustainable development first of all requires the change of consciousness on ethical principles, which will promote cooperation and mutual understanding.

2. Sustainable development of the region is a dynamic process that can save and accumulate the potential of the region, increase its long-term competitiveness, and balance development indicators in the region. This is achieved through the definition and adherence to development priorities in the environmental, social and economic spheres, effective management of the resource base of the region, ensuring social justice and social responsibility, coherence of interests of all stakeholders, etc.

3. In recent years, many regions in Ukraine, while defining regional policy priorities, choose regional sustainable development as a tool for the effective management of the territory. However, most often the definition of the priorities of sustainable development of the region is declarative, they are not supported by scientific substantiation and methodological development of regional policy guidelines.

5 Suggestions for practice

However, for the effective implementation of certain strategic priorities, a change in the public consciousness is needed in order to understand the importance of regional sustainable development, as well as the definition of clear criteria for the achievement of sustainable development and continuous monitoring of its level. We consider education as one of the most effective tools, that will allow to form a system of knowledge, skills and competences necessary for the formation of citizens responsible for the future of the country.

References

1. The Sustainable Development Goals [online], Available at: https://www.un.org/sustainabledevelopment/sustainable-development-goals/ (2018)
2. G. Likhonosova, I. Trunina, O. Sushchenko, The territory recreation ensuring management system design. *Economic theory and practice*, **1**, 635-647, (2018)
3. An introduction to EU Cohesion Policy 2014-2020 [online], Available at: http://ec.europa.eu/regional_policy/sources/docgener/informat/basic/basic_2014_en.pdf (2018)
4. I. Z. Dénes, Adopting the European Model versus National Egoism: The Task of Surpassing Political Hysteria. *European Review*, **20**(4), 514-525, (2012)

5. United Nations Department of Economic and Social Affairs, *World population projected to reach 9.8 billion in 2050, and 11.2 billion in 2100* [online], Available at: https://www.un.org/development/desa/en/news/population/world-population-prospects-2017.html (2017)
6. United Nations' Food and Agriculture Organization (FAO) [online], Available at: www.fao.org/home/en/ (2018)
7. UN World Food Programme [online], Available at: https://www.wfp.org/ (2018)
8. X. Liang, D. Si, X. Zhang, Regional Sustainable Development Analysis Based on Information Entropy—Sichuan Province as an Example. *International Journal of Environmental Research and Public Health*, **14**, 1219, (2017)
9. T. Owens, Higher education in the sustainable development goals framework. *European Journal of Education*, **52**, 414-420, (2017)
10. M. Szabó, M. Csete, T. Pálvölgyi, Resilient Regions from Sustainable Development Perspective. *European Journal of Sustainable Development*, **7**(1) 395-411, (2018)
11. Y. Choi, Regional Cooperation for the Sustainable Development and Management in Northeast Asia. *Sustainability 2018*, **10**, 548, (2018)
12. Z. Wysokińska, Implementing the Main Circular Economy Principles within the Concept of Sustainable Development in the Global and European economy, with Particular Emphasis on Central and Eastern Europe – the Case of Poland and the Region of Lodz. *Comparative Economic Research*, **21**(3), 75-93, (2018)
13. R. Gogonea, A. Baltalunga, A. Nedelcu, D. Dumitrescu, Tourism Pressure at the Regional Level in the Context of Sustainable Development in Romania. *Sustainability 2017*, **9**, 698, (2017)
14. K. Mally, Regional differences in slovenia from the viewpoint of achieving europe's sustainable development. *Acta geographica Slovenica*, **58**(2), 31-46, (2018)
15. M. Cerf, The Sustainable Development Goals: Contextualizing Africa's Economic and Health Landscape. *Global Challenges*, **2**, (2018)
16. D. Pearce, G. Atkinson, Capital theory and the measurement of sustainable development: an indicator of "weak" sustainability". *Ecological Economics*, **8**, 103-108, (1993)
17. R. Costanza, H. Daly, Natural Capital and Sustainable Development [online], Available at: https://doi.org/10.1046/j.1523-1739.1992.610037.x (1992)
18. D. North, Institutions. *Journal of Economic Perspectives*, **5**(1), 97-112, (1991)
19. The official site of the United Nations [online], Available at: http://www.un.org/en/sections/about-un/overview/ (2018)
20. The official site of UNESCO, *UN Decade of ESD* [online], Available at: https://en.unesco.org/themes/education-sustainable-development/what-is-esd/un-decade-of-esd (2018)
21. Treaty of Amsterdam, *Amending the treaty on European Union, the treaties establishing the European Communities and certain related acts* [online], Available at: http://www.europarl.europa.eu/topics/treaty/pdf/amst-en.pdf (2018)
22. Millennium Development Goals: Ukraine. *National Consultations*, **6**, (2016)
23. J. Cygler, W. Sroka, M. Solesvik, K. Dębkowska, Benefits and Drawbacks of Coopetition: The Roles of Scope and Durability in Coopetitive Relationships. *Sustainability 2018*, **10**, 2688, (2018)

24. I. Trunina, O. Vartanova, O. Sushchenko, O. Onyshchenko, Introducing ERP System as a Condition of Information Security and Accounting System Transformation. *International Journal of Engineering & Technology*, **7**(4.3), 530-536, (2018)
25. D. Zagirniak, I. Khovrak, V. Perevozniuk, The Convergence of the Systems of Education at Scottish and Ukrainian Universities in the Context of Ethic Leadership. *International Journal of Engineering & Technology*, **7**(4.3), 550-556, (2018)
26. Education for Sustainable Development Goals, *Learning Objectives, UNESCO* [online]. Available at: http://creativecommons.org/licenses/by-sa/3.0/igo/ (2018)
27. O. B. Vartanova, Professional competence of the personnel: modern requirements and approaches to formation. Problems of a Systematic Approach in the Economy: Sb. Sciences Works of the National Aviation University [Професійна компетентність персоналу: сучасні вимоги та підходи до формування. *Проблеми системного підходу в економіці: зб. наук. праць Національного авіаційного університету*], [online], Available at: http://psae-jrnl.nau.in.ua/journal/5_61_2017_ukr/14.pdf (2017)
28. I. Trunina, Development of entrepreneurship entity competitive strategy using competence-based approach. *Actual problems of economics*, **11**(173), 206-213, (2015)
29. I. Buzko, Y. Dyachenko, M. Petrova, N. Nenkov, D. Tuleninova, K. Koeva, Artificial Intelligence technologies in human resource development. *Computer Modelling & New Technologies*, **20**(2), 26-29, (2016)

Randomness vs. fuzziness in managerial decision-making

Simona Hašková[1,*]

[1]The Institute of Technology and Business in České Budějovice, School of Expertness and Valuation, Okružní 517/10, 37001 České Budějovice, Czech Republic

Abstract. Managers often deal with uncertainty of a different nature in their decision processes. They can encounter uncertainty in terms of randomness or fuzziness (i.e., mist, obscurity, inaccuracy or vagueness). In the first case (randomness), it can be described, for example, by probability distribution, in the second case (fuzziness) it cannot be characterized in such a way. The methodological part of the paper presents basic tools for dealing with the uncertainty of both of these types, which are techniques of probability theory and fuzzy approach technique. The original contribution of the theoretical part is the interpretation of these different techniques based on the existence of fundamental analogies between them. These techniques are then applied to the problem of the project valuation with its „internal" value. In the first case, the solution is the point value of the statistical E[PV], in the second case the triangular fuzzy number of the subjective E[PV]. The comparison of the results of both techniques shows that the fuzzy approach extends the standard outcome of a series useful information. This informative „superstructure" of the fuzzy approach compared to the standard solution is another original benefit of the paper.
Key words: uncertainty, probability, fuzziness, expected value, fuzzy number

1 Introduction

In the real world, much of the decision-making takes place in an environment where the consequences of the decisions cannot be predicted with certainty. This is due to the existence of internal and external factors that the decision-maker is able to influence only partially or not at all. In practice, however, it is in many cases quietly assumed that uncertainty in the sense of fuzziness (i.e. in the sense of mist, obscurity, inaccuracy or vagueness), whatever its nature, can be considered as randomness, which is a doubtful assumption.

We can only talk about randomness in conjunction with the elements of the universe whose objectively identified basic statistical characteristics are known. If nothing is known about it, it is uncertainty (see [1]). One of the reasons why it is necessary to distinguish between randomness and uncertainty is that they cannot be treated equally. Uncertainty is the main source of vagueness in many decision-making processes. We perceive it as a type of uncertainty that is associated with the use of intuitive (or only generally or linguistically

* Corresponding author: haskovas@post.cz

defined) terms (for details see [2]). Mathematics models these concepts by means of fuzzy sets, i.e. by the classes in which there is no sharp transition between membership and non-membership (this occurs when the characteristic function of the class is more than two-valued).

Decision-making in a fuzzy environment means a decision-making process in which goals and / or constraints, but not necessarily the whole system, are of fuzzy nature. This means that targets and / or limitations are classes that are not strictly defined. An example of a fuzzy constraint is: „The price of product A should be considerably higher than β", where β is a specified constant. Similarly, an example of a fuzzy target: „y is located near y_0", where y_0 is a constant. The underlined words in these examples are the source of vagueness (see [3]). Fuzzy targets and fuzzy constraints can be defined as fuzzy sets in space of alternatives. Fuzzy decisions can be understood as intersection of given goals and constraints. The optimal decision is the set of points in the space of alternatives where the membership function of fuzzy decision achieves the best possible value.

In the case of randomness, the uncertainty concerns the membership or non-membership of the object in a non-fuzzy set. An example of randomness is the statement „The probability that Z will be profitable this year is 0.7", which is an expression of uncertainty about Z's membership in a non-fuzzy set of profitable companies.

The difference between randomness and uncertainty is reflected in various mathematical techniques. In the first case, the techniques of probability theory help to cope with the existence of randomness; in the case of fuzziness, the technique of fuzzy approach is applied.

Fuzzy techniques are simpler than probability techniques in many respects because the concept of probability in probability theory corresponds to a simpler concept of membership function in fuzzy approach. In addition, for operations a + b, or respectively a · b, where a, b are the real numbers, the simpler operations max{a,b} and min{a,b} apply (more in [4]).

The aim of the paper is to put into context fuzzy approach and probabilistic approach, i.e. to specify the main principles and differences of both the techniques and to identify analogies between them. This is in the theoretical level the subject of the methodological part. The application part compares both approaches within the problem of quantifying the „internal" value of the reality close project of a developer company. The results of the approaches within the mini-study are discussed and interpreted. The conclusion summarizes the essentials, including the introduction of the original contributions of the paper.

2 Methodological approach to random and vague phenomena in management

Statistical procedures of *probability theory* in investment decision making are based on calculations of the expected values of future flows generated by the investment. The most commonly used criterion is the expected current value $E[PV]$, which is the present value of the expected annual cash flows $E[CF_i]$ in years $i = 1, 2$ to n – see relation (1).

$$E[PV] = \sum_{i=1}^{n} E[CF_i] \,/\, \prod_{j=1}^{i}(1 + r_j), \tag{1}$$

where CF_i, $i > 0$ are net cash flows (positive or negative) generated by the project in i- year of operation, r_j is the positive discount rate per annum valid in the j-year of the project life, E is the symbol of the weighted average.

Probabilistic analysis of the cash flow can be used if the probabilities of possible outcomes are known, i.e. if the distribution of the frequency of possible outcomes is known. If it is not known, as it is typical in the case of demand for new products, most decision-makers rely on the knowledge and opinions of experts in predicting the time cash flow of the investment at the stage of its assessment.

In the literature, we find a number of authors who have dealt with the approach of evaluating investments in terms of the expected present value and its alternatives and the variance of the present value of probabilistic cash flows in different time ranges. Let us note, for example, Zinn et al. [5] who processed the formulas for the expected net present value, variance and semi-variance of net present values for different cash flow profiles at random time. In [6] Tufekci and Young presented the moments of net present value of probabilistic investment alternatives. The publication [7] compares two discount methods for the evaluation of multi-time stochastic income flows, which are identical and time independent.

The fuzzy method is based on the theory of *fuzzy sets* and represents an alternative approach in the case of uncertain variables characterized by the absence of any statistical description in the structure of the CF_i values and / or uncertain values of the discount rates r_j.

The theory of fuzzy sets has its foundations in various versions of fuzzy logic that were created by adapting the binary numerical characteristics of propositional operators to the interval $\langle 0,1\rangle$. For example, if $|A| \in \{0,1\}$, or respectively $|B| \in \{0,1\}$ is the truth value of A, or respectively B, we can classify the classical truth tables for negation, conjunction, disjunction, and implication by the numerical characteristics of these operators: $|\neg A| = 1 - |A|$, $|A \wedge B| = \min\{|A|,|B|\}$, $|A \vee B| = \max\{|A|,|B|\}$ and $|A \rightarrow B| = 1$ for $|A| \leq |B|$, $|A \rightarrow B| = 1 - (|A| - |B|)$, otherwise. Moving from the set $\{0,1\}$ to the interval $\langle 0,1\rangle$, we get the semantic form of Lukasiewicz's fuzzy propositional calculus, originally formulated by Lukasiewicz and Tarski ([8]). Fuzzy logic serves here only as a tool for the exact handling of fuzzy sets, the theory of which was published by L. A. Zadeh in 1965 – see [9].

In brief, the *fuzzy set* is a class of arranged pairs in which the first member is an element of the given universe of consideration, the second number comes from the interval $\langle 0,1\rangle$, which assigns it a degree of membership to a subset of the universe (i.e. to the support of the fuzzy set). This membership degree corresponds to the extent to which the element is compatible with the fuzzy set support. More specifically:

Let the set U be a field of reasoning or discussion (universe), let $\mu_{\underline{A}}: U \rightarrow \langle 0,1\rangle$ be a membership function and let $\underline{A} = \{(y, \mu_{\underline{A}}(y)): y \in U\}$ be the set of all pairs $(y, \mu_{\underline{A}}(y))$, in which the numbers $0 \leq \mu_{\underline{A}}(y) \leq 1$ state to the given $y \in U$ the degree of membership of the pair $(y, \mu_{\underline{A}}(y))$ to the set $\underline{A}$. Then $\underline{A}$ is a *fuzzy subset* on the universe U. The significant characteristic of the fuzzy subset $\underline{A}$ is its support $U_{\underline{A}} = \{y: 0 < \mu_{\underline{A}}(y) \leq 1, y \in U\} \subset U$. In terms of fuzzy logic is $\mu_{\underline{A}}(y) = |y \in U_{\underline{A}}|$, where $|y \in U_{\underline{A}}|$ denotes the degree of veracity of the proposition that y is the element of the support of the fuzzy set $\underline{A}$. The element $y \in U$ with $\mu_{\underline{A}}(y) = 0.5$ is called the *crossover point* in $\underline{A}$. At values greater than 0.5, the element y rather belongs to $U_{\underline{A}}$, at the values smaller it rather does not.

The fuzzy subset $\underline{A}$, whose support $U_{\underline{A}} \subset U \subset R$, where R is the set of real numbers, and its function $\mu_{\underline{A}}$ is gifted by the property of normality and convexity, i.e. at least in the case of one element $x \in U_{\underline{A}}$, it applies $\mu_{\underline{A}}(x) = 1$, and $\mu_{\underline{A}}(x') \geq \min\{\mu_{\underline{A}}(x_1), \mu_{\underline{A}}(x_2)\}$ for all $x' \in \langle x_1, x_2\rangle \subset U_{\underline{A}}$, we call the fuzzy number. Theoretically, there may be different shapes of membership function $\mu_{\underline{A}}$ of fuzzy numbers: triangular, trapezoidal, bell, sinusoidal, cosinusoidal (see e.g. [10]). In such a way defined fuzzy numbers can be used to the formal representation of *uncertain* variables.

There is a significant analogy between the function $f(x)$, which is the probability density of a *random* variable x and the function $\mu_{\underline{A}}(x)$, which is the degree of the membership of the element x to the support of the *uncertain* variable (a fuzzy number $\underline{A}$). For example, a similar role that in the case of the *random* variable x plays an average value or expected value E[x], which is the horizontal coordinate of the centre of gravity of the area under the function $f(x)$ in its definition field, in the case of the *uncertain* variable this role is played by the horizontal coordinate of the centre of gravity under the course of the function $\mu_{\underline{A}}(x)$ above the interval defined by fuzzy support $\underline{A}$.

This analogy can be used to solve a number of problems with variables that are out of any statistical description. A trusted spot estimate, which can hardly be obtained in such a case, can be replaced by the corresponding coordinate of the position of the centre of gravity of an adequate fuzzy number with a support matching to the set of all possible results. In practice, we often use this method to measure something that is difficult to quantify (such as the growth of quality of life) with something relatively close to what is easily measurable (e.g. GDP growth - see [11]). Let $\underline{A} = (A_L, A, A_R)$ and $\underline{B} = (B_L, B, B_R)$ be triangular fuzzy numbers in which the indices L and R mark the left and right edges of their supports. Let the middle numbers without an index be their *subjectively* expected values in which $\mu_{\underline{A}}(A) = \mu_{\underline{B}}(B) = 1$ can be reasonably assumed (subjectively expected values are usually at the centre of the fuzzy number supports and in the case of the symmetrical probability densities they coincide with the *statistically* expected values).

Through the algebraic operations of the (+), (–), (·) and (/) of the calculus of triangular fuzzy numbers developed in [12], from which let us mention $\underline{A}$ (+) $\underline{B}$ = $(A_L+B_L, A+B, A_R+B_R)$, $\underline{A}$ (–) $\underline{B}$ = $(A_L-B_R, A-B, A_R-B_L)$, k (·) $\underline{A}$ = $(k\cdot A_L, k\cdot A, k\cdot A_R)$ and $\underline{A}$ (/) $\underline{B}$ = $(A_L/B_R, A/B, A_R/B_L)$, it is possible to formulate the fuzzy number $\underline{PV} = (PV_L, PV, PV_R)$ in the case of uncertain cash flows (the fuzzy numbers $\underline{CF_i}$) and uncertain discount rates (the fuzzy numbers $\underline{r_i}$) as follows ([13]):

$$PV_L = \sum_{i=1}^{n}[\max\{CF_{iL},0\} / \prod_{j=1}^{i}(1+r_{jR}) + \min\{CF_{iL},0\} / \prod_{j=1}^{i}(1+r_{jL})],$$

$$PV = \sum_{i=1}^{n} CF_i / \prod_{j=1}^{i}(1+r_j), \qquad (2)$$

$$PV_R = \sum_{i=1}^{n}[\max\{CF_{iR},0\} / \prod_{j=1}^{i}(1+r_{jL}) + \min\{CF_{iR},0\} / \prod_{j=1}^{i}(1+r_{jR})].$$

3 Application of probabilistic and fuzzy approach

This probability and fuzzy approach will be applied to a reality close managerial decision-making problem. Basic input information and task data are:

- The developer company decides to invest in a construction project of a modern residential building in the outskirts of the capital.
- Construction completion including the inspection of apartments is planned within two years from the construction launch. If there is a delay in the plan for a year, the company chooses between the option to complete the project with an annual delay or to sell the project in the third year at an estimated price of EUR 23 millions. The possibility of construction completion according to schedule in time or with an annual delay is equally probable.
- In case of completion of the construction, the apartments will be sold to personal ownership in the next year. The amount of budgeted income from the sale of the apartments (net income is the difference between the sum of revenues of the apartment sales and operating costs, paid fixed costs associated with an investment and income tax) depends on the size of the demand for apartments.

- In the case of the apartment sale in the third year strong demand is estimated with 70 % probability, weak demand with 30 % probability; in the case of the one-year later sale, strong demand is estimated with a probability of 60 % and weak demand with 40 % probability.
- The discount rate of the project *r* is equal to the average project capital costs of 15 %. As the company does not intend to change the structure of its long-term funding sources in the near future, it can be considered a constant.
- The prediction of net incomes N_{31} and N_{32}, or respectively N_{41} and N_{42} from the sale of apartments in the third, or respectively in the fourth year of the project is captured in Tab. 1.

Table 1. Total budgeted net project income

Demand in the third year	**Total net income from sale in the third year N_{31} and N_{32}**
Strong 0.7	Eur 40 mil.
Weak 0.3	Eur 25 mil.
Demand in the fourth year	**Total net income from sale in the fourth year N_{41} a N_{42}**
Strong 0.6	Eur 38 mil.
Weak 0.4	Eur 23 mil.

Source: Own processing.

The task of the manager is to evaluate the project with its maximum „internal" value using the E[PV] criterion and to determine the highest possible investment cost that is acceptable for a project with an adequate risk rate if it is not to be loss-making. The aim of the study is to assess whether the criterion in question is the mainstay of the manager's decision in this case.

3.1 Model of sequential decision-making situation

Data of Tab. 1 are captured in the leaves of the decision tree shown in Fig. 1. The ellipses show situational nodes characterizing the state of the project implementation after the second year of its lifetime and the estimated level of demand for apartments in the third and fourth year. Rectangles represent the decision nodes (the decision whether to invest in the project and the decision whether to sell it in the event of its delay or to finish it).

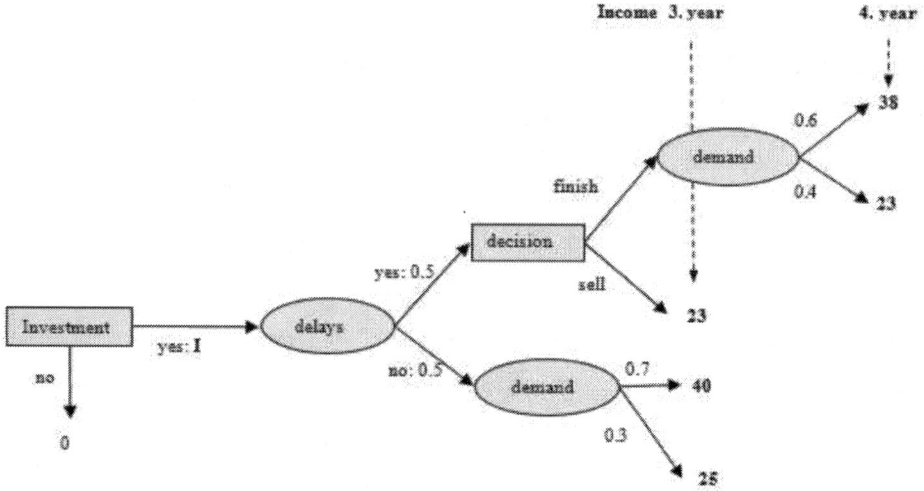

Fig. 1. Decision tree of the project of the construction of a modern residential building at current prices of 3rd and 4th year in millions of Eur

Source: Own processing.

3.1.1 Probabilistic evaluation based on data distribution and budgeted net income

The model of the decision task in Fig. 1 is a sequential decision tree illustrating the statistical solution of the problem based on the unplanned forecasts of the point estimates of the random variables listed in Tab. 1. Construction completion time and demand level are also considered random variables described by an even distribution of probability density in the case of a delay, and in case of demand by some probability distribution according to the opinion of the financial manager of the company.

Given that there are generally positive numbers in the tree sheets with the exception of the root node, therefore E[PV] > 0. The question of whether to invest asked in the root decision node can be reduced to the question of how much to invest (I = ?). In the second decision node, in addition to the amount of EUR 23 million for the sale of the project, the amount calculated by the statistical approach $(0.6 \cdot 38 + 0.4 \cdot 23) / 1.15 = 27.8$[†] is taken into account. This amount is greater than 23 – the decision is then known in advance and therefore this decision node can be omitted. The given step allows us to simplify the tree structure of Fig. 1 as shown in Fig. 2.

[†] The dot separating the numerical order refers to a number that is not of monetary significance, the comma separating the numerical order refers to the number, which is expressed in millions of Euros.

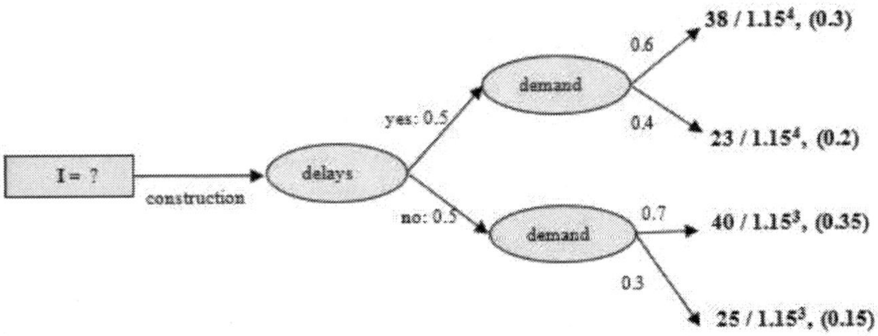

Fig. 2. The simplified tree of Fig. 1

Source: Own processing.

In it, four paths leading from the root to the sheets show four possible scenarios of the project run. Each of the scenarios is rated in its sheet by its current value $PV_{ij} = N_{ij} / (1 + r)^i$ to the time t = 0 of the project run (the tree root), where N_{ij} are the net incomes from Tab. 1. The brackets following the PV_{ij} values are the estimated occurrence probabilities of the relevant scenario.

It applies: $E[PV] = 0.3 \cdot PV_{41} + 0.2 \cdot PV_{42} + 0.35 \cdot PV_{31} + 0.15 \cdot PV_{32} = 0.5 \cdot (0.6 \cdot PV_{41} + 0.4 \cdot PV_{42}) + 0.5 \cdot (0.7 \cdot PV_{31} + 0.3 \cdot PV_{32}) = 0.5 \cdot (E[PV_4] + E[PV_3])$.

The equality $E[PV] = 0.5 \cdot (E[PV_4] + E[PV_3])$ allows to reduce the original four scenarios to two scenarios and to redraw the tree from Fig. 2 to Fig. 3, showing that $E[PV_4] = 32 / 1.15^4$ and $E[PV_3] = 35{,}5/1.15^3$ with the same probability of occurrence of both of the scenarios.

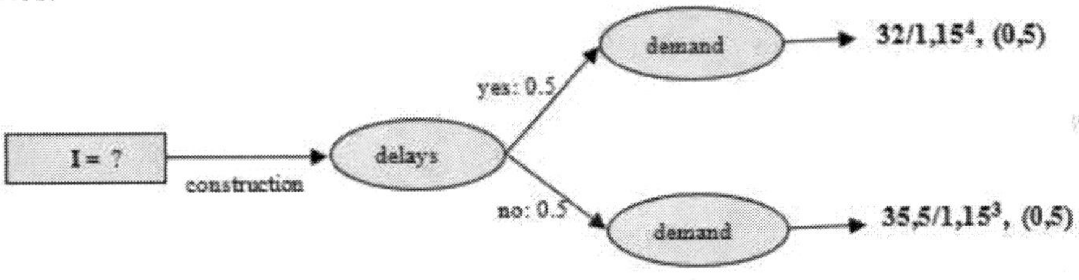

Fig. 3. Stochastic tree of the project after the reduction of four scenarios to two

Source: Own processing.

The solution sought is $E[PV] = 0.5 \cdot (E[PV_4] + E[PV_3]) = 0.5 \cdot (32 / 1.15^4 + 35{,}5 / 1.15^3) = 20{,}82$ million EUR, which is simultaneously the „internal" value of the project. project. It also defines the highest possible investment cost for the lossless project.

3.1.2 Evaluation of the project within fuzzy approach

The basic method of considering data uncertainty in the fuzzy approach is the replacement of the estimated point values of the variables with triangular fuzzy numbers of type (L; S; P), where L is the smallest considered value, P is the largest, and S is the centre between them. This is an application of the general principle of indifference, which states that if there are more alternative results, and, at the same time, there is no relevant reason to prefer one or the other, it is appropriate to act as if all were equally probable. Then S is the *subjectively* expected value and (unlike the statistically expected value that is objective as being derived from the observation) it does not converge to any of the boundaries of the interval of possible values. In our model, the uncertain variable is the demand for apartments, from which the net income is also derived. In Fig. 1 the subjectively expected value of the fuzzy approach corresponds then to 30,5 / 1.15 = 26,5, which is compared with the amount of EUR 23 million for the sale of the project (part 3.3.1, the second decision node). Similarly as in the case of the comparison of a statistically expected value, this subjective value is higher than 23, therefore this decision node can be omitted and the model can be reduced in analogy to Fig. 3 on the fuzzy decision tree model in Fig. 4, wherein the second subjective value is 32,5.

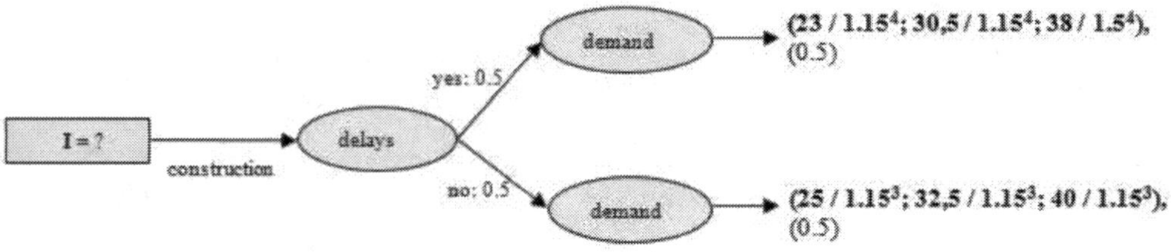

Fig. 4. Model of the considered project from a fuzzy approach perspective

Source: Own processing.

By the application of the tools of interval calculus mentioned in the methodological part, we obtain the fuzzy number of the sought solution in the form:

$(E[PV]_L, E^*[PV], E[PV]_R) = (0.5 \cdot 23 / 1.15^4 + 0.5 \cdot 25 / 1.15^3;\ 0.5 \cdot 30{,}5 / 1.15^4 + 0.5 \cdot 32{,}5 / 1.15^3;\ 0.5 \cdot 38 / 1.15^4 + 0.5 \cdot 40 / 1.15^3) = \mathbf{(14{,}8;\ 19{,}4^*;\ 24)}$,

in which the subjectively expected present value of potential net incomes from the sale of apartments is marked with the asterisk.

The fuzzy number (14,8; 19,4*; 24) represents the interval of possible „internal“ values of the project in which the left number can be understood as the value of the pessimistic scenario, the right number as the optimistic scenario and the middle number as the subjectively expected real value. At the same time, the interval of maximum investment costs is defined.

3.2 Procedures analysis and results discussion

Comparing the results of the standard and fuzzy analysis, we see that the fuzzy approach extends the standard result for a number of other useful information. Firstly, the „internal“ value of the E[PV] project as a purely statistical variable is based on the weighted average of the relevant scenario values. In contrast, the „internal“ fuzzy value of the project is a triangular fuzzy number $\underline{E[PV]}$ = $(E[PV]_L, E^*[PV], E[PV]_R)$, whose extreme values define the lowest and highest possible present values with an average value as the subjectively expected value. The interval $(E[PV]_L, E[PV]_R)$ can be understood as the span between the result of the pessimistic and optimistic development of the project.

Furthermore, the fact that $E^*[PV] = 19{,}4$ is lower than $E[PV] = 20{,}82$ only confirms the opinions of some authors e.g. in [14] that managers tend to overestimate their prognosis of positive outcomes – in the cash flow forecasts of projects the managers exaggerate positive flows and reduce negative flows. This trend corresponds to the probability distribution of the demand for apartments in the 3rd, or respectively 4th year. As a result the most of the large investment projects are completed late, they significantly exceed the budget and many of them will never meet the expectations of investors [15], [16]).

The manager, who derives from E[PV] = 20,82 that the investment of € 20 million in the project earns, for example, a minimal required gain of € 0,5 million, is wrong. From the fuzzy analysis, it is clear that to reach this target, the initial investment should not exceed

14,3 million Euro (i.e. for 0,5 less than PV_L); the question is whether the realization of the project would be feasible under these circumstances at all.

From the stated it can be seen that knowing the boundaries of the intervals of possible E[PV], which the fuzzy approach provides, can be very significant for decision-makers.

4 Conclusion

In the world of management, most of the decision-making processes take place in an environment where the consequences of decisions made cannot be estimated with certainty. There is uncertainty of different nature that plays its role – it can be uncertainty in terms of randomness or uncertainty in the sense of fuzziness. The first mentioned can be expressed, for example, by probability distribution based on historically known data of the same or similar phenomena, in the latter case, it is not possible to describe it in such a way as the frequency of occurrence of the phenomenon is unknown. The theory of probability theory helps with the existence of uncertainty of the randomness type, in the case of fuzziness the technique of fuzzy approach can be successfully applied.

The most commonly used probability criterion in managerial decision making is the expected present value E[PV], which is the present value of expected annual cash flows $E[CF_i]$ in years $i = 1, 2$ to n. In the fuzzy approach, the decision leans on the fuzzy number $\underline{E[PV]}$ = ($E[PV]_L$, E*[PV], $E[PV]_R$) of uncertain cash flows (fuzzy numbers $\underline{CF_i}$) and uncertain discount rates (fuzzy numbers $\underline{r_i}$), where index L, or respectively R stands for left, respectively right border of the support of the respective fuzzy number.

The differences and analogies of approaches have been demonstrated to address the valuation of the project of the construction and sale of apartment units by its „internal" value. Comparing the results of probability and fuzzy analysis, it has emerged that the fuzzy approach extends the standard result for a number of useful information. In particular, E[PV] is a weighted average whose calculation erases all boundary values of the project scenarios; the result is a number that carries less information than information of „non-wiped out" scenarios. In contrast, the fuzzy approach generates limits of possible values of fuzzy number supports in which the middle value is a subjectively expected value not burdened by excessive optimism or scepticism of a manager regarding the occurrence of CF flows within individual scenarios. Knowledge of these limits is beneficial for investors with a different risk sensitivity relationship.

The described information advantages of fuzzy approach compared to the standard solution are the original benefits of the application part; the theoretical superstructure of the article is based on identifying the analogy between probabilistic approach and fuzzy approach.

References

1. R. E. Bellman, L. A. Zadeh, Decision-Making in a Fuzzy Environment. *Management Science,* **17**(4), B-141, (1970)
2. J. M. Mendel, *Uncertain rule-based fuzzy logic systems: introduction and new directions.* Springer International Publishing, (2001)
3. L. A. Zadeh, The role of fuzzy logic in the management of uncertainty in expert systems. *Fuzzy sets and systems*, **11**, 199-227, (1983)
4. L. A. Zadeh, J. Kacprzyk, *Fuzzy logic for the management of uncertainty.* Wiley-Interscience, (1992)

5. C. D. Zinn, W. G. Lesso, B. Motazed, A probabilistic Approach to Risk Analysis in Capital Investment Projects. *The Engineering Economist*, **22**(4), 239-260, (1977)
6. S.Tufekci, D. B. Young, Moments of the Present Worths of General Probabilistic Cash Flows Under Random Timing. *The Engineering Economist*, **32**(4), 303, (1987)
7. V. Benzion, J. Yagil, On the Discounting Formula for a Stream of Independent Risky Cash Flows. *The Engineering Economist*, **32**(4), 337, (1987)
8. T. J. Ross, *Fuzzy logic with engineering application*. Wiley, (2009)
9. L. A. Zadeh, *Fuzzy Sets, Fuzzy Logic, And Fuzzy Systems*. World Scientific Pub Co Inc, (1996)
10. C. Kahraman, *Fuzzy multi-criteria decision making: theory and applications with recent developments*. Springer US, (2008)
11. R. L. Ackoff, From data to wisdom. *Journal of applied systems analysis*, **16**(1), 3-9, (1989)
12. C. Kahraman, Fuzzy versus probabilistic benefitúcost ratio analysis for public work projects. *International Journal Applied Mathematics and Computer Sci*ence, **11**(3), 705-718, (2001)
13. S. Haskova, Contribution to Economic Efficiency Evaluation of Projects in Terms of Uncertainty. *35th International Conference on Mathematical Methods in Economics 2017 (MME 2016)*, 213-218, (2017)
14. D. Kahneman, D. Lovallo, Timid Choices and Bold Forecats: A cognitive perspective on Risk Taking. *Management Science*, **39**(1), 17-31, (1993)
15. B. Flyvbjerg, Curbing Optimism Bias and Strategic Misrepresentation in Planning: Refernce Class Forecasting in Practice. *European Planning Studies*, **16**(1), 3-21, (2008)
16. M. Vochozka, Using neural Networks to Determine the Financial Plan. *Innovation Management, Entrepreneurship and Corporate Sustainability (IMECS 2016)*, 742-755, (2016)

The fuzzy approach for estimating the creditworthiness and credit risk of issuers of corporate bonds

Simona Hašková[1,*]

[1]The Institute of Technology and Business in České Budějovice, School of Expertness and Valuation, Okružní 517/10, 37001 České Budějovice, Czech Republic

Abstract. The uncertainty in the financial market is often perceived as a risk of deviation from expected results. However, uncertainty is associated with vagueness in the sense of ambiguity or obscurity, which, unlike the risk, is not describable in the form of deterministic or stochastic models. Given the vagueness of the data entering the credit risk assessment models, the fuzzy decision-making process is introduced in the theoretical part and implemented in the application part as an effective alternative to standard models. The fuzzy approach is used to address the problem of the creditworthiness of the corporate bonds issuers and the credit risk. Data sources are of a qualitative and quantitative nature based on the ratings of Standard and Poor's and Moody's and financial indicators relevant for the creditor. The resulting creditworthiness and credit risk values rectify the credit ratings of rating agencies and are the original contribution to the topic solved. In conclusion, the impact of company ownership on the creditor´s subjective perception of creditworthiness, or respectively credit risk is outlined.
Key words: fuzziness, creditworthiness, credit risk, ROA, rating

1 Introduction

A typical feature of the financial market is uncertainty. This uncertainty is standardly conceived in terms of risk and described by probabilistic and statistical models. However, there is often vagueness in terms of inaccuracy, ambiguity, obscurity behind uncertainty, which, unlike uncertainty understood as risk, is not describable in the form of deterministic or stochastic models on which an objective prediction can be established. In financial practice, however, it is often silently assumed that the uncertainty in the sense of vagueness, regardless of its nature, can be considered a random phenomenon that can be handled, for example, by a distribution function.

For the purpose of credit risk analysis the structural models are used which, as known, overestimate corporate bonds. Structural models use the contingent claims approach to evaluate the put options in the agreement between creditors and equity holders. In practice, these models are applied despite their unsatisfactory explanation of corporate bonds current prices and spreads ([1]).

* Corresponding author: haskovas@post.cz

The reduced models based on rating evaluation belong to other frequently used instruments for credit risk assessment. These models lean on the assumption that bonds grouped according to the rating values are homogeneous in terms of risk. For each risk, group estimates of several characteristics are made such as the spot yield curve, the probability of default and the rate of return. These estimates are then used to calculate the theoretical price of each bond in the group. In addition to selected financial indicators and other information, whose exact set is not known, credit rating agencies derive their ratings, to a certain extent, from subjective factors and variables that are not directly related to a particular firm. Additionally, the method used to allocate the output estimates based on input variables is unknown. A practical consideration of credit rating is the credit spread, i.e. the credit spread between the corporate debt and the benchmark level, such as government bonds, of corresponding yield curve.

Elton et al. in [2] identified bond and rating characteristics beyond the rating categories of S & P and Moody's, which provide important information for the valuation of corporate bonds. These include, for example, the differences between S & P and Moody's ratings, the rating of the bond and the rating of the company issuing the bond or whether the bond is new or has been traded for more than a year.

The risk of a badly rated bond can mask excessive credit risk, which, in the extreme case, can result in the inability of an issuer company to pay the debts to which it has been committed by issuing the bond. Credit risk is closely related to the company's creditworthiness, which expresses the credibility of an economic entity to repay its debts (see [3]). The highest financial creditworthiness is usually attributed to sovereign bonds due to the state repayment guarantee. On the other hand, some corporate bonds may have a low credit score, the lowest financial creditworthiness is usually linked with junk bonds (more in [4], [5]) - see Tab. 1 matching the risk of debt default and bond quality.

Table 1. Ratings of Standard and Poor´s a Moody´s and the interpretation

Ratings	Interpretation	Grade ratings	Risk rate
AAA-Aa	Highest-high quality	Investment	None-low risk
A-Baa	Strong-adequate payment capacity	Investment	Medium risk
Ba-C	Likely to fulfil obligation-high risk	Speculative	High risk

Source: Own processing according to [6].

Taking into account the uncertainty of the data entering the standard models for the credit risk assessment, the fuzzy multi-criterial decision approach will be further discussed and applied as an effective alternative tool for reducing inaccurate outputs of deterministic and stochastic models, and as a means of refining official ratings.

In the field of financial market the fuzzy approach is used, for example, for evaluating the real options (see also [7], [8], [9]), the asset current value assessment (e.g. [10]), investment projects evaluation (e.g. [11]). Its use can be found in many management areas such as subcontracting (e.g. [12]), energy planning (e.g. [13]), analysis of human reliability (e.g. [14]), etc.

The article has the following structure: the methodological part is devoted to the basics of the fuzzy approach based on which a model of fuzzy multi-criteria evaluation is constructed. In the application section this model is used to assess the creditworthiness of the selected bond issuers and their credit risk. Unlike many previous works, we do not entirely rely on the rating value itself, which is, by its very nature, a vague characteristic, but we also take into account important factors of economic nature from the creditor point of view.

2 Methodological approach

In the tasks of expert analysis or expert control of vaguely defined systems vague terms are usually presented by linguistic variables and their linguistic values (the so-called terms), where the word „linguistic“ is used here in the meaning of „*in natural language communicated*“.

Desirable relationship between combinations of the terms of the input linguistic variables and the output linguistic variable are given by a set of rules. The tool of mathematical solution to this type of tasks is the fuzzy logic on which the fuzzy approach is based (more in [15]). The fuzzy approach purposefully modifies the assumptions and thus the conclusions of the so-called „principle of extension“ of fuzzy logic. Its individual phases can be briefly summarized as follows (the scheme of this procedure described is illustrated in Fig. 1):

- In the first phase of the fuzzy approach, which is the fuzzification of the numerical vector $x = (x_1,...,x_n)$ entering the relation K, the vector x selects the inferential rules relevant to its multi-criteria evaluation. These are the rules $((A_{1i},...,A_{nj}), B_k) \in P$, in which formalizations of terms $A_{1i},...,A_{nj}$ on the left side, all numbers $\mu_{\underline{A}1i}(x_1),..., \mu_{\underline{A}nj}(x_n)$ are nonzero.
- Moving from the symbolic notation $((A_{1i},...,A_{nj}), B_k)$ of the selected inference rule to its on x dependent logic notation in the form $(\min\{\mu_{\underline{A}1i}(x_1),..., \mu_{\underline{A}nj}(x_n)\}, \mu_{\underline{B}k}) = (\mu_{\underline{A}^*}(x), \mu_{\underline{B}k})$ we obtain a number $\mu_{\underline{A}^*}(x)$ indicating not only the degree of truthfulness that the situation characterized by the vector x actually occurred, and therefore the degree of eligibility of the rule selection $((A_{1i},...,A_{nj}), B_k)$ to its solution, but also the degree of significance of the term B_k in the multi-criteria evaluation of vector x. $\mu_{\underline{B}k}$: $Y \rightarrow \langle 0,1\rangle$ is the standard membership function of the elements $y \in Y$ to the fuzzy set $\underline{B}_k$. On the other hand $\mu_{\underline{B}k}^*: Y \rightarrow \langle 0,1\rangle$ in the form $\mu_{\underline{B}k}^*(y) = \min\{\mu_{\underline{A}^*}(x), \mu_{\underline{B}k}(y)\}$ limits from above the function $\mu_{\underline{B}k}$ with the number $\mu_{\underline{A}^*}(x)$.
- The components of the relation K will then be the elements of the set $Px = \{B_k: \exists y:(\min\{\mu_{\underline{A}^*}(x),\mu_{\underline{B}k}(y)\} > 0), ((A_{1i},...,A_{nj}), B_k) \in P, y \in Y\}$. These are the partial intermediate results whose aggregation (logical sum) we obtain the formalized term $\underline{B}_{agg} = \{(y, \mu_{agg}(y)): y \in Y\}$ on Y, in which it applies
$$\mu_{agg}(y) = max\left\{min\{\mu_{\underline{A}^*}(x), \mu_{\underline{B}k}(y)\}: \left((A_{1i}, ..., A_{nj}), B_k\right) \in P, y \in Y\right\}.$$
- The final phase of the fuzzy approach is defuzzification, in which last step we obtain the output value y_x of the alternative $(a_1,...,a_c) \in A$ as the mean value of the elements $y \in \underline{B}_{agg}$ weighted by values $\mu_{agg}(y)$ of their significance. Therefore
$$y_x = \int y \cdot \mu_{agg}(y)dy / \int \mu_{agg}(y)dy \quad (1)$$
where $\int$ is the symbol of a definite integral over the universe Y.

The process of the multi-criteria assessment of creditworthiness and credit risk of corporate bond issuers is illustrated in a solution to the specific task whose input parameters are of qualitative and quantitative nature.

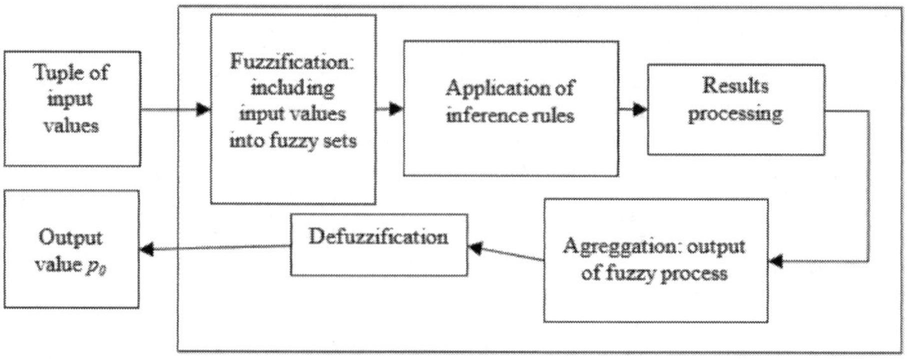

Fig. 1. Schematic representation of the multicriteria problem solving process within the fuzzy approach

Source: Own processing.

3 Procedures and results of the application of fuzzy approach for evaluating the creditworthiness and credit risk of bond issuers

Creditworthiness of the bond issuer expresses the credibility of the company in the financial market. It is closely related to the company's economical results, its solvency, the ability and willingness to meet its obligations and ability to valorise the capital invested. In this context, it includes information on, for example, the company´s payment discipline, the current status of commitments towards different creditors and the financial situation. In a wider context, it also relates to the company's stability in the market, the reputation or the life cycle of the company. Creditworthiness is often mistakenly identified with the rating - the ability to repay debts, which is one of the components of creditworthiness.

In the following analysis of creditworthiness and credit risk, three characteristics are selected, which co-determine the company creditworthiness – the Standard and Poor's and Moody's ratings, the firm production power and the equity to total asset ratio. Credit risk is given as the reciprocal value of creditworthiness [16].

3.1 Input parameters of the fuzzy model

The index values of components of the issuer bond quality, which are the official 2017 Rating assessment (R), the Production power (ROA) and the Equity to asset ratio (VA), are input parameters to the fuzzy multi-criteria model to assess the creditworthiness and credit risk of bonds issuers ČEZ, a.s., Letiště Praha, a.s. and Česká telekomunikační infrastruktura a.s. (see Table 1).

Rating assessment (R) is the resultant of ratings of Standard and Poor's and Moody's; it expresses the credibility of the debtor in terms of solvency. As mentioned in the introductory section, index construction includes qualitative and quantitative inputs. The resulting values gain linguistic forms (see Table 1).

Index Production power (ROA) given by the ratio of EBIT to total assets is an indicator of the company's production ability. It is given by the ROA average of the three consecutive past periods in relative terms in the interval 0 - 0.5; the higher the ROA value is the higher the production power. It is obvious that ROA is lower in the case of companies with a high proportion of assets, which is the case of all companies analysed. Compared to the average ROA within the industry in 2016, ČEZ, a.s. and Česká telekomunikační infrastruktura, a.s. are slightly below average, Letiště Praha, a.s., on the contrary, shows an highly above average result.

The Equity to asset ratio is based on the financial statements of three consecutive past periods, of which the average is calculated. The index is expressed in relative terms in the interval 0 - 0.8, with a higher value representing a better result from the creditor point of view. Compared to the 2016 industry average, ČEZ, a.s reaches below average value, Letiště Praha, a.s. above-average value and Česká telekomunikační infrastruktura, a.s. corresponds to the industry average.

Table 2. The input parameters of indexes Rating assessment, Production power and Equity / Asset ratio

Company/Index	Rating assessment 2017 (R)	Production power (ROA) ⟨0,0.5⟩ / Sector Average 2016 ⟨0,0.5⟩	Equity / Asset ratio (VA) ⟨0,0.8⟩ / Sector Average 2016 ⟨0,1⟩
ČEZ, a.s. (X)	Low to medium risk A-, Baa1	0.043 0.0597	0.42 0.70
Letiště Praha, a.s. (Y)	Low A1	0.29 0.0331	0.8 0.42
Česká telekomunikační infrastruktura, a. s. (Z)	Medium Baa2	0.047 0.0567	0.47 0.43

Source: Justice.cz, Veřejný rejstřík (2018) [on-line] https://portal.justice.cz/Justice2/Uvod/uvod.aspx, MPO, Finanční analýza podnikové sféry za rok 2016, (2017) [on-line] https://www.mpo.cz/cz/rozcestnik/analyticke-materialy-a-statistiky/analyticke-materialy/financni-analyza-podnikove-sfery-za-rok-2016--228985/ (Own processing).

As noted, there is a direct relationship between the values of these indices, the bonds issuers and the credit risk, which is defined as the reciprocal value of creditworthiness. This provides us with the possibility to solve the assigned task in two steps: the first step is to estimate the creditworthiness of the bond issuer; the second step is to utilize this estimate for the inclusion of the company in the relevant credit risk category.

3.2 Fuzzy model of the task

The ROA and VA indices are assigned with numerical data in the interval ⟨0,0.5⟩ and ⟨0,0.8⟩. Index R is of a qualitative nature. In the set of monitored companies, it is the most favourable for Letiště Praha, a.s., less favourable for ČEZ, a.s. and the least favourable for Česká telekomunikační infrastruktura, a.s.

At first glance, the elements of a given set of companies are separated into classes (categories) defined by a trio of classifiers (individual indices). The non-standard character of the task is that both the class and the classifier function are defined vaguely in the case of the Rating assessment. Therefore, the task as a whole is solved within the fuzzy approach here introduced.

For this purpose, we first normalize the values of inputs by specifying them in % within the range of their universe. If the universe *U* of ROA variables is the interval 0 to 0.5, then ROA_X is ≈ 8.6 %, ROA_Y ≈ 58 %, and ROA_Z ≈ 9.4 %. Analogously, for VA values in the *U* universe within the interval ⟨0,0.8⟩ applies VA_X = 52 %, VA_Y = 100 % and VA_Z = 58 %. The estimate of the values of the R parameter depends on the distribution of points *a* to *d* in the definition of the trapezoid fuzzy numbers (see Figure 2) and from it resulting crossover points. If expert opinion, that would help us with the distribution of points *a* to *d*, is unavailable, it is best to assume that they are distributed evenly, i.e. $a = 20$ %, $b = 40$ %, $c = 60$ % and $d = 80$ %. In addition, it is necessary to normalize R, expressed as the risk of non-payment of the obligations, as the rate of solvency of the bond issuer, i.e. as the opposite value of the risk. If the low, low to medium and medium risk is 20 %, 35 % and 50 % of the maximum risk, i.e. 100 %, then $R_X \approx 65$ %, $R_Y \approx 80$ % and $R_Z \approx 50$ % are the solvency rates of the bond issuers. The resulting values of normalized fuzzy inputs are presented in Table 3.

Table 3. Values of normalized fuzzy inputs in percentage terms

Company/Index	Rating assessment (R)	Production power (ROA)	Equity / Asset ratio (VA)
ČEZ, a.s. (X)	65 %	8.6 %	52 %
Letiště Praha, a.s. (Y)	80 %	58 %	100 %
Česká telekomunikační infrastruktura, a.s. (Z)	50 %	9.4 %	58 %

Source: Own processing.

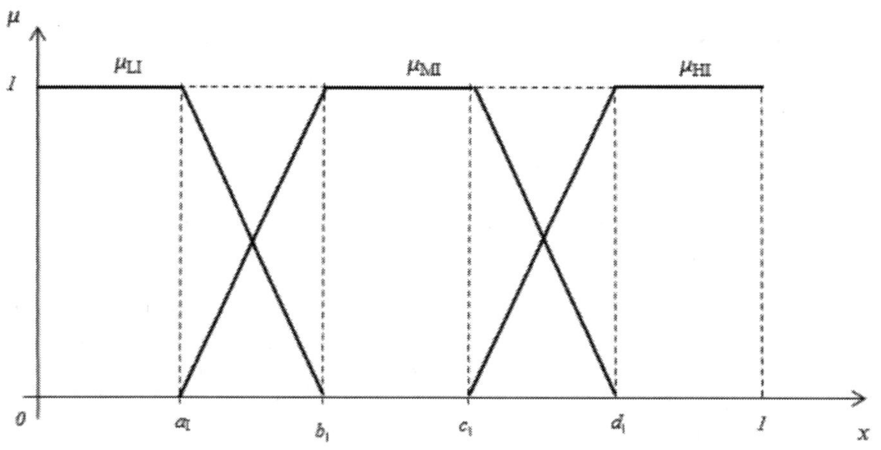

Fig. 2. The course of the membership function *L*, *M* and *H* of linguistic variables R, ROA, VA, and the output function *B*, where *L* stands for the low, *M* for the mean and *H* for the high value of the variable

Source: Own processing.

The following applies:

L) $\mu_{LI}(x) = 1$ for $0 \le x \le a_I$,
$\mu_{LI}(x) = (b_I - x) / (b_I - a_I)$ for $a_I \le x \le b_I$,
$\mu_{LI}(x) = 0$ for $x \ge b_I$

M) $\mu_{MI}(x) = (x - a_I) / (b_I - a_I)$ for $a_I \le x \le b_I$,
$\mu_{MI}(x) = 1$ for $b_I \le x \le c_I$,
$\mu_{MI}(x) = (d_I - x) / (d_I - c_I)$ for $c_I \le x \le d_I$,
$\mu_{MI}(x) = 0$ otherwise

H) $\mu_{HI}(x) = 0$ for $0 \le x \le c_I$,
$\mu_{HI}(x) = (x - c_I) / (d_I - c_I)$ for $c_I \le x \le d_I$,
$\mu_{HI}(x) = 1$ for $x \ge d_I$

The set of inference rules $P = \{((\underline{ROA}, \underline{VA}, \underline{R}), \underline{B}): \underline{ROA}, \underline{VA}, \underline{R}, \underline{B} \in \{\underline{L}, \underline{M}, \underline{H}\}\}$, where L, M and H stands for the low, the medium and the high value of the parameter, has a total of 27 elements in which the selection of the right sides (without knowledge of the matter and the expert's help) is carried out by the „predominant element" method in the left side vector (if neither of them agrees, then $\underline{B} = \underline{M}$), see Table 4.

Table 4. A set of twenty-seven inferential rules

LLL→L	LLH→L	LMH→M	MMM→M	MLL→L	MHM→M	HHH→H	HLH→H	HHM→H
LLM→L	LHL→L	LMM→M	MLM→M	MLH→M	MHH→H	HLL→L	HMM→M	HLM→M
LML→L	LHM→M	LHH→H	MML→M	MMH→M	MHL→M	HHL→H	HMH→H	HML→M

Source: Own processing.

4 Results and discussion

For ČEZ, a.s. company we get: $P^* = \{(1, \mu_{\underline{L}}), (0.75, \mu_{\underline{M}}), (1, \mu_{\underline{M}}), (0.25, \mu_{\underline{H}})\}$, $B = \{\min\{1, \mu_{\underline{L}}\}, \min\{0.75, \mu_{\underline{M}}\}, \min\{1, \mu_{\underline{M}}\}, \min\{0.25, \mu_{\underline{M}}\}\}$, $\mu_{agg} = \max\{\min\{0.75, \mu_{\underline{M}}\}, \min\{0.25, \mu_{\underline{M}}\}\}$, therefore $\mu_{agg}(y) = 0$ pro $y < 20$, $\mu_{agg}(y) = (y - 20) / 20$ for $20 \leq y < 35$; $\mu_{agg}(y) = 0.75$ for $35 \leq y < 75$; $\mu_{agg}(y) = (80 - y) / 20$ for $75 \leq y < 80$; $\mu_{agg}(y) = 0$ for $y \geq 80$.

By means of the numerical approximation of the integrals values in the formula for calculating the horizontal coordinate of the centre of gravity position, see Part 2 relation (1), the value p_0 representing the value of the bond issuer creditworthiness can be estimated as

$$p_0 = (0.5 \cdot 20 + 0.75 \cdot (30 + 40 + 50 + 60 + 70) + 0.5 \cdot 80) / (5 \cdot 0.75 + 2 \cdot 0.5) \approx 50\ \%.$$

Analogously, for Letiště Praha, a.s. company: $P^* = \{(1, \mu_{\underline{M}}), (1, \mu_{\underline{H}}), (1, \mu_{\underline{H}})\}$, $B = \{\min\{1, \mu_{\underline{M}}\}, \min\{1, \mu_{\underline{H}}\}, \min\{1, \mu_{\underline{H}}\}\}$, $\mu_{agg} = \max\{\min\{1, \mu_{\underline{H}}\}\}$, therefore $\mu_{agg}(y) = 0$ for $y < 60$; $\mu_{agg}(y) = (y - 60) / 20$ for $60 \leq y < 80$; $\mu_{agg}(y) = 1$ for $80 \leq y \leq 100$. Then

$$p_0 = (90 + 100) / 2 \approx 95\ \%.$$

For Česká telekomunikační infrastruktura, a.s. company applies: $P^* = \{(1, \mu_{\underline{L}}), (1, \mu_{\underline{M}}), (1, \mu_{\underline{M}})\}$, $B = \{\min\{1, \mu_{\underline{L}}\}, \min\{1, M_{\underline{L}}\}, \min\{1, \mu_{\underline{M}}\}\}$, $\mu_{agg} = \max\{\min\{1, _{\underline{M}}\}\}$, therefore $\mu_{agg}(y) = 0$ for $y < 20$; $\mu_{agg}(y) = (y - 20) / 20$ for $20 \leq y < 40$; $\mu_{agg}(y) = 1$ for $40 \leq y < 60$; $\mu_{agg}(y) = (80 - y) / 20$ for $60 \leq y < 80$; $\mu_{agg}(y) = 0$ for $y \geq 80$. Then

$$p_0 = (50 + 60) / 2 \approx 55\ \%.$$

The creditworthiness results can be within the intervals in Fig. 2 interpreted as follows: $0 \leq x \leq a_I$ low; $a_I \leq x \leq b_I$ low to medium; $b_I \leq x \leq c_I$ medium; $c_I \leq x \leq d_I$ medium to high; $x \geq d_I$ high. From it follows that a high grade of creditworthiness belongs to Letiště Praha, a. s., the middle grade to the other two companies.

As mentioned in the last paragraph of section 3.1, the second step is to utilize the bond issuer creditworthiness for inclusion of the company in the appropriate credit risk category. There is a direct relationship between bond issuer creditworthiness and credit risk – creditworthiness is the reciprocal value of credit risk (the overall results are shown in Table 5).

As a result of this relationship the set of basic values of variables breaks down into credit risk classes analogously in the previous sense: $\frac{1}{100} \leq x < \frac{1}{d_I}$ low, $\frac{1}{d_I} \leq x < \frac{1}{c_I}$ low to medium, $\frac{1}{c_I} \leq x < \frac{1}{b_I}$ medium, $\frac{1}{b_I} \leq x < \frac{1}{a}$ medium to high, $\frac{1}{a_I} \leq x < \infty$ high.

Within these intervals Letiště Praha, a.s. shows low credit risk value, the remaining companies medium credit risk value.

Table 5. Input parameters of the fuzzy model and resulting output values of creditworthiness and credit risk

Company/Index	Rating assessment (R)	Production power (ROA)	Equity / Asset ratio (VA)	Bond issuer creditworthiness	Credit risk
ČEZ, a.s. (X)	65 %	8.6 %	52 %	50 % Medium	2.00 % Medium
Letiště Praha, a.s. (Y)	80 %	58 %	100 %	95 % High	1.05 % Low
Česká telekomunikační infrastruktura a.s. (Z)	50 %	9.4 %	58 %	55 % Medium	1.81 % Medium

Source: Own processing.

The resulting values of creditworthiness, or respectively credit risk could be corrected upwards or respectively downwards due to the fact that a major shareholder of ČEZ, a.s. company is the state (the 70 % share); similarly, Letiště Praha, a.s. is owned by Český Aeroholding, a.s. company whose only shareholder is the Ministry of Finance of the Czech Republic. This fact, in the eyes of investors, can be a guarantee for the repayment of creditors´ claims. In the case of Česká telekomunikační infrastruktura, a.s. company the 100 % owner is the PPF investment group, which has a stable position on the Czech financial market. This also contributes to the growth of subjective credibility in creditors´ mind.

5 Conclusion

In view of uncertain input data of vague nature entering the standard credit risk rating models, the fuzzy approach has been introduced to handle this uncertainty. In practice, credit risk is often assessed on the basis of reduced models leaning on rating assessments. In addition to selected financial indicators and other information whose exact set is not known, credit rating agencies base their ratings to a certain extent on subjective factors and variables that are not directly related to a particular creditor firm.

The problem of credit risk is here understood as an analysis of a vaguely defined system whose vague concepts are presented by linguistic variables and their linguistic values, the so-called terms. Desirable relationships between combinations of the terms of the input linguistic together with numerical variables, and the terms of output linguistic and numerical variable are specified by a set of rules. The mathematical tool of this type of a task is the fuzzy logic whose individual phases are shown in the diagram in Fig. 1.

The credit risk is closely related to the company's creditworthiness, which expresses the credibility of the economic entity to repay its debts; the reciprocal value of creditworthiness is credit risk. Input parameters for creditworthiness and credit risk estimate of bond issuers ČEZ, a.s., Letiště Praha, a.s. and Česká telekomunikační infrastruktura, a.s. are the index values of the sub-components of the „quality" of the bond from the viewpoint of the creditors; they are the rating agencies assessments and relevant financial indicators.

The resulting values of the creditworthiness and credit risk of bond issuers qualitatively and quantitatively improve the standard ratings of rating agencies. The fuzzy approach can be considered as an appropriate alternative not only to reduced models but also to deterministic and stochastic methods.

References

1. S. M. Schaefer, I. A. Strebulaev, Structural models of credit risk are useful : Evidence from hedge ratios on corporate bonds. *Journal of Financial Economics,* **90**(1), 1-19, (2008)
2. E. J. Elton, M. J. Gruber, D. Agrawal, C. Mann, Factors affecting the valuation of corporate bonds. *Journal of Banking & Finance,* **28**(11), 2747-2767, (2004)
3. M. Vochozka, V. Machová, Enterprise Value Generators in the Building Industry. *SHS Web of Conferences, EDP Sciences – Innovative Economic Symposium 2017*, **39**, (2017)
4. V. Stehel, M. Vochozka, The analysis of the Economical Value Added in transport. *Naše more: znanstveno-stručni časopis za more i pomorstvo,* **63**(3 Special Issue), 185-188, (2016)
5. K. Ayotte, Bankruptcy and enterpreneurship : The value of a fresh start. *The Journal of Law, Economics, & Organization,* **23**(1), 161-185, (2006)
6. G. E. Pinches, K. A. Mingo, A multivariate analysiss of industrial bond ratings. *The Journal of Finance,* **28**(1), 1-18, (1973)
7. C. Carlsson, R. Fullér, A fuzzy approach to real option valuation. *Fuzzy sets and systems,* **139**(2), 297-312, (2003)
8. M. Collan, R. Fullér, J. Mezei, A fuzzy pay-off method for real option valuation. *Advances in Decision Sciences,* **2009,** 1-16, (2009)
9. Z. Zmeškal, Application of the fuzzy-stochastic methodology to appraising the firm value as a European call option. *European Journal of Operational Research,* **135**(2), 303-310, (2001)
10. J. S. Yao, M. S. Chen, H. W. Lin, Valuation by using a fuzzy discounted cash flow model. *Expert Systems with Applications,* **28**(2), 209-222, (2005)
11. S. H. Ho, S. H. Liao, A fuzzy real option approach for investment project valuation. *Expert Systems with Applications,* **38**(12), 15296-15302, (2011)
12. W. Ho, X. Xu, P. K. Dey, Muti-criteria decision making approaches for supplier evaluation and selection : A literature review. *European Journal of operational research,* **202**(1), 16-24, (2010)
13. S. D. Pohekar, M. Ramachandran, Application of multi-criteria decision making to sustainable energy planning – A review. *Renewable and Sustainable Energy Reviews,* **8**(4), 365-381, (2004)
14. M. Grabisch, The application of fuzzy integrals in multicriteria desicion making. *European Journal of Operational Research,* **89**(3), 445-456, (1996)
15. S. Hašková, P. Volf, The contribution of foreign direct investments to the convergence of regions in the Czech Republic. *Littera Scripta,* **2**, 23-33, (2017)
16. A.Saunders, L. Allen, *Credit risk management in and out of the financial crisis: new approaches to value at risk and other paradigms.* **528,** John Wiley & Sons, New Jersey, (2010)

Factors Affecting Innovative Activities of SMEs in Slovak Regions

Katarina Haviernikouá[1,*] and *Marcel* Kordoš[1]

[1]Alexander Dubček University in Trenčín, Študentská 2, 911 50 Trenčín, Slovak Republic

Abstract. Innovations and innovation policies and strategies implementations are one of the most important aspects of current regional economics and business development issues. It is the key not only to creating more jobs, building a greener society and improving quality of life, but also to maintaining companies' competitiveness on regional markets and states'/economies' competitiveness enhancement within the international economics system. The goal of this paper is oriented on Slovak SMEs that carrying out their activities in eight administration regions and have knowledge or experience with cluster cooperation. Questionnaire research was executed in 8 Slovak regions where respondents were owners or managers of SMEs. For test of homogeneity of variance (homoscedasticity) the Levene's test was used. Innovations as individual stakeholders as well as whole cluster have an impact on clusters' growing development and success and ultimately contribute to the development of whole region.
Key words: SMEs, innovative activities, technology

1 Introduction and Theoretical Background

Small and medium-sized enterprises (SMEs) are considered to be the most important elements of national economies [1]. Cluster includes manufacturing companies, universities, scientific workplaces, organizations in the economy. For small and medium-sized businesses cluster offer a wide variety of benefits and make an environment be suitable for innovation and knowledge creation [2]. For this reason, regions with strong clusters and SMEs are considered to be innovation leaders. Globalization further deepens on these trends. Despite this fact, the level of awareness about benefits of cluster cooperation is still low. Some organizations, such as SMEs are not able to cope with difficulties or, on the contrary, or to use the benefits globalization brings to world markets. The process of globalization requires the adoption of highly flexible innovative solutions, which, in particular, are often not able to be implemented by SMEs [3, 4].

Innovation provides real benefits for us as citizens, consumers, and workers. It speeds up and improves the way we conceive, develop, produce and access new products, industrial processes and services [5]. It is the key not only to creating more jobs, building a greener society and improving our quality of life, but also to maintaining our competitiveness on global market [6]. Innovation policy is about helping companies to

* Corresponding author: katarina.haviernikova@tnuni.sk

perform better and contributing to wider social objectives such as growth, jobs and sustainability [7].

Technology offers opportunities to business organizations to increase their profits and growth through the introduction of new and improved goods and services and through changes to their production processes. Technology also helps firms to restructure their global and regional patterns of production through investment in low cost locations or by sub-contracting to cheaper suppliers [8]. However, technology can also pose threats and challenges for firms particularly if they allow themselves to fall behind their competitors. Technological advance, because it involves change in products or production processes, is a risky business particularly for firms being not able to manage change very well [9, 27].

The rapid internationalization of technology means that firms need to monitor both their domestic and their foreign technological environments. For many industries, technology is of the utmost importance and can determine whether firms prosper or fall by the wayside [10]. That also applies not only to states but also to particular regions. Only economy with strong economic regions as an active member of international division of labor, producing high-tech products based on innovations is able to achieve sustainable economic growth [11, 28]. Failing to do so, economy will be caught in the world economy periphery having no sources to secure economic development for sufficient quality of life. Thus issues such as innovation and technology in international and regional business environment, technological change and innovation by means of clusters and spillovers as the globalization engines and the role of innovations in regional economy are the crucial issues to achieve regional competitiveness enhancement [12, 13].

Technology has become internationalized. Thus, transnational companies take opportunities offered by the international environment to develop global research strategies, carrying out research and development both at home and abroad. Technology is a double-edged sword for business offering many opportunities but also challenges [14]. On one hand, it opens up a variety of opportunities for business in terms of new products, processes, and markets [15]. On the other hand, it lets firms more open to a range of competitive threats such as takeover, increased competition and even to the theft of their technologies.

Joseph Schumpeter set technological change, specifically innovation – the creation and diffusion of new ways of doing things – at the very nears of the processes of economic growth and development [16]. Technological change is a fundamental force in shaping the patterns of transformation of the economy [17]. Therefore, it is one of the most important processes underlying the globalization of economic activity. Technological change, therefore, is a socially and institutionally embedded process. The ways in which technologies are used – even their very creation – are conditioned by their social and their economic context [18, 29]. Nowadays this means primarily the values and motivations of capitalist business enterprises, operating within an intensely competitive system. Choices and uses of technologies are influenced by the drive for profit, capital accumulation and investment, increased market share and so on. Secondly, technology should be seen as, essentially, an enabling or a facilitating agent - asserting new structures, new organizational and geographical arrangements of economic activities, new products and new processes [19, 20, 21]. On the other hand, in a highly competitive environment, once a particular technology is in use by one firm, then its adoption by others may become virtually essential to ensure competitive survival.

The idea underlying the notion of national innovation systems is that the specific combination of social, cultural, political, legal, educational and economic institutions and practices varies systematically among national contexts [22, 30]. Such nationally differentiated characteristics help to influence the kind of technology system that develops there together with its subsequent development trajectory.

National and regional systems of innovation, in fact, consist of aggregations of localized knowledge clusters, sometimes termed technology districts or technopoles [23, 30]. Many of them are associated with major metropolitan areas, although some have developed outside the metropolitan sphere in rather less urbanized areas. Most of them are the outcome of historical process of cumulative, path-dependent growth processes although some are the deliberate creations of national technology policy. These technological agglomerations form one of the most significant features of current global economy [24, 31].

Technological change within the innovation processes are the main factors of current global economy development [25, 32]. In this paper we focus only on those aspects of technological change that specifically influence the economic activities of SMEs in Slovak regions. We will try to identify some of those features of technological change that are most important in the regionalization of economic activity. Technological change is the dynamic core of economic growth and development; it is fundamental to the evolution of a global economic system. We focus on the geography of innovation, on the different scales – national and local – at which innovation processes operate.

2 Goal and Research Methodology

The focus of this paper is oriented on Slovak SMEs that carry out their activities in eight administration regions and have knowledge or experience with cluster cooperation. Within the frame of the project VEGA 1/0953/16 "The evaluation of clusters´ impact measurement on regional development of the Slovak Republic", several questionnaire surveys were realized in years 2016-2017. One part of them also covered the issues of SMEs' innovative activities. Innovation within individual stakeholders as well as whole cluster have an impact on clusters' growing development and success and ultimately contribute to the development of whole region. Into research that has connection with innovative activities were involved 320 SMEs of which 40 enterprises are from each region (BA – Bratislava, TT – Trnava, TN – Trenčín, ZA – Žilina, NR – Nitra, BB – Banská Bystrica, PO – Prešov, KE – Košice). Respondents were owners or managers of SMEs. SMEs operate in one of the following economic sectors: Industry, Agriculture, Energy, Tourism. Slovak clusters also operate in these economic sectors. Table 1 presents the main characteristics of respondents. The biggest category size of respondents in each region except Žilina region was presented by category of medium sized enterprises (50-249 employees). The most respondents according to economic sectors belonged to economic sector of Industry (47.19%), the least were in the category of Energy (5.00%). Within the category of medium sized enterprises (50 – 249 employees), the most respondents also belong to SMEs that carry out their activities in Industry sector (29.69%), at least in sector of Energy (1.25%). Within the category of small enterprises (10-49 employees), it was the same situation (Industry – 11.25%, Energy – 1.25%). The different composition of respondents by economic sector we can see in the last category – micro enterprises (with 0-9 employees), where the most respondents belonged to sector of Agriculture (10.94%) and the least, again to sector of Energy (2.50%).

Table 1. Main characteristics of the respondents (%)

Region	Number of employees			Total	Economic sector	Number of employees			Total
	0-9	10-49	50-249			0-9	10-49	50-249	
BA	3.75	3.44	5.31	12.50	**Industry**	6.25	11.25	29.69	47.19
TT	4.06	3.13	5.31	12.50	**Agriculture**	10.94	10.00	7.50	28.44
TN	3.75	3.44	5.31	12.50	**Energy**	2.50	1.25	1.25	5.00
ZA	0.63	7.81	4.06	12.50	**Tourism**	6.25	8.44	4.69	19.38
BB	4.06	3.44	5.00	12.50					
NR	3.75	3.13	5.63	12.50					
PO	3.75	1.88	6.88	12.50	**Total**	**25.94**	**30.94**	**43.13**	**-**
KE	2.19	4.69	5.63	12.50					
Total	**25.94**	**30.94**	**43.13**	-					

Source: Own research.

To meet the main goal of this paper two working hypotheses have been stated:

WH1 The assessment of factors that have impact on innovative activities of SMEs depends on size category of SMEs.

WH2 The assessment of reasons that lead SMEs toward cluster cooperation in the context of innovation activities depends on size category of SMEs.

To each working hypothesis, we set null and alternative hypothesis due to the statistical method being used.

For working hypothesis WH1 the set of eight factors that can have impact on innovative activities of SMEs were set: F1. Taxation, laws, regulations. F2. Low own innovation potential. F3. Customers' requirements. F4. Structure of the company. F5. Support from enterprise management. F6. Collaboration with suppliers. F7. Functioning work teams. F8. Cooperation in research and development, production and marketing.

Respondents could rate each factor by points from value 0 to 5, while value 0 means that particular factor most embarrass the innovative activities of SMEs and value of 5 means that particular factor is most conducive to innovation activities of SMEs.

Using the analysis of variance for particular rated factors can be determined by whether their perception in the size categories is identical. For evaluation we used statistical methods (Levene's test, Lilliefors test and non-parametric multi-selective Kruskal-Wallis Test), the calculations were realized in statistical software Statistica [26].

For working hypotheses WH2 respondent could choose one from the six reasons that could mainly lead them towards cluster cooperation from the point of view of innovation. Particular reasons are: R1. Strategy of common business activities, R2. Management of common business activities, R3. Risk management, R4. Cost sharing, R5. New business impulses, R6. Optimizing the innovative idea. This W2 was evaluated by using Chi-square test.

3 Results and discussion

The necessary information for the analysis of variance in case of WH1 is shown in Table 2. The basic statistical characteristics are: the respondent evaluation average, the respondents' evaluation range measured by standard deviation.

Table 2. The statistical characteristics of the evaluated factors

Factor	μ	σ2	SD
F1	1.80	0.94	0.97
F2	2.48	1.35	1.16
F3	3.70	1.67	1.29
F4	3.08	1.15	1.07
F5	3.53	1.77	1.33
F6	3.70	1.40	1.18
F7	3.89	1.06	1.03
F8	3.53	1.47	1.21

Source: Own research.

The respondents in terms of size category of enterprise perceived the factor F7 being the most important factor that have influence on innovative activities of SMEs - Functioning work teams (3.89±1.03 points). The less important is the factor F1. Taxation, laws, regulations (1,80±0,97 points).

The percentage of respondents rating in total represents Table 3. As we can see, the highest rated value of 5 was obtained from respondents by two factors: F3 Customers' requirements (28.40%) and F7 Functioning work teams (28.40%). The least of respondents' responds to this value obtained the factor F1 Taxation, laws, regulations (2.47%). It means that this factor is not so important for SMEs in case of obtaining benefits from cluster cooperation towards innovation activities.

Table 3. Respondents' rating of given factors (%)

Factor	0	1	2	3	4	5
F1	2.47	44.44	29.63	19.75	1.23	2.47
F2	3.70	19.75	20.99	40.74	9.88	4.94
F3	3.70	6.17	4.94	16.05	40.74	28.40
F4	3.70	3.70	14.81	45.68	24.69	7.41
F5	3.70	6.17	9.88	22.22	30.86	27.16
F6	2.47	2.47	11.11	19.75	37.04	27.16
F7	1.23	2.47	4.94	17.28	45.68	28.40
F8	2.47	3.70	9.88	30.86	28.40	24.69

Source: Own research.

For analysis of the variance, the parametric or nonparametric tests could be used. Parametric tests should be used when two following conditions are met: homogeneity of variance and normality.

For test of homogeneity of variance (homoscedasticity) the Levene's test was used, by which authors tested the null hypothesis that the population variances are equal. If the resulting p-value of Levene's test is less than significance level (0.05), the null hypothesis of equal variances is rejected and we can conclude that there is a difference among the variances in the population. Results of this test are in Table 4.

Table 4. Levene's test of homoscedasticity

Factor	F	p
F1	3.78279	0.023793
F2	1.80602	0.165993
F3	10.95737	0.000025
F4	0.04446	0.956523
F5	2.98917	0.051750
F6	6.45246	0.001792
F7	0.08403	0.919421
F8	28.89852	0.000000

Source: Own calculation, Significance level of $p < 0.05000$.

The results showed that null hypothesis about homoscedasticity is confirmed only for factors: F2. Low own innovation potential, F4. Structure of the company and F7. Functioning work teams.

To verify the normality of rated factors we used the Lilierfors' test. Results presents Table5 We test null hypothesis that the results of rated factors came from normal break-up values, against alternative hypothesis that results of rated factors came from basic set with different distribution. The results in Table 5, showed that at the significance level of 0.05 we rejected the null hypothesis of normality for each factor.

Table 5. Lilliefors test of normality

Factor	Lilliefors	S-W	p
F1	$p < 0.01$	0.832198	0.000000
F2	$p < 0.01$	0.914241	0.000000
F3	$p < 0.01$	0.819456	0.000000
F4	$p < 0.01$	0.884363	0.000000
F5	$p < 0.01$	0.873099	0.000000
F6	$p < 0.01$	0.860933	0.000000
F7	$p < 0.01$	0.817839	0.000000
F8	$p < 0.01$	0.886303	0.000000

Source: Own research.

The results of Levene's and Lilliefors test are prerequisites to use a non-parametric Kruskal – Wallis test of variance. The results of variance analysis of particular rated factors done with this test are summarized in Table 6.

Table 6. The results of Kruskal-Wallis Test

Factor	Kruskal-Wallis Test
F1	*H (2, N= 320) =10.81229 p =0.0045*
F2	*H (2, N= 320) =37.44249 p =0.0000*
F3	H (2, N= 320) =4.827061 p =0.0895
F4	H (2, N= 320) =4.595731 p =0.1005
F5	H (2, N= 320) =2.776082 p =0.2496
F6	H (2, N= 320) =3.592882 p =0.1659
F7	H (2, N= 320) =2.366810 p =0.3062
F8	*H (2, N= 320) =6.573793 p =0.0374*

Source: Own research.

The grouping variable was the size category of SMEs. The authors tested the null hypothesis: The difference in mean values of rated factors among the size categories of SMEs is not statistically significant. The calculated p-value of variance intensity analysis of rated factors is higher than 0.05 for factors: F3. Customers' requirements, F4. Structure of the company. F5. Support from enterprise management, F6. Collaboration with suppliers. F7. Functioning work teams. The null hypothesis is not rejected. For the rest of factors, the null hypothesis is rejected. It means that there are differences in respondents rating of particular factors. This has been confirmed by results of the multiple comparison performed by using the p-level. We can conclude, that at significance level of 0.05 the responses of respondents vary among the individual size categories of SMEs. The level of p-values was in each cases lower than 0.05.

At the end we can conclude, that working hypothesis was confirmed only in several cases of rated factors: F1. Taxation, laws, regulations, F2. Low own innovation potential and F8. Cooperation in research and development, production and marketing. Within evaluation of this factors, the size category of SMEs plays the significant role.

For working hypothesis WH2, the basic information about respondents' decision is summarized in Table 7.

Table 7. Percentage statements of the respondents' answers (%)

Reason	Number of employees			Total
	0-9	10-49	50-249	
R1	10.63	10.94	19.69	**41.25**
R2	6.56	3.44	5.94	**15.94**
R3	0.31	2.50	1.56	**4.38**
R4	0.94	1.25	1.56	**3.75**
R5	5.00	10.00	10.00	**25.00**
R6	2.50	2.81	4.38	**9.69**
Total	**25.94**	**30.94**	**43.13**	**100.00**

Source: Own research.

The reason R1 has been marked by the most respondents (41.25%). Strategy of common business activities as the most important from the point of view of innovation activities of SMEs in case of cluster cooperation. Reason R4. Cost sharing obtained only 3.75% of respondents´ votes.

If we evaluate the choice of reasons by the size category of enterprises, we can see a similar situation as was in the overall assessment, except in the category of micro enterprises, where the least respondents´ votes was obtained by reason R3. Risk management.
To verify the working hypothesis WH2 we used statistical method Chi square test. We stated null hypothesis: Within the size categories of SMEs there is not difference in choosing the most important reason. The alternative hypothesis means opposite. The low level of p value <0,05 means that the H0 is rejected and we accept the alternative hypothesis H1. The result of Chi-square test (Pearson chi-sq: 16. 5885, df=10, p=0.083999) showed, that null hypothesis is not rejected. It means that evaluation of particular reasons doesn't depend on size category of SMEs. As the most important reasons for cluster cooperation in case of innovative activities are R1. Strategy of common business activities (41.25%), R5. New business impulses (25.00%) and R2. Management of common business activities (15,94%).

4 Conclusion

The innovation capacity of SMEs is currently considered to be one of the conditions for the competitiveness of enterprises. In a changing business environment the innovation represent the tool for their managing. This paper presents the perception of selected factors that can have impact on innovation activities of SMEs: F1. Taxation, laws, regulations. F2. Low own innovation potential. F3. Customers' requirements. F4. Structure of the company. F5. Support from enterprise management. F6. Collaboration with suppliers. F7. Functioning work teams. F8. Cooperation in research and development, production and marketing. The results showed the differences in perception of these factors by SMEs. As the most important factor that has impact on innovation activities of SMEs was marked the F7. Functioning work teams. This factor was evaluated by 147 respondents by value 4 within the used scale. The least important factor according to SMEs is F1. Taxation, laws, regulations. This factor was marked in 143 cases by value 1. The statistical significant differences in evaluation of SMEs for these two factors were also confirmed by using the Kruskal –Wallis test.

The innovation potential of SMEs is its internal characteristic and it is one of the basic elements for the development of this type of business entities. One way how to provide the transfer of innovation into SMEs is their connection into cluster cooperation, where subjects from main three areas are met: research and education, public administration, business environment. There are many reasons that lead the SMEs towards connection into cluster cooperation from the point of view of innovation. This paper provides evaluation results of six particular reasons by SMEs: R1. Strategy of common business activities, R2. Management of common business activities, R3. Risk management, R4. Cost sharing, R5. New business impulses, R6. Optimizing the innovative idea. As the most important reason was marked by 41,25% of respondents R1. Strategy of common business activities. By using Chi –square statistics we verified if there are differences in evaluation of selected reason in dependence of size category of business entity. Results showed, that evaluation of given reasons doesn't depend on the size category of SMEs.

The needs of innovation for the business environment of SMEs are the essence for their survival. The innovation activities contribute to their improving and gaining the competitive advantage. SMEs need to become aware that their interaction with the other regional subjects that represent the holder of innovation in form of cluster cooperation, gives them just what could move them forward and solve their everyday problems.

Acknowledgements

This research was financially supported by the Slovak Ministry of Education's Scientific grant agency VEGA: "The evaluation of clusters´ impact measurement on regional development of the Slovak Republic". Project registration number: [Reg. No.: 1/0953/16].

References

1. E. Krajňáková, A. Navikaite, V. Navickas, Paradigm Shift of Small and Medium-Sized Enterprises Competitive Advantage to Management of Customer Satisfaction. *Inzinerine Ekonomika-Engineering Economics*, **26**(3), 327-332, (2015)
2. S. Brakman, H. Garretsen, Ch. Van Marrewijk, A. Van Witteloostuijn, *Nations and firms in the global economy - an introduction to international economics and business.* Cambridge University Press, (2006)
3. A. Malakauskaite, V. Navickas, Contribution of Clusters to the Competitiveness of Companies: Revelation and Evaluation. *Inzinerine Ekonomika-Engineering Economics*, **22**(1), 50-57, (2011)
4. S. Vojtovic, The Impact of the Structural Funds on Competitiveness of Small and Medium-Sized Enterprises. *Journal of Competitiveness*, **8**(4), 30-45, (2016)
5. L. Hamilton, P. Wepster, *The International Business Environment.* OUP Oxford, (2009)
6. P. Balaz, A. Hamara, Export Dependency of Slovakia on German's Economy. *Politicka Ekonomie*, **64**(5), 573-590, (2016)
7. T. De Castro, P. Hnát, Czech FDI Performance: Between Global Value Chains and Domestic Reforms. *Foreign Direct Investment in Central and Eastern Europe: Post-Crisis Perspectives,* 51-75, (2017)
8. P. Burda, J. Abrham, Z. Horváthová, Factors Influencing Online Civic Participation in Mid-Sized Czech Towns. *Transformations in Business & Economics*, **16** (2B), 607-618, (2017)
9. P. Drulák, R. Druláková, The richness of the liberal tradition in international relations: Karl Deutsch on political community and the European integration. *International Relations*, **28**(3), 333-349, (2014)
10. L. Fojtíková, Trends in the revealed comparative advantages of the EU member states. *Economic Annals-XXI*, **161**, 9-10, 7-11, (2016)
11. E. Ivanová, J. Masárová, Performance evaluation of the Visegrad Group countries. *Economic Research-Ekonomska Istrazivanja*, **31**(1), 270-289, (2018)
12. L. Lipková, D. Braga, Measuring Commercialization Success of Innovations in the EU. *Marketing and Management of Innovations*, **4**, 15-30, (2016)
13. M. Okręglicka, M. Mynarzová, R. Kaňa, Business process maturity in small and medium-sized enterprises [Dojrzałość procesów biznesowych w małych i średnich przedsiębiorstwach]". *Polish Journal of Management Studies*, **12**(1), 121-131, (2015)
14. S. Vojtovic, V. Navickas, V. Gruzauskas, Sustainable Business Development Process: The Case of the Food and Beverage Industry. *Advancing Research in Entrepreneurship in the Global Context,* **68**, 225-239, (2016)
15. H. Machková, A. Sato, Analysis of Competitiveness of Belgian Sugar Industry. *Listy Cukrovarnicke a Reparske*, **133**(12), 390-392, (2017)
16. J. H. Dunning, S. M. Lundan, *Multinational Enterprises and the Global Economy.* Edward Elgar Publishing, (2008)

17. J. Mruškovičová, F. Dano, Optimizing the Number of Stops within the Logistic Routes Using the Mathematical Model. *AD ALTA-Journal of Interdisciplinary Research*, **7**(1), 113-115, (2017)
18. J. Taušer, M. Arltová, P. Žamberský, Czech exports and german GDP: A closer look. *Prague Economic Papers*, **24**(1), 17-37, (2015)
19. L. Mura, M. Orlíková, Social Entrepreneurship and Social Enterprises: The Case of Slovakia. *Innovation Management Entrepreneurship and Corporate Sustainability*, 495-505, (2016)
20. K. A. Reinert, *An Introduction to International Economics*. Cambridge University Press, (2012)
21. S. M. Obadi, M. Korček, Competitive Advantage of International Trade of the EU-28 and USA. *Proceedings of 8th Annual International Scientific Conference: Competition*, (2016)
22. M. Okręglicka, A. Lemańska-Majdzik, Cooperative with Competitive Environment as a Determinant of Competitive Position of Microenterprises in Poland. *International Scientific Conference Knowledge for Market Use 2017: People in Economics - Decisions, Behavior and Normative Models,* (2017)
23. T. Sadílek, D. Zadražilová, Current Trends in German Sugar Industry. *Listy Cukrovarnicke a Reparske*, **132**(12), 390-393, (2016)
24. J. Taušer, R. Čajka, Hedging techniques in commodity risk management. *Agricultural Economics (Czech Republic)*, **60**(4), 174-182, (2014)
25. F. Varadzin, Global Public Goods and Integration. *Proceedings of the 3rd International Conference on European Integration,* 1052-1059, (2016)
26. J. Betáková, R Zeman., J. Dvorský, T. Pavlenko, Analysis Of Selected Risks Arising From Human Activities. *WIT Transactions on Ecology and the Environment,* **199**, 71-81, (2015)
27. J. Sejkora, Poor Economics: A Radical Rethinking of the Way to Fight Global Poverty. *Ekonomicky Casopis*, **62**(4), (2014)
28. S. Zemanová, R. Druláková, Making Global Goals Local Business in V4 Countries: V4 Entrepreneurs And The UN Global Compact. *Globalization and its Socio-Economic Consequences,* 2490-2497, (2016)
29. E. Haňuláková, F. Daňo, Circular Economy as a New Managerial Approach. *AD ALTA-Journal of Interdisciplinary Research*, **8**(1), 95-98, (2018)
30. S. Žák, M. Kuchta, A. Miklošík, Monetising Content through Delivery of Advertisements: The Case of Ad Blockers. *AD ALTA-Journal of Interdisciplinary Research*, **8**(1), 175-179, (2018)
31. J. Sejkora, Comparative advantage, economic structure and growth: The case of Senegal. *South African Journal of Economic and Management Sciences*, **20**(1), 1-9, (2017)
32. K. Krajčo, Analysis of foreign direct investment in the industry of the Slovak Republic. *Proceedings of the 2nd international conference on European integration 2014,* (2014)

Using artificial intelligence to analyse businesses in agriculture industry

Jakub Horák[1,*]

[1]University of Žilina, Faculty of Operation and Economics of Transport and Communications, Univerzitná 1, 01026 Žilina, Slovak Republic

Abstract. Artificial intelligence is largely used in many technical applications and allows you to provide various solutions in problem estimation, regression, or optimization. Artificial intelligence, specifically artificial neural networks, extend to the area of economics and finance. They are used primarily for operations that can´t be identified analytically. Neural networks are suitable for modelling very complex strategic decisions, for large sets of data, and so on. The main advantage is the ability to learn and then to capture hidden and strongly non-linear dependencies. In this paper they are used for the analysis of agricultural businesses. The aim is to analyse the state of the agricultural sector through the use of Kohonen networks and then to assess its future development. On the basis of the analysis, significant and large clusters of businesses are depicted, and the most significant clusters are analysed. It is possible to estimate the number of businesses that will be successful, those that will stagnate and those that will fail in the following period. Application of Kohonen networks is rather complex, but they have great potential and the results are very interesting.
Key words: Kohonen networks, cluster analysis, prediction, agricultural businesses, sector development

1 Introduction

Artificial neural networks (ANNs) try to copy processes in the human brain and the nervous system through computer devices. The concept of an artificial neural network first appeared in biology and psychology. According to Kliešstik [1], ANNs are computational models that are inspired by biological neural networks, namely neuronal behavior. The use of ANNs, however, crosses over to other areas as well. Currently, these networks are widely used to address possible future problems, to predict values [2]. Altun, Bilgil and Fidan [3] claim that ANNs are extensively used in many technical applications and provide various solutions in problem estimation, regression, or optimization. ANNs are used for various purposes and it is appropriate to use them for demanding operations that can´t be identified analytically. Guresen and Kayakutl [4] state that they are therefore mainly used to model very complex strategic decisions. According to Sánchez and Melina [5], ANNs can be applied to function approximations, classification, and prediction of time series. They help us understand functions that exist in model groups and environments that are constantly changing. ANNs

* Corresponding author: horak@mail.vstecb.cz

are able to offer solutions to complex models of improved artificial intelligence. Beiranvand et al. [6] maintain that the results of ANNs are very promising, and their performance and accuracy when solving a company's key performance indicators are considerably higher than those of traditional statistical techniques.

The advantage of ANNs is the ability to work with large data, accuracy of results, or simpler use of the acquired neural network [7]. ANNs are a group of intelligent data analysis technologies that differ from other classical techniques [8]. According to Santin [9], ANNs have many advantages over conventional methods. They are able to analyse complex patterns very quickly and with high precision and are flexible in their own use. Sayadi et al. [10] contend that some of the main advantages of neural network methods for predicting key company indicators are the ability to learn, the ability to generalize, etc. The disadvantage of ANNs is their demand for large sample data, because to create such an amount of data, many test trials are needed, which is not ideal for the user. Another disadvantage is the process of optimizing topology of hidden layers, which is time consuming, complicating the computation process [11]. Rowland and Vrbka [12] consider ANNs to be particularly disadvantageous in that they require high data quality and architecture definition, the possibility of illogical network behavior.

One of the ANN models is the Kohonen Network, which can be used to group sets of data into different groups. The data is grouped so that the records within the group tend to be similar to each other, and the records in the different groups are different. Many experimental results show that Kohonen's networks are very effective for assessing companies [13]. The Kohonen network consists of an input layer that is completely interconnected with the output layer and learns by self-organization. This network has a wide range of uses because it is an alternative network that is usable for most of the ANNs calculations. It is used primarily for speech processing, audio editing, security applications, photos, videos, and allows the projection of high-dimensional data into lower-dimension data [14]. Many experts even claim that Kohonen's maps are a highly versatile tool, and if one does not know what method to use, he can use the Kohonen network with a quiet conscience. There is no need for the author to know about the information he processes and the network will categorize it by itself [15].

The Kohonen networks are used in this paper to analyse agricultural businesses. Like every branch of the national economy, agriculture also has its own specificities. These consist mainly in satisfying basic human needs. Agriculture is a source of food of both animal and plant origin, which is necessary for human nutrition. It produces many raw materials for other industries. According to Waheed et al. [16], agriculture is primarily characterized by its maximum dependence on natural conditions, weather. Agricultural land is therefore the main production means for agricultural production. Agricultural production systems use sophisticated techniques to correlate human, natural, industrial and economic resources. This is done to meet the demand for food in today's highly competitive and demanding market in terms of environmental and social sustainability [17].

The aim of this article is to carry out a cluster analysis of enterprises operating in the area of agriculture.

2 Data and methods

A data set of 4213 enterprises having operated in the area of agriculture in 2016 is made up for purposes of this article. The data set of an enterprise is generated from Albertina database. Comprehensive financial statements of these companies (all financial data are provided in thousands CZK) are available. The analysis deals only with some of the items:

1. Total assets: the size of the enterprise, volume of assets (fixed, current and other assets) in particular

2. Fixed assets: this item refers to the degree of the participation of fixed assets in an enterprise and informs about the technology used in the production. The higher the participation of fixed assets is, the higher the supposed automated production should be.
3. Current assets: this item refers to the amount of financial resources, claims and inventories of the enterprise. They mostly concern working capital, i.e. assets that change their nature within a period shorter than one year and that immediately participate in main activities of the enterprise.
4. Equity: they refer to the degree of possible business risk that owners of enterprises take; they mostly concern the amount of equity that may be lost in case of insolvency.
5. Borrowed capital: they refer to the capital that is not associated with the participation in the commercial management of the company. They provide an external view of a potential success of the enterprise.
6. Performance: they refer to main activities of the enterprise – production.
7. Added value: it refers to a value added to the capital investment generated by the enterprise through its main activity.
8. Operating profit: it informs about the success rate of the enterprise in its main activity.
9. ROE: income of equity means the evaluation of the equity of the enterprise, i.e. the capital provided to the entrepreneur.
10. Economic result before taxation: it refers to a success or failure of the enterprise in its entire activity.

In this way, these items rather refer to characteristics, not value generators, which may have, though indirect, influence on the enterprise efficiency.

The data set is subsequently subjected to a cluster analysis, Kohonen networks in particular. The cluster analysis uses Statistica software Dell, version 12. Module Data Mining is applied as a tool for neural networks. Here, neural networks without teacher forcing are chosen – Kohonen networks. The analysis data are amassed; they all concern continuous predictors. The data set is divided into three parts:

1. Training data set: it involves 70% of enterprises from the data set. This data set is demonstrated by Kohonen network.
2. Testing data set: it involves 15% of enterprises from the data set. This data set helps verify parameters achieved from Kohonen network.
3. Validation data set: it involves 15% of enterprises from the data set. This data set also helps verify achieved Kohonen network – its applicability in particular.

The topological length and topological width of Kohonen map is determined to 10. The number of repetitions – iterations – is determined to 10,000. All the same, we should take into account that the error level is the most decisive factor. Unless the parameters of Kohonen network improve with individual iterations, the training will be terminated before 10,000th iteration. In case of a tangible improvement of parameters in 10,000th iteration, the whole procedure has to be repeated and a higher value of required iterations has to be generated in order to achieve the best possible result.

3 Results

According to the methodological part of the article, the data were divided into three data sets – training, testing and validation. Defining characteristics of all three data sets and the original data set are suggested in Tab. 1.

Table 1. Descriptive characteristics of the dataset

Samples	Total assets	Fixed assets	Current assets	Equity	Borrowed capital	Perform.	Added value	Operating profit	ROE	Economic result before taxation
Minimum (Training)	0	-325	-731	-517381	-3629	-3744	-21623	-33385	-4764.50	-27561
Maximum (Training)	703378	62152700	8122509	67905791	2295046	11844244	7831937	6356727	14086.50	6493751
Average (Training)	78961	54510	23916	55566	23176	31617	9518	4933	10.87	4709
Standard deviation (Training)	1302210	1148973	156285	1255363	68527	233653	145402	117554	321.28	120263
Minimum (Testing)	-2009	-1772	-2009	-24105	-8980	0	-9606	-19093	-1421.78	-44599
Maximum (Testing)	153017	539774	1117112	1274713	311733	1310205	384836	196452	1685.50	130947
Average (Testing)	54844	33103	21190	33664	20995	26175	6805	2744	4.64	3129
Standard deviation (Testing)	108531	62538	55404	82012	37082	68332	20271	9836	105.17	9642
Minimum (Valid.)	-460	-77	-3071	-76429	-623	-446	-12964	-27325	-3862.00	-31939
Maximum (Valid.)	896402	6843702	2105131	7744524	1209418	2248884	1439845	355045	2600.00	115722
Average (Validation)	72759	47430	24695	48527	23985	30873	9580	3103	0.03	2295
Standard deviation (Validation)	130054	85382	51325	102234	41875	66905	18555	10115	321.85	11109
Minimum (Overall)	-2009	-1772	-3071	-517381	-8980	-3744	-21623	-33385	-4764.50	-44599
Maximum (Overall)	703378	62152700	8122509	67905791	2295046	11844244	7831937	6356727	14086.50	6493751
Average (Overall)	74414	50238	23624	51225	22970	30689	9120	4330	8.31	4110
Standard deviation (Overall)	1099644	967693	137071	1057887	63981	201369	124025	98614	281.92	100738

Source: Author.

Ideally, the characteristics of individual data sets would be approximately identical. Nevertheless, regarding the fact that enterprises are randomly divided, that is not the situation. All the same, the random division of data does not necessarily have to have a negative impact on the result.

The calculation of Kohonen network was then carried out. The network is referred to as SOFM 10-100 further in the text. Enterprises were divided into separate clusters in a topological grid 10 x 10. The frequency of separate clusters is apparent from the three-dimensional diagram in Figure 1. The picture suggests that the highest number of enterprises in individual squares of the topological grid is situated in (1, 8); then squares (3, 9) follow and then (3, 10), (4, 10) and (5, 10). On the whole we might say that the highest number of 5,000 analysed enterprises is situated in sector (1, 7), (5, 7) and (1, 10), (5, 10).

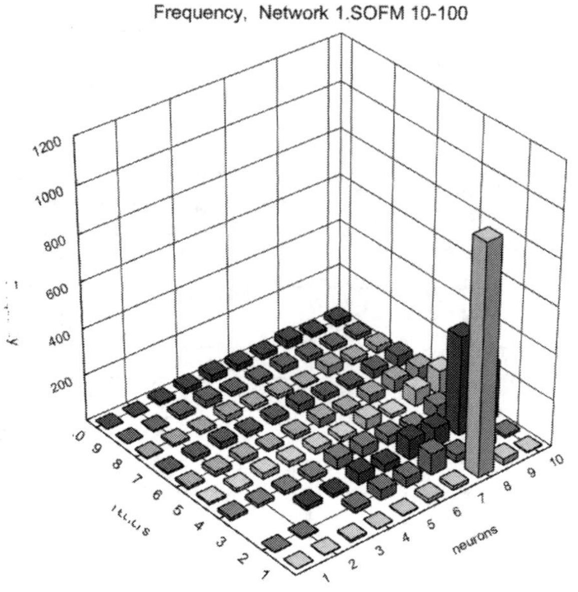

Fig. 1. Diagram of the frequency of enterprises in the topological grid

Source: Author.

For more detailed results see Table 2.

Table 2. Frequencies of network SOFM 10-100

	Network: 1.SOFM 10-100 / Samples: Training, Testing, Validation									
	1	2	3	4	5	6	7	8	9	10
1	1	3	9	2	4	17	16	990	26	4
2	2	7	0	20	44	42	90	30	29	11
3	0	0	18	14	48	32	86	82	427	249
4	13	13	15	19	46	45	37	22	64	150
5	9	14	26	22	27	22	47	15	72	103
6	10	21	24	13	19	37	21	39	55	56
7	6	17	28	17	25	36	23	17	37	48
8	6	10	12	28	17	16	7	40	23	30
9	2	5	19	31	18	19	28	23	18	18
10	1	2	12	24	31	26	21	34	14	26

Source: Author.

The table suggests that our attention should be drawn to two clusters – (1, 8) and (3, 9) in particular. In regard to the calculating operations, these are very remarkable data sets, which represent more than one third of all enterprises in the data set; the dominance is therefore possible and may be evident from Kohonen diagram in Figure 2.

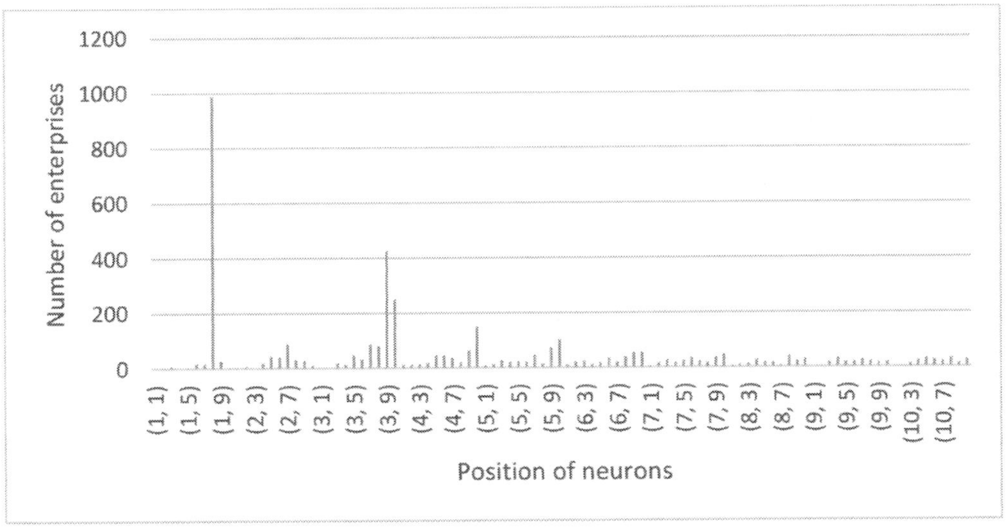

Fig. 2. Kohonen diagram for selected network cells SOFM 10-100

Source: Author.

Although the data set is relatively heterogeneous, the cluster analysis suggests certain similarities in some characteristics of the enterprises in question.

Cluster (1, 8)

Cluster (1, 8) of the topological grid contains 990 enterprises. Its individual clusters constitute the biggest data set; the following average values might be derived from the input parameters:

1. Total assets: 1,284.2 mil. CZK,
2. Fixed assets: 687.8 mil. CZK,
3. Current assets: 583.1 mil. CZK,
4. Equity: 1,052 mil. CZK,
5. Borrowed capital: 225.2 mil. CZK,
6. Performances: 516 mil. CZK,
7. Added value: 138.2 tis. CZK,
8. Operating profit: 11.1 mil. CZK,
9. ROE: 27.6%.
10. Economic result before taxation: 290 mil. CZK.

As a matter of fact, a typical enterprise can be derived from average values of this cluster of enterprises. A small enterprise is here to be dealt with, whose operating profit is minimal. Despite this fact, its existence is not jeopardized. The degree of enterprise automation is relatively low. The enterprise holds a large amount of current assets (claims and inventories).

Provided a homogeneous cluster of enterprises is to be dealt with, the validity of its basic logical link needs to be assessed. Of interest may also be a close link between total assets and the amount of the operating profit (see Figure 3 for more details).

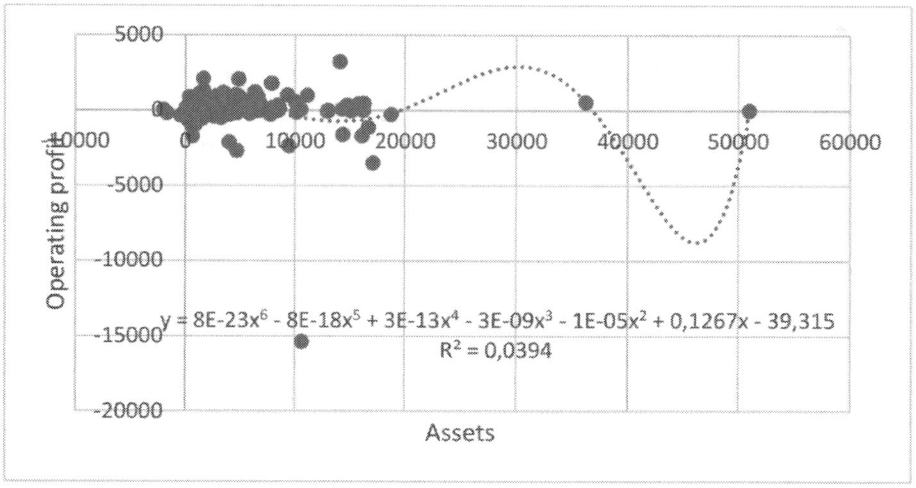

Fig. 3. Link of total assets and the operating profit of the cluster (1, 8)

Note: All values in graphs are in thousands of CZK.

Source: Author.

Total assets and operating profit correlate at more than 0.2; any significant correlation of both quantities is therefore ruled out. However, the data are provided with a regression curve – 6th degree polynom in particular. Based on this polynom the development of the operating profit can be assessed considering the size of the enterprise determined by the amount of total assets.

Figure 4 suggests a relationship between borrowed capital and operating profit.

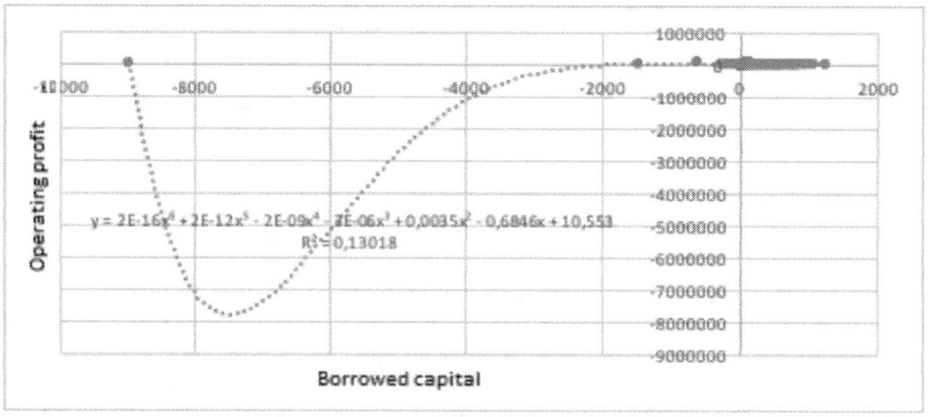

Fig. 4. Borrowed capital and operating profit of cluster (1, 8)

Source: Author.

The figure shows that the data suggest an extreme value of a negative foreign capital and a very negative operational result at the same time; this extreme needs to be removed from the source data used in the diagram for further assessments of the entire data set. Zoo Tábor, plc. is here to be dealt with. Figure 5 illustrates the relationship between borrowed capital and operating profit without the extreme value.

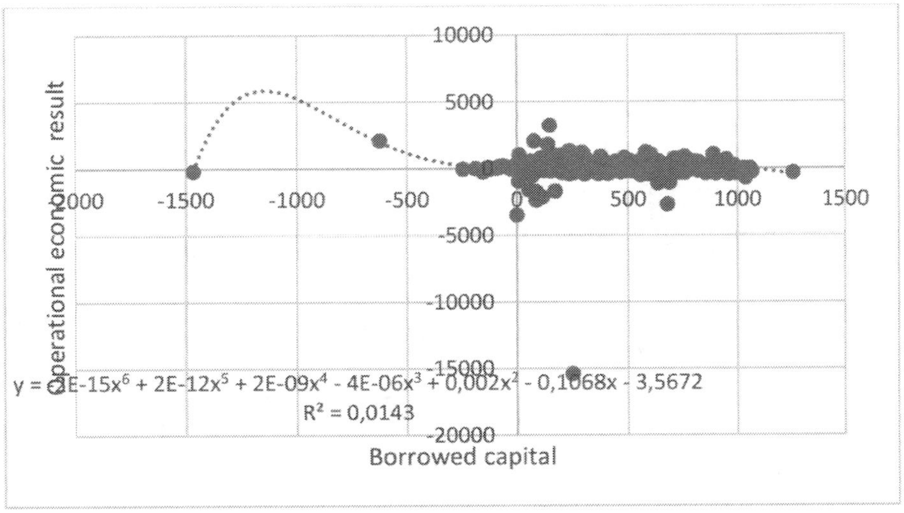

Fig. 5. Borrowed capital and operating profits of cluster (1, 8) without the identified extreme value

Source: Author.

The relationship can show us financial leveraging in case of a relatively high correlation between both quantities; however, it amounts only to 0.36. Despite this fact, the diagram items are provided with 6th degree polynom. This polynom suggests that the optimal degree of enterprise debts situated in cluster (1, 8) is up to 1 mill. CZK.

Cluster (3, 9)

Cluster (3, 9) of the topological grid is ranked second as far as the frequency is concerned. It comprises 427 enterprises. In a view of input parameters, typical values of this cluster are as follows:

1. Total assets: 1,244.6 mil. CZK,
2. Fixed assets: 497.4 mil. CZK,
3. Current assets: 728.4 mil. CZK,
4. Equity: 518 mil. CZK,
5. Borrowed capital: 707.2 mil. CZK,
6. Performances: 1,174.8 mil. CZK,
7. Added value: 391 mil. CZK,
8. Operating profit: 85.4 mil. CZK,
9. ROE: 20.82%.
10. Economic result before taxation: 107.9 mil. CZK.

The typical enterprise of this cluster is twice as large as cluster (1, 8) (in a view of the amount of total assets). This enterprise is slightly less automated, but it possesses a higher amount of current and total assets. The biggest difference lies in the amount of the operating profit. The enterprise of cluster (3, 9) generates almost by 7% lower income of the equity – 20 in particular which means 82%. All the same, even this case is worth analysing of some of its variables.

Figure 6 suggests the relationship between total assets and the operating profit.

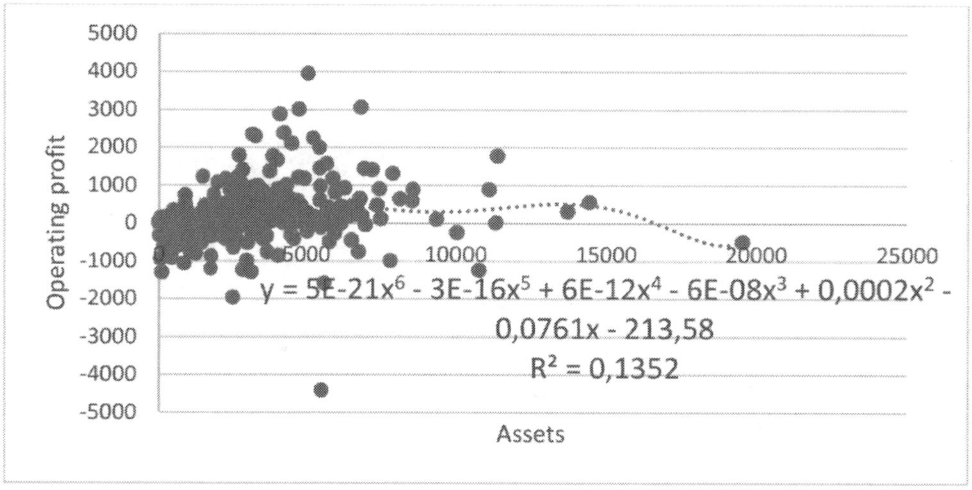

Fig. 6. Relationship between total assets and the operating profit of cluster (3, 9)

Source: Author.

The correlation coefficient of both quantities amounts to 0.38, which doesn't show any correlation between the two variables. However, this cluster manifests that a higher amount of assets means a better operating profit; the connecting line has again a shape of 6th degree polynom.

Figure 7 depicts a relationship between borrowed capital and operating profit.

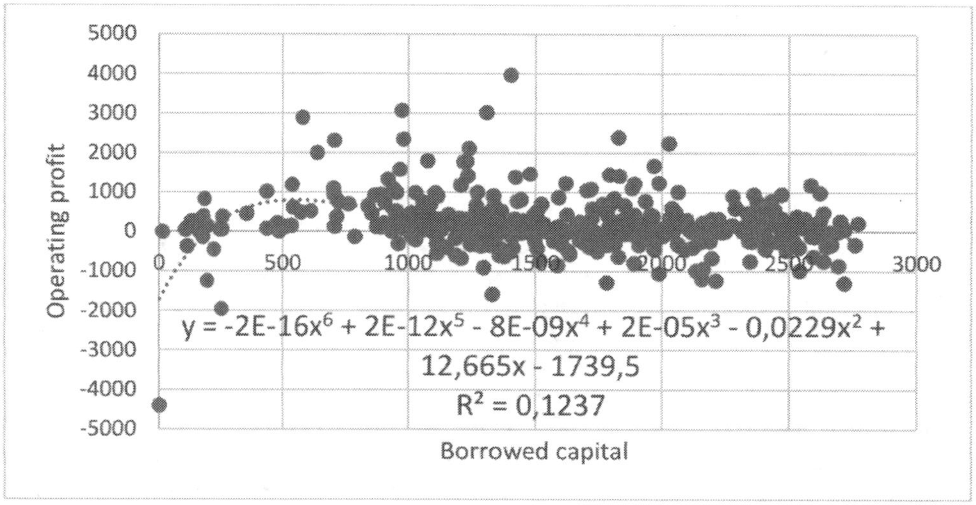

Fig. 7. Borrowed capital and operating profit of cluster (3, 9)

Source: Author.

Neither this case shows a high correlation coefficient. Nevertheless, even this example demonstrates that higher debts do not mean a better operating profit; on the whole, the total debt should not exceed 2 mil. CZK.

4 Conclusion

The aim of this article was to carry out a cluster analysis of enterprises operating in the area of agriculture. The specified aim was achieved only partially.

A cluster analysis using neural networks without teacher forcing – Kohonen networks – was carried out. According to the cluster analysis, enterprises were divided into clusters in Kohonen map (10 x 10 clusters). In regard to the number of enterprises, some of the clusters are of a major significance. Clusters (1, 8) and (3, 9) contain hugest numbers of enterprises. These clusters were subjected to a further analysis. Both clusters were provided with typical examples of these clusters. They differed in their size and the amount-generated operating profit. Enterprises of cluster (1, 8) were on the verge of profitability in 2016. The average operating profit for an enterprise amounted to 10 thousand CZK, which means minimal debts in the ideal world. Of interest might also be its degree of automated production. A typical enterprise of cluster (3, 9) is more than twice larger than that of cluster (1, 8). The participation of fixed assets was lower than in that of cluster (1, 8); this enterprise, on the contrary, holds a larger number of current assets. The average ROE of this enterprise was 20.82%. On the whole, the analysis pointed to two reasonable conclusions:

1. Larger enterprises (i.e. an enterprise with a larger number of assets) generate a better operating profit on average.
2. Financial leverage has a more positive effect in larger enterprises; i.e. the bigger the amount of borrowed capital is, the higher the operating profit and the income of equity are.

References

1. T. Kliеštik, Models of autoregression conditional heteroskedasticity garch and arch as a tool for modeling the volatility of financial time series. *Ekonomicko-manažerské spektrum*, **7**(1), 2-10, (2013)
2. H. Pao, A comparison of neural network and multiple regression analysis in modeling capital structure. *Expert Systems with Applications*, **35**(3), 720-727, (2008)
3. H. Altun, A. Bilgil, B. C. Fidan, Treatment of multi-dimensional data to enhance neural network estimators in regression problems. *Expert Systems with Applications*, **32**(2), 599-605, (2007)
4. E. Guresen, G. Kayakutlu, Definition of artificial neural networks with comparison to other networks. *Procedia Computer Science*, **3**, 426-433, (2011)
5. D. Sánchez, P. Melin, Modular Neural Networks for Time Series Prediction Using Type-1 Fuzzy Logic Integration. In: P. Melin, O. Castillo, J. Kacprzyk. *Design of Intelligent Systems based on Fuzzy Logic, Neural Networks and Nature-Inspired Optimization*, 141-145, (2015)
6. V. Beiranvand, A. Abu Bakar, Z. Othman, A Comparative Survey of Three AI Techniques (NN, PSO, and GA) in Financial Domain. *Proceedings of the 7th International Conference on Computing and Convergence Technology*, 332-337, (2012)
7. P. Šuleř, Using Kohonen´s neural networks to identify the bankruptcy of enterprises: Case study based on construction companies in South Bohemian region. *Proceedings of the 5th International Conference Innovation Management, Entrepreneurship and Sustainability*, Prague, Czech Republic, pp 985-995, (2017)

8. A. Hogenboom, W. Ketter, J. Van Dalen, U. Kaymak, J. Collins, A. Gupta, Adaptive Tactical Pricing in Multi-Agent Supply Chain Markets Using Economic Regimes. *Decision Sciences*, **46**(4), 791-818, (2015)
9. D. Santin, On the approximation of production functions: a comparison of artificial neural networks frontiers and efficiency techniques. *Applied Economics Letters*, **15**(8), 597-600, (2008)
10. A. R. Sayadi, S. M. M. Tavassoli, M. Monjezi, M. Rezaei, Application of neural networks to predict net present value in mining projects. *Arabian Journal of Geosciences*, **7**(3), 1067-1072, (2012)
11. M. S. Hossain, Z. C. Ong, Z. Ismail, S. Noroozi, S. Y. Khoo, Artificial neural networks for vibration based inverse parametric identifications: A review. *Applied Soft Computing*, **52**, 203-219, (2017)
12. Z. Rowland, J. Vrbka, Using artificial neural networks for prediction of key indicators of a company in global world. *Proceedings of the 16th International Scientific Conference Globalization and Its Socio-Economic Consequences*, Rajecké Teplice, Slovakia, pp. 1896-1903, (2016)
13. X. Han, L. Wang, Stock Company Comprehensive Assessment Model Based on Kohonen Network. *2008 Second International Conference on Genetic and Evolutionary Computing*, 185-188, (2008)
14. V. Konečný, O. Trenz, Classification of companies with the assistance of self-learning neural networks. *Agricultural Economics*, **56**(2), 51-58, (2010)
15. M. Vochozka, Formation of complex company evaluation method through neural networks based on the example of construction companie´s collection. *AD ALTA – Journal of Interdisciplinary Research*, **7**(2), 232-239, (2017)
16. R. Waheed, D. Chang, S. Sarwar, W. Chen, Forest, agriculture, renewable energy, and CO 2 emission. *Journal of Cleaner Production*, **172**, 4231-4238, (2018)
17. J. Bronnmann, F. Asche, The Value of Product Attributes, Brands and Private Labels: An Analysis of Frozen Seafood in Germany. *Journal of Agricultural Economics*, **67**(1), 231-244, (2016)

Comparison of exponential time series alignment and time series alignment using artificial neural networks by example of prediction of future development of stock prices of a specific company

Jakub Horák[1,*], *Tomáš* Krulický[1]

[1]Institute of Technology and Business in České Budějovice, School of Expertness and Valuation, Okružní 517/10, 370 01 České Budějovice, Czech Republic

Abstract. Accurate stock price prediction is very difficult in today's economy. Accurate prediction plays an important role in helping investors improve return on equity. As a result, a number of new approaches and technologies have logically evolved in recent years to predict stock prices. One is also the method of artificial neural networks, which have many advantages over conventional methods. The aim of this paper is to compare a method of exponential time series alignment and time series alignment using artificial neural networks as tools for predicting future stock price developments on the example of the company Unipetrol. Time series alignment is performed using artificial neural networks, exponential alignment of time series, and then a comparison of time series of predictions of future stock price trends predicted using the most successful neural network and price prediction calculated by exponential time series alignment is performed. Predictions for 62 business days were obtained. The realistic picture of further possible development is surprisingly given based on the exponential alignment of time series.
Key words: prediction, stock price, tome series, exponential alignment, artificial neural networks

1 Introduction

Time series are defined as sequences of spatially and de facto comparable observations which are organized based on time [1]. According to De Baets and Harvey [2], time series can be defined as ordered sequences of values of a variable at equidistant time intervals. According to the authors, time series analysis can be applied, for example, in process and quality control, in sales predictions, in census analysis, etc.

Currently, the application of time series is used in many different disciplines. Significant uses of time series include financial time or economic series that are very specific. Economic time series are often volatile, especially because over time, economic operators face different

* Corresponding author: horak@mail.vstecb.cz

circumstances that are constantly changing. Time series are routinely monitored at monthly, quarterly or annual intervals. However, the financial time series are monitored in the daily, sometimes hourly, timeframe [3].

One of the most important tasks of time series analysis is prediction. When predicting time series, it is considered a basic step to specify a given prediction problem. With prediction, you need to get the most accurate idea of the predictive variables, all available data and their nature, determine the prediction method, and set the prediction horizon. It is also important to decide whether data will be adjusted for the calculation, and if so, how. Because the prediction is needed for further decision making, it is best to know the system as much as possible [4].

Accurate stock price prediction is very difficult. Unpredictability is particularly noticeable as a result of the global financial crisis [5]. Accurate prediction of stock price movements plays a significant role in helping investors improve the return on their purchased stocks [6]. According to Vochozka [7], traditional stock price prediction methods can´t be analysed and applied well due to problems such as forecast inaccuracies, slow training rates, etc.

According to Ghiassi et al. [8], the analysis of stock price prediction includes many artificial intelligence approaches, especially artificial neural networks. According to Zhang (2004), a wide range of applications and new technologies have been developed over the last few years to predict stock prices such as the ARIMA method, artificial neural networks, etc.

Artificial neural networks are currently considered a very effective method for analysing, collecting and predicting different data, so their use is possible in solving complex problems or in various complex situations [10].

It depends very much on the purpose of time series and on the nature of the data. As for the process of time series and neural networks, neural networks capture time series behavior and try to predict individual data points as best as they can. However, in order for neural networks to work best with the time series, it is very important to learn the neural networks correctly [11]. The efficiency of the proposed time series and learning process seems to be a good tool for predicting a very complex non-linear time series [12].

Artificial neural networks have many advantages over conventional methods. They are flexible in their own use, they can quickly and with high precision analyse complex patterns etc. [13]. According to Šuleř [14] their advantage lies in the accuracy of the results, the ability to work with large data, etc. According to Vochozka [15], the disadvantage of neural networks is their demand for large input data. To create such amount of data, many test observations are needed, which is demanding for users.

The ARIMA model is currently a widely used class of time series prediction models [16]. This model allows you to predict the behavior of time series in the future and understand their properties [17]. According to Mélard and Pasteels [18], the ARIMA model is mainly used for a short-term prognosis when data is not available or the model has poor prediction characteristics. We regard the prediction of future stock market values as one of the most needed financial applications. At present, the ARIMA model is often used to predict future stock price developments [19].

Over the past twenty years, exponential alignment has been successful in many empirical studies, and is currently well established as an accurate prediction method. Exponential alignment is one of the many approaches to time series analysis. It is usually used to model a trendline component of time series (but can also be used for other time series components) that are not time-stable and are rapidly changing. If this is the case, normal decomposition can´t be used and other modelling methods such as ARIMA or other stochastic models [20] are sought. Applying exponential alignment to prognostic time series usually depends on three basic methods: simple exponential alignment, exponential alignment corrected by a trend and its seasonal variations. A common approach to selecting a method appropriate for a specific time series is based on the validation of prediction on the retained portion of a sample using criteria such as a mean absolute percentage error [21].

The aim of the paper is to compare the method of exponential alignment of time series and time series alignment using artificial neural networks as tools for predicting future price development of the company on the example of Unipetrol.

2 Data and methods

Unipetrol is one of the most prominent business entities in the Czech Republic. Its activity, structure, vision and activity is characterized by the following [22]:

"Unipetrol is the most important refining and petrochemical group in the Czech Republic and one of the main players in Central Europe. In the Czech Republic, it is the largest oil processor, one of the most important plastic producers and the owner of the broadest petrol station network under the Benzina brand. Revenues in the group in 2016 amounted to 87.8 billion CZK. The Unipetrol Group has been part of the largest Central European refining and petrochemical group PKN Orlen, based in Poland, since 2005. The group is divided into two business segments: downstream (combining refinery and petro-chemistry) and retail fuel distribution. Within the downstream segment, the company operates refineries in Litvínov and Kralupy nad Vltavou. The group is number one on the Czech wholesale fuel market. In its Litvínov locality, the group operates an ethylene unit with subsequent polymer production. In 2016, Benzina's petrol station network with 363 petrol stations (as of 31 December 2016) and the estimated market share of 17.6% of the retail fuel sales was the leader in the Czech market (company estimate according to the Czech Statistical Office by October 2016). Within the subsidiaries PETROTRANS, s.r.o. and UNIPETROL DOPRAVA, s.r.o. the group operates a wide range of logistics and transportation services. Unipetrol R & D Center, a.s. in Ústí nad Labem and the Polymer Institute Brno branch are engaged in research and development in the field of petro-chemistry. As of 31 December 2016, the group employed more than 4,500 workers from various professions. The parent company of the group is UNIPETROL, a.s. "

Data on stock prices is available from January 2, 2006 to April 30, 2018, with a total of 3,076 entries. The data comes from the Prague Stock Exchange. These are the final prices of each day on which the stocks of the company were traded during that period. Data statistics are listed in Table 1.

Table 1. Data input statistics for the data file

Statistic	Stock price (Output)
Minimum (Training)	105.5
Maximum (Training)	385
Average (Training)	198.86
Standard Deviation (Training)	64.58
Minimum (Testing)	107.5
Maximum (Testing)	380
Average (Testing)	195.96
Standard Deviation (Testing)	62.86
Minimum (Validation)	106
Maximum (Validation)	381
Average (Validation)	199.49
Standard deviation (Validation)	59.57
Minimum (Overall)	105.5
Maximum (Overall)	385
Average (Overall)	198.51
Standard Deviation (Overall)	64.49

Source: Authors.

2.1 Time series alignment using artificial neural networks

For data processing, DELL's Statistica version 12 will be used. The data mining tool of the neural network will be used. In particular, we will use time series (by regression).

We will generate Multilayer Perceptron Networks (MLP) and Neural Networks of Basic Radial Functions (RBF). Time will be the independent variable. We will determine the company's stock price as the dependent variable. We divide the time series into three sets - training, testing, and validation. The first group will be 70% of input data. Based on the training set of data, we generate neural structures. In the remaining two sets of data, we always leave 15% of the input information. Both groups will serve us to verify the reliability of the found neural structure or found model. The delay of the time series will be 1. We will generate 10,000 neural networks. Out of those we will preserve 5 with the best characteristics[†]. In the hidden layer, we will have at least two neurons, at most 20. In the case of the radial core function, at least 21 neurons, at most 30, will be in the hidden layer. For the multilayer perceptron network, we will consider these activation functions in the hidden layer and in the output layer:

- Linear,
- Logistic,
- Atanh,
- Exponential,
- Sinus.

Other settings are left by default (ANS – automated neural network). The results, if the outputs are not adequate, can then be corrected by adjusting the weights of individual neurons in the structure using the VNS tool (own neural networks).

As soon as we generate neural networks, we will evaluate their validity on an expert basis, not just by statistical characteristics. Because of neural network shortcomings (black box,

[†] We will proceed using the smallest square method. We will terminate network generation if there is no improvement, ie the reduction of the sum of squares value. Thus, we will preserve those neural structures whose sum of squares of residues to actual gold development will be as low as possible (ideally zero).

neural network overflow, etc.), neural networks may have excellent statistical parameters but won't be usable for realistic predictions. We will confront the predictions of the most successful networks between each other. Price developments will be predicted for the next 62 days on which the stocks will be traded. However, it should be noted that the prediction of the course of the price development is likely to be inaccurate. We will expect greater precision when predicting the price for the next one day after the input data set.

2.2 Exponential alignment of time series

DELL's Statistica software will also be used for processing. Specifically, it will be a module of advanced models. In them, we select time series / prediction. We then choose Exponential alignment and Prediction. The analysis settings will be as follows:-

1. We will use the model with exponential,
2. A seasonal component will not be established,
3. Alpha and Gamma will be both set to 0.1,
4. We will predict for the next 62 business days,
5. Set the maximum number of iterations to 50,
6. The convergence criterium will be 0.0001,
7. The compliance rate indicator will be the mean average square of the error,
8. Keep the other settings as default.

2.3 Comparison of time series

Finally, we compare the time series, i.e. the prediction of future stock price developments predicted by the most successful neural network and the prediction of prices calculated by exponential time series alignment.

3 Results

3.1 Time series alignment using artificial neural networks

Table 2 provides 5 neural networks with the best characteristics out of 10,000 generated structures.

Table 2. Overview of preserved neural networks

	Network name	Training perform.	Testing perform.	Validation perform.	Training error	Testing error	Validation error	Training algorithm	Error function	Activ. of hidden lyr.	Output act. funct.
1	RBF 1-25-1	0.987764	0.986103	0.989493	50.66039	54.57406	45.63646	RBFT	Sum of sq.	Gauss	Identity
2	MLP 1-20-1	0.989723	0.987887	0.990214	42.60973	47.6501	42.41196	BFGS (Quasi-Newton) 733	Sum of sq.	Tanh	Logistic
3	MLP 1-20-1	0.988754	0.987263	0.989342	46.7116	49.95894	45.79425	BFGS (Quasi-Newton) 584	Sum of sq.	Logistic	Logistic
4	MLP 1-20-1	0.990416	0.988148	0.991166	40.30469	46.76308	38.14231	BFGS (Quasi-Newton) 1206	Sum of sq.	Tanh	Logistic
5	MLP 1-17-1	0.989041	0.986828	0.989956	45.40605	51.71415	43.3042	BFGS (Quasi-Newton) 523	Sum of sq.	Logistic	Logistic

Source: Authors.

The first neural network differs from others. It is a neural network of the basic radial function. The RBFT was used as a training algorithm. As with other preserved neural networks, the error function was determined by the sum of the smallest squares. For the activation of neurons in a hidden layer, Gauss function is used, for the activation of the output layer of neurons, identity is used. Other preserved neural networks are multilayer perceptron neural networks (MLPs). All were created using the Quasi-Newton algorithm (always using a different version). To activate the hidden layer of neurons, they use two functions – namely hyperbolic tangent and logistic functions. To activate neurons the outer layer uses a single function – logistic. The performance of the neural network is described by the value of the correlation coefficient. Correlation coefficients of all preserved networks and all data sets are shown in Table 3.

Table 3. Performance of preserved neural networks

Neural network	Training perform.	Testing perform.	Validation perform.
RBF 1-25-1	0.987764	0.986103	0.989493
MLP 1-20-1	0.989723	0.987887	0.990214
MLP 1-20-1	0.988754	0.987263	0.989342
MLP 1-20-1	0.990416	0.988148	0.991166
MLP 1-17-1	0.989041	0.986828	0.989956

Source: Authors.

We are ideally searching for such a neural network, which has a correlation coefficient ideally the closest to 1. The performance of all three sets should be ideally similar. This means that the structure created by the training set of data is valid and validated on the other two data sets. Of course, we must not forget that the neural network should have minimal errors in all three sets of data. The value of the correlation coefficients of all neuronal structures and data sets is always higher than 0.986. Differences between individual neural networks are minimal. This will be important in analysis of prediction statistics (see Table 4 below).

Table 4. Prediction statistics of individual neural structures

Statistic	1. RBF 1-25-1	2. MLP 1-20-1	3. MLP 1-20-1	4. MLP 1-20-1	5. MLP 1-17-1
Minimum prediction (Training)	120,1925	123,5535	119,5583	120,9355	119,4795
Maximum prediction (Training)	387,5554	374,8945	384,771	385	381,3817
Minimum prediction (Testing)	120,1919	123,5555	119,5586	120,9518	119,5251
Maximum prediction (Testing)	387,5525	374,8302	384,6541	385	380,9729
Minimum prediction (Validation)	120,2007	123,5533	119,5627	120,9459	119,4791
Maximum prediction (Validation)	387,5476	374,894	384,7621	385	381,347
Minimum residues (Training)	-54,031	-45,0864	-49,1901	-39,3348	-45,535
Maximum residues (Training)	37,6793	49,0984	40,8764	38,1443	39,5881
Minimum residues (Testing)	-56,01	-39,976	-43,0483	-38,7561	-43,9389
Maximum residues (Testing)	30,1897	33,3395	38,5448	31,2415	37,5442
Minimum residues (Validation)	-34,6119	-42,2136	-32,397	-35,0735	-32,8213
Maximum residues (Validation)	36,4955	48,7342	45,0962	33,3061	44,0382
Minimum standard residues (Train.)	-7,5912	-6,907	-7,1972	-6,1958	-6,7575
Maximum standard residues (Train.)	5,2938	7,5216	5,9808	6,0083	5,875
Minimum standard residues (Testing)	-7,5818	-5,7912	-6,0904	-5,6675	-6,1101
Maximum standard residues (Testing)	4,0866	4,8298	5,4533	4,5686	5,2208
Minimum standard residues (Valid.)	-5,1235	-6,482	-4,7874	-5,679	-4,9876
Maximum standard residues (Valid.)	5,4024	7,4832	6,664	5,3929	6,6921

Source: Authors.

If we monitor the statistics of the predictions of individual neural networks, we necessarily conclude that the differences between the networks are absolutely minimal for all statistics.

The graphical development of prices and predictions can also tell us the right outcome. Figure 1 provides a graphical comparison of Unipetrol's actual stock prices and predictions calculated using all neural networks.

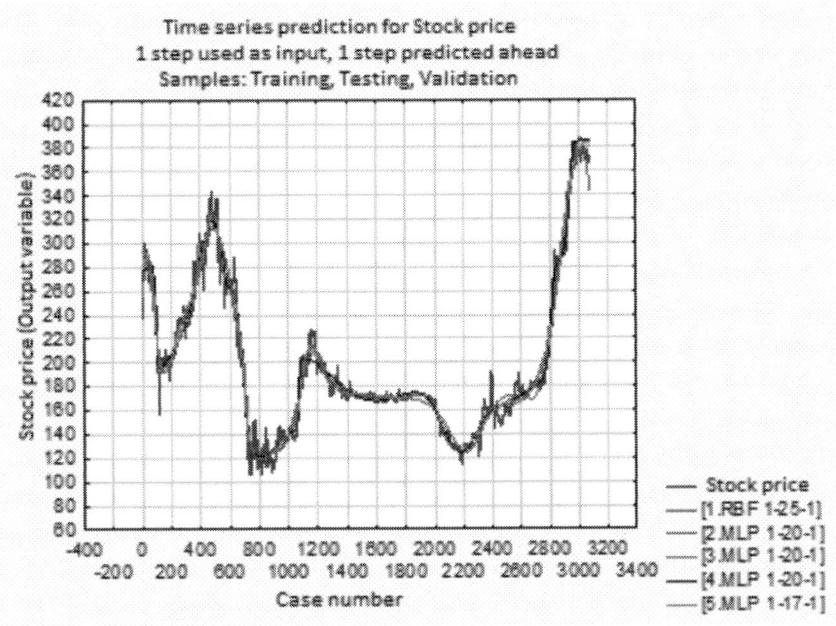

Fig. 1. Time series with prediction for 62 trading days

Source: Authors.

It can be seen from the picture that all networks have more or less accurately copied the real price movements in past data. However, at some moments, for example on the 100th trading day, the stock price fluctuated downwards. These fluctuations, but in both directions, are repeated several times. It is a question of how far they are the result of a turbulent environment that none of the preserved neural networks has been able to reliably describe. However, despite this fact, we can assume that all preserved neural networks are applicable in practice.

After network training, prediction was performed over the next 62 business days. Figure 2 gives you a close look at these 62 business days.

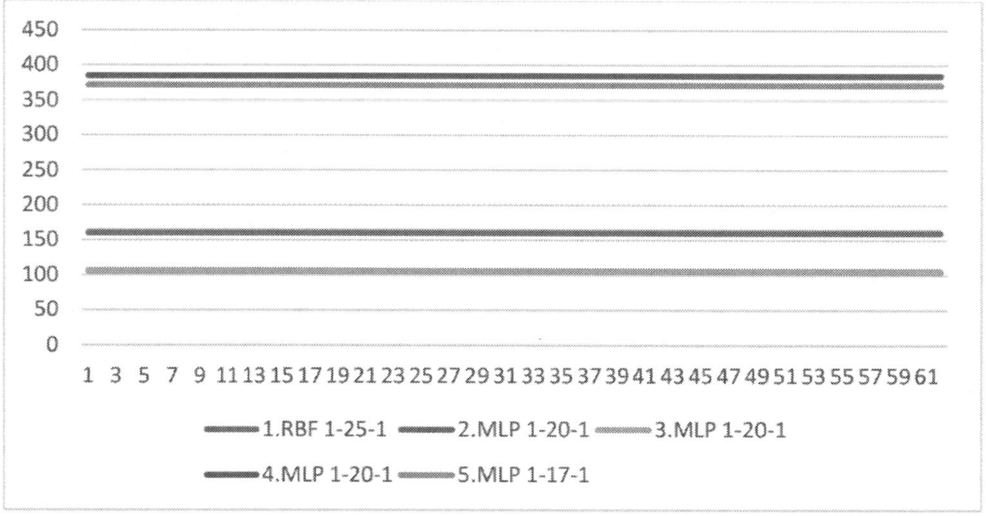

Fig. 2. Prediction for the next 62 business days

Source: Authors.

It is clear from the graph that networks are always (or almost always) predicting a constant development of Unipetrol's stock price over the review period. The only network that does not predict the constant development of stock prices over the projection period is Network Number 5, MLP 1-17-1. However, the price differences between days vary at the level of one thousand CZK. There is a significant difference between the predictions of individual networks. At first glance, and in view of price trends so far, the constant price seems unlikely to be in the observed period of 62 days.

Table 5 provides a detailed view of selected prediction cases. Specifically, this is about every tenth case of prediction.

Table 5. Values of selected predictions by individual neural networks

Case	1.RBF 1-25-1	2.MLP 1-20-1	3.MLP 1-20-1	4.MLP 1-20-1	5.MLP 1-17-1	Maximum	Minimum	Difference	% of max. price
10	160.364	105.5	105.5	385	371.766	385	105.5	279.5	72.60 %
20	160.364	105.5	105.5	385	371.7664	385	105.5	279.5	72.60 %
30	160.364	105.5	105.5	385	371.7667	385	105.5	279.5	72.60 %
40	160.364	105.5	105.5	385	371.767	385	105.5	279.5	72.60 %
50	160.364	105.5	105.5	385	371.7673	385	105.5	279.5	72.60 %
60	160.364	105.5	105.5	385	371.7676	385	105.5	279.5	72.60 %

Source: Authors.

The table shows that the 2nd and 3rd neural networks have exactly the same predictions. This is always the minimum amount of predictions at a given time. Other predictions are higher. The table below also offers a summary of the maximum and minimum predictions of the case. It also calculates the difference between the minimum and maximum predictions. Subsequently, this difference is compared to the maximum predicted and expressed as a percentage of the difference to the maximum prediction. We find that the difference between the minimum and maximum predictions is 72.6%. Therefore, it is clear that the interval between the lowest and the highest predictions is too large for the future estimate. This means that the actual price may move at this interval, but the value of the prediction is zero for the investor. However, if we expertly assess predictions for the next day, we would choose the network number 5, i.e. MLP 1-17-1 as the most successful. On the imaginary second place, we would put network number 4, MLP 1-20-1.

3.2 Exponential alignment of time series

Figure 3 shows a comparison of the actual time series (blue curve) and the aligned time series (orange curve).

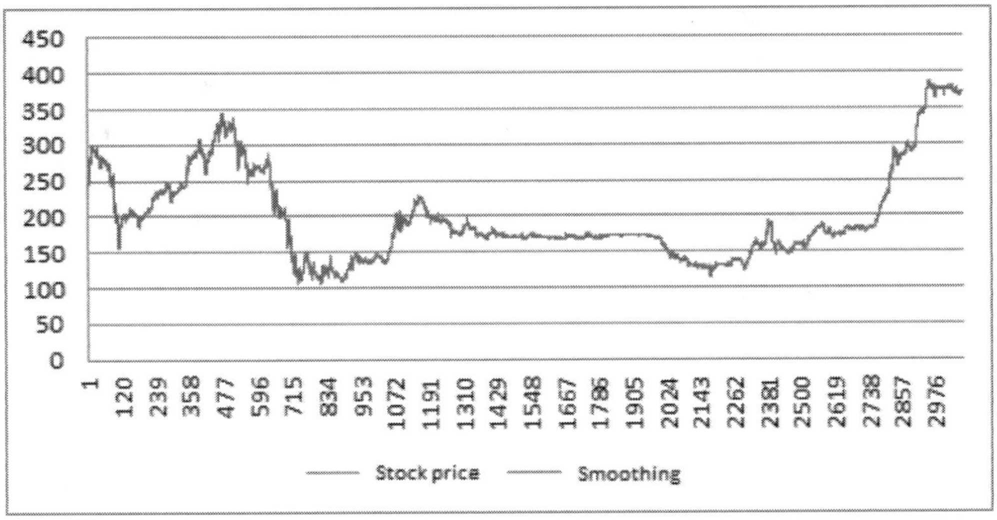

Fig. 3. Applied exponential time series alignment method

Source: Authors.

It may seem optically that the course of both curves is exactly the same. Perhaps the observer will notice minor differences. Even more interesting will be to enrich the comparison by adding residues which will precisely define the differences between the two curves. Such a comparison is the subject of Figure 4.

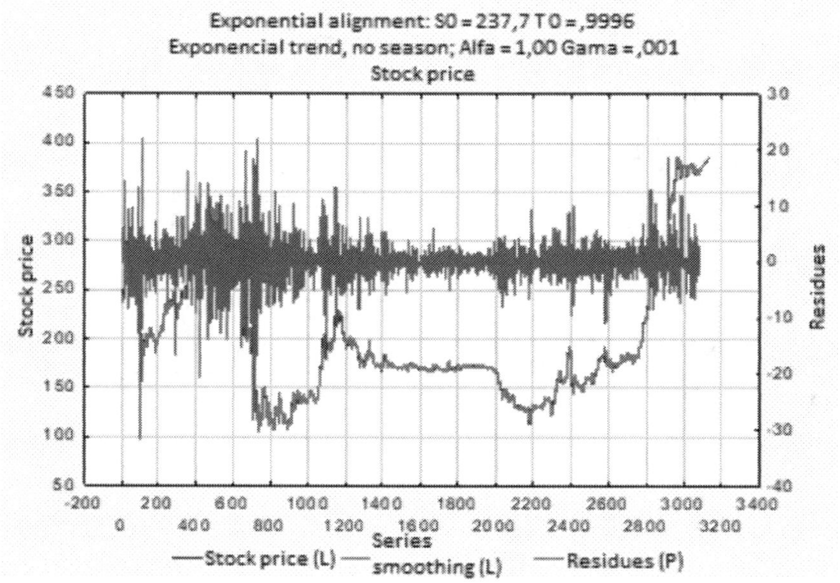

Fig. 4. Exponential time series alignment residues

Source: Authors.

In essence, this is a graph that matches the figure 3. However, this is a box graph that provides the second variable for comparison. In addition, residues of an exponentially aligned time curve are displayed. The colour of the exponentially aligned curve, which is now red, has changed as opposed to the graph in figure 3. We can now follow the curve of residues, which are presented in green colour. The x-axis for the residue graph is on the right side of the image. The residual value ranges from -32 to 22 CZK per stock. In relative terms, we range from approximately - 15% to the average stock price to 10%. Because these are extreme values, we can boldly state that the model appears at first glance as very interesting.

3.3 Comparison of methods

In order to better compare the use of both methods, i.e. artificial neural networks and exponential time series alignment, we will also use the graph (see Figure 5).

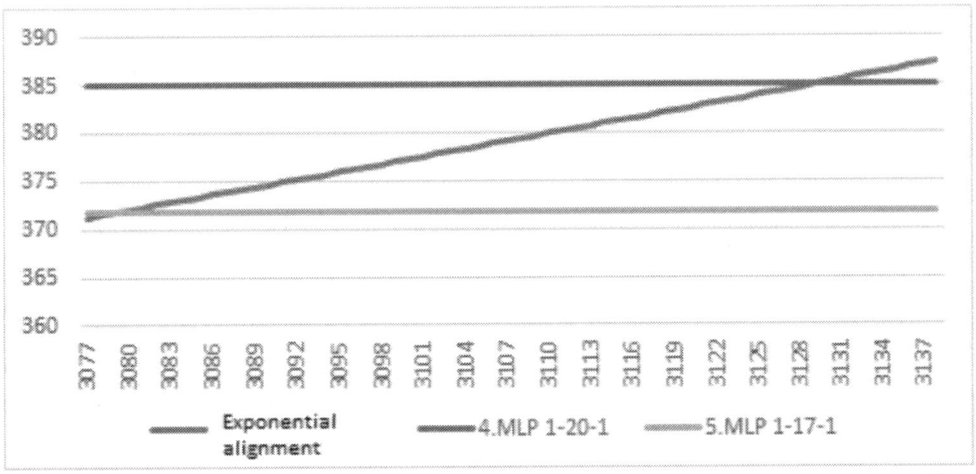

Fig. 5. Comparison of predictions calculated by neural networks and exponential time series alignment

Source: Authors.

For comparison, we used the neural network number 4 and 5, in particular MLP 1-20-1 and MLP 1-17-1, in addition to prediction using exponential time series alignment. Both neural networks predict constant or near-price development. However, this may seem unlikely, as we predict the price for 62 business days (i.e. for more than 2 months). Looking at Unipetrol's stock price so far, we find that there is no analogy in the past where the price of stocks remains constant. Logically, the development predicted by the exponential alignment of time series appears more suitable. Over 62 business days, the company's stock price is expected to grow from approximately 371 CZK to 387 CZK. The change in the 16 CZK is realistic (in relative terms it is a change of less than 3.7% of the price).

4 Conclusion

The aim of the paper was to compare the method of exponential alignment of time series and time series alignment using artificial neural networks as tools for predicting future price development of a company, as exemplified on Unipetrol.

First, the data file was analysed. Subsequently, neural networks were generated, of which five with the best characteristics were preserved. Using statistically interpreted results, it has been found that it is possible to use up to two neural networks in practice: 4 (MLP 1-20-1) and 5 (MLP 1-17-1).

Then, the exponential alignment of time series was used. A prediction was obtained for 62 business days, which was then compared to the results of two preserved neural networks. At the time of the calculations it was not possible to determine which prediction is or is not correct. Based on a simple argumentation, we have inclined to the fact that, a realistic picture of further development is given rather by a prediction based on an exponential alignment of time series.

Generally, experts have found that artificial neural networks show better results than conventional statistical methods. In this case, however, the result was different. But the question remains why. We can believe that this has happened due to the possible turbulent development of Unipetrol's stock prices or by using an inappropriate software tool – Statistica for this task. Or we can conclude that the prediction for 62 business days is too long. It will therefore be appropriate to devote more effort to this research. We will test other prediction tools such as ARIMA. We will find further developments in the stock prices and verify their interpretation with subsequent real stock price developments.

References

1. M. Sheikhan, N. Mohammadi, Time series prediction using PSO-optimized neural network and hybrid feature selection algorithm for IEEE load data: revue littéraire mensuelle. *Neural Computing and Applications*, **23**(3-4), 1185-1194, (2013)
2. S. De Baets, N. Harvey, Forecasting from time series subject to sporadic perturbations: Effectiveness of different types of forecasting support. *International Journal of Forecasting*, **34**(2), 163-180, (2018)
3. J. Jaramillo, J. D. Velasquez, C. J. Franco, Research in Financial Time Series Forecasting with SVM: Contributions from Literature. *IEEE Latin America Transactions*, **15**(1), 145-153, (2017)
4. P. Rostan, A. Rostan, The versatility of spectrum analysis for forecasting financial time series. *Journal of Forecasting*, **37**(3), 327-339, (2018)
5. S. Hašková, Prediction of Future Development of Share Prices Using Neural Networks Based on Daily and Monthly Share Price Data. *Mladá veda*, **5**(9), 33-48, (2017)
6. Z. Qun, L. Xu, G. Zhang, LSTM Neural Network with Emotional Analysis for Prediction of Stock Price. *Engineering Letters*, **25**(2), 167-175, (2017)
7. M. Vochozka, Inventory management using artificial neural networks in a concrete case. *Proceedings of the 5th International Conference Innovation Management, Entrepreneurship and Sustainability*, Prague, Czech Republic, 1084-1094, (2017)
8. M. Ghiassi, H. Saidane, D. K. Zimbra, A dynamic artificial neural network model for forecasting time series events. *International Journal of Forecasting*, **21**(2), 341-362, (2005)
9. X. Zhang, Evolution of ARMA demand in supply chains. *Manufacturing and Service Operations Management*, **6**(2), 195-198, (2004)
10. J. Vrbka, Z. Rowland, Stock price development forecasting using neural networks. *Proceedings of the International conference Innovative Economic Symposium 2017, Strategic Partnerships in International Trade, Web of Conferences*, **39**, (2017)
11. Y. H. Hu, J. Hwang, *Handbook of neural network signal processing*. Boca Raton: CRC Press, (2002)
12. Y. Chen, B. Yang, J. Dong, Time-series prediction using a local linear wavelet neural network. *Neurocomputing*, **69**(4-6), 449-465, (2006)
13. D. Santin, On the approximation of production functions: a comparison of artificial neural networks frontiers and efficiency techniques. *Applied Economics Letters*, **15**(8), 597-600, (2008)
14. P. Šuleř, Using Kohonen´s neural networks to identify the bankruptcy of enterprises: Case study based on construction companies in South Bohemian region. *Proceedings of the 5th International Conference Innovation Management, Entrepreneurship and Sustainability*, Prague, Czech Republic, 985-995, (2017)
15. M. Vochozka, Formation of complex company evaluation method through neural networks based on the example of construction companies collection. *AD ALTA: Journal of Interdisciplinary Research*, 7(2), 232-239, (2018)
16. P. H. B. F. Franses, The Econometric Modelling of Financial Time Series. *International Journal of Forecasting*, **16**(3), 426-427, (2000)

17. A. B. Sánchez, C. Ordóñez, F. S. Lasheras, F. J. De Cos Juez, J. Roca-Pardiñas, Forecasting SO 2 Pollution Incidents by means of Elman Artificial Neural Networks and ARIMA Models. *Abstract and Applied Analysis*, 1-6, (2013)
18. G. Mélard, J. M. Pasteels, Automatic ARIMA modeling including interventions, using time series expert software. *International Journal of Forecasting*, **16**(4), 497-508, (2000)
19. M. F. Anaghi, Y. Norouzi, A Model for Stock Price Forecasting Based on ARMA Systems. *2nd International Conference on Advances in Computational Tools for Engineering Applications*, Beirut, Lebanon, 265-268, (2012)
20. L. Marek, M. Vrabec, ARIMA models and exponential smoothing. *33rd International Conference on Mathematical Methods in Economics*, Czech Republic, 502-507, (2015)
21. B. Billah, M. L. King, R. D. Snyder, A. B. Koehler, Exponential smoothing model selection for forecasting. *International Journal of Forecasting*, **22**(2), 239-247, (2006)
22. Unipetrol, *Unipetrol company in brief* [online], Available at: http://www.unipetrol.cz/cs/VztahySInvestory/Stranky/unipetrol-ve-zkratce.aspx, (2018)

Customer evaluation of loyalty programs in selected companies of the South Bohemian region

Tsolmon Jambal[1], *Jaroslav* Stuchlý[1,*]

[1]Institute of Technology and Business in České Budějovice, School of Corporate Strategy, Okružní 517/10, 37001 České Budějovice, Czech Republic

Abstract. Satisfied and constantly returning customers is a tremendous benefit for the business. There is also a well-thought loyalty program to help customers return. The aim of the article is to present preferences within the framework of possible tools used in loyalty programs and their comparison with what is actually offered to customers in the South Bohemian region. The attention is focused on younger and middle age customers, which is more open to new technologies and trends. The article is based on partial research outputs focusing on customer loyalty preferences. The data were obtained through a questionnaire survey (e-questionnaire) in 2018. The obtained data are evaluated using basic descriptive statistics. Also, there is the independent two-sample t-test used to average the mean values of two groups of data in terms of their determination of significant differences.
Key words: loyalty programs, benefits of loyalty programs, point system, forms of loyalty programs of South Bohemian customers

1 Introduction

At the moment, it is the main effort of all companies to keep a customer who is making profits for the company, and this approach is called the CRM strategy. According to Kotler and Kellner, the CRM system is able to acquire important customer information, "Who are our customer?", "What they need?" and "How they behave" [1].

Reviews three different perspectives on loyalty, and relates these to a framework for understanding customer loyalty that encompasses customer brand commitment, customer brand acceptance and customer brand buying [2].

Yuping Liu [3] points to diverse responses across consumers suggest a need to consider consumer idiosyncrasies when assessing the impact of loyalty programs. By nature, loyalty programs are one-to-one programs. How much a consumer can benefit from such a program depends on his or her "investment" in the relationship with the firm.

Loyalty programs are a classic tool for sales promotion. The customer can get involved in these programs, where they can get a variety of discounts on their purchases, gifts, or trips.

* Corresponding author: stuchly@mail.vstecb.cz

Typical are, for example, retail chains where loyalty cards can be obtained, at banks where they offer bonus programs, but these are typical for many other organizations today. They represent a very effective way to persuade customers to make regular purchases. Thanks to these programs, it is very efficient to build databases with customer contacts [4].

The magic of loyalty programs lies not only in the possibility of getting a discount or an advantage for a particular program operator and his partner, but also the possibility of combining different loyalty programs. New trends point to such a combination if the benefit is drawn by a customer card and the payment for the goods or service is made by means of a payment card, where the turnover is returned back to the card account.

2 Data and Methods

This research was developed within the framework of an internal grant on the topic "Supporting the subjects taught by linking the theoretical knowledge with the practice of companies by analysing and evaluating the level of customer relationship management in a selected sample of enterprises in the South Bohemian Region". The data were obtained through a questionnaire survey (e-questionnaire) in 2018. The select sample is composed of 269 respondents.

Data analysis is performed by basic methods of descriptive and inductive statistics including correlation, simple and multiple linear regression. Numerical calculations and statistical analyses are required for Excel, R system in version R-2.15.3 and statistical program SPSS [5].

In addition, quantitative responses are evaluated using the minimum and maximum values, mod, median, mean and sampling standard deviations. Scattering analysis allows you to determine the dependence of a numerical (quantitative) variable on a categorical (qualitative) variable. One-factor scattering analysis was used in the work.

The questionnaire contained 5 demographic and 13 evaluation questions. Demographic responses are evaluated by the distribution of absolute and relative frequencies. In the South Bohemian Region, in 2017 there were 639,180 inhabitants, which is 6% of the total population of the Czech Republic. It is the sixth most populated region. The South Bohemian region ranks according to the population density of 63.4 inhabitants per km2 to the least populous region [6].

3 Results and Discussion

The research was attended by 116 male respondents (43.45%) and 151 female respondents (56.55%). The age of respondents was assessed as a qualitative variable in the interval (up to 29), (30-39), (40-49), (50-59) and over 60 years. The average age of respondents is 34.44 years and a median of 35 years. According to the nature of the online questionnaire, the majority of respondents (71.6%) are under the age of 39 (see Table 1).

Table 1. Structure of respondents by age

Age	up to 29	30-39	40-49	50-59	over 60 years
Relative Frequency (%)	46.1	25.5	19.1	6.7	2.6
Count	123	69	51	19	7

Source: Authors.

The structure of the respondents according to their educational attainment is clearly dominated by graduates with a school-leaving examination, which corresponds to the age structure of the inhabitants of the South Bohemian Region (Fig. 1).

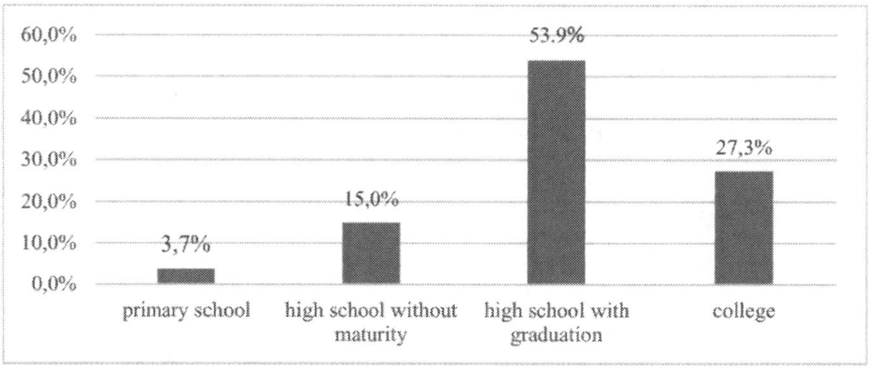

Fig. 1. Composition of the respondent by education

Source: Authors.

Fig. 2 shows that 6% of respondents (160 persons) are living in the municipality from 1,000 to 50,000 inhabitants in the South Bohemian Region.

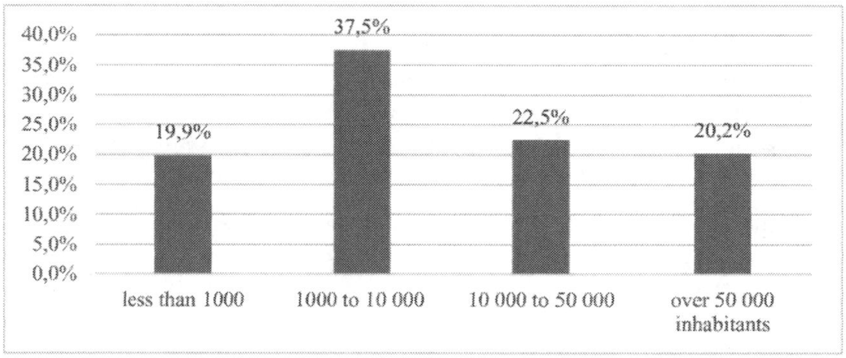

Fig. 2. Structure of correspondents according to the number of inhabitants in the municipality

Source: Authors.

Asked the answer, "what kind of discount do you like?" Respondents answered that they like the loyalty program in the form of a percentage discount (39.70%) and a combination of the percentage discount and the crown discount (32.96%), see Fig. 3.

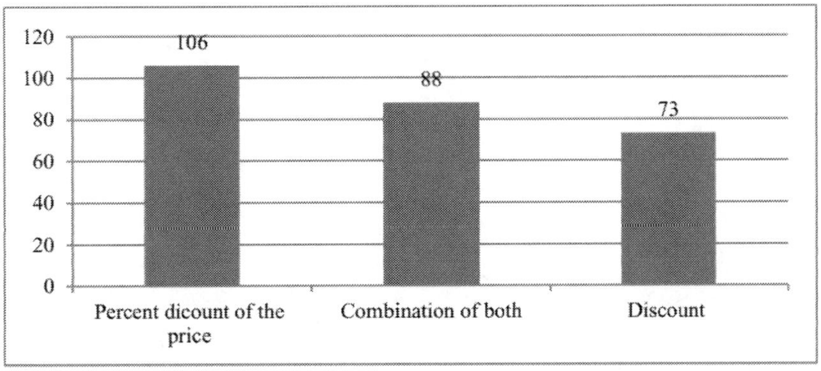

Fig. 3. Structure of respondents according to the form of the loyalty program

Source: Authors.

To the question "How many loyalty programs do you use?" 14.78% of respondents answered 2 loyalty programs. Respondents use an average of 6.2 loyalty programs. The maximum number of loyalty programs is 30 and the minimum value is 0. Median 4 and the modus was measured by 2. The standard deviation is 10.52, see Fig. 4.

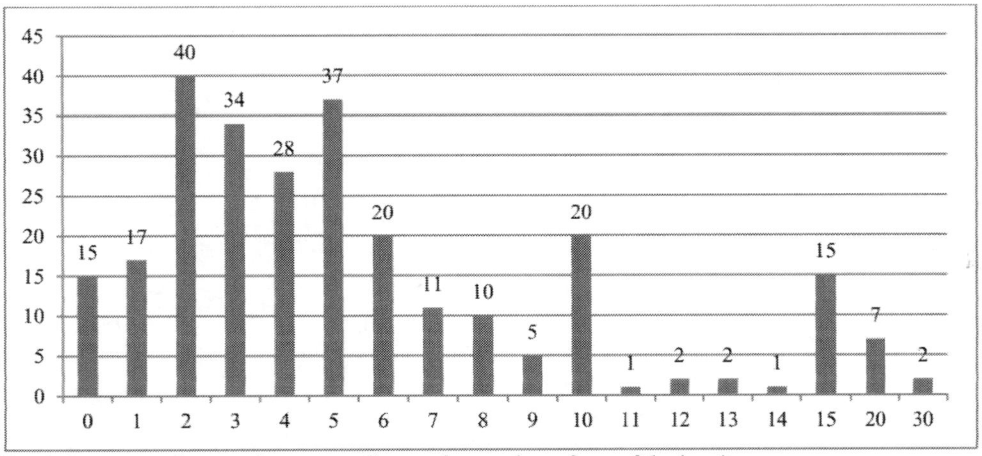

Fig. 4. Structure of respondents according to the number of use of the loyalty program

Source: Authors.

To the question "How many percentage discounts do you expect to participate in loyalty programs?" responded in the range of 0 to 100%. Although to the question, "How many percentage discounts do you get for participating in loyalty programs?" in the range of 0 to 50% (see Table 2).

Table 2. Numerical characteristics of the actual and expected increment

Parameter	Expected increment	Actual increment
Average value	22.78	10.02
Median	20	10
Modus	10	10
Standard deviation	17.27	9.1
Oblique	1.6	1.98
Spikiness	3.79	5.56
Quantum span	16	18
Minimum	0	0
Maximum	100	50

Source: Authors.

We see from the box diagram (Fig. 5) that the level and variability of the expectation variable is higher than the actual discount.

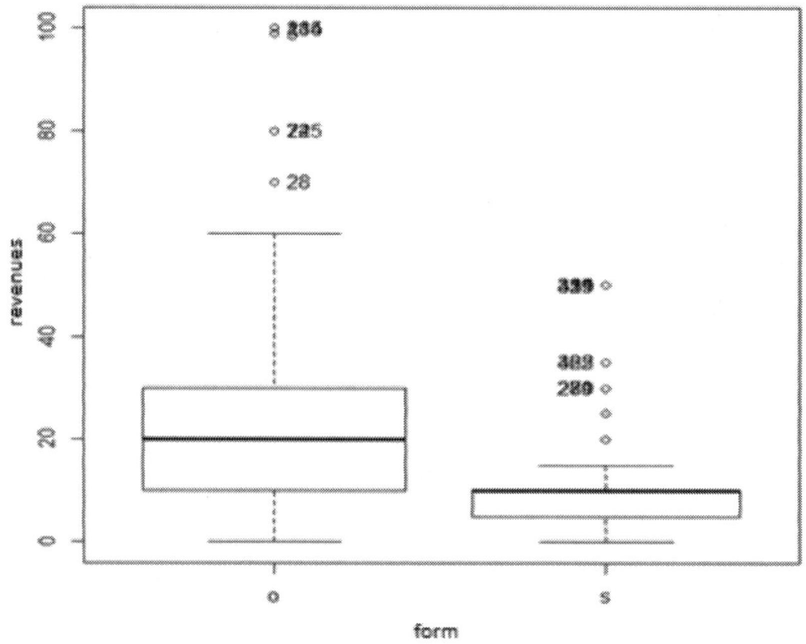

Fig. 5. Box diagram for expected (o) and actual (s) discount

Source: Authors.

People surveyed from the South Bohemian Region are not worried about providing information in the area of the mailing address, name and surname, age, contact address, title, telephone number and date of birth. However they are concerned about household income.

Regarding the permanent residence and the number of the family member they have a neutral opinion. Fig. 6 shows that 215 respondents (80.59%) disagree with the provision of family income information.

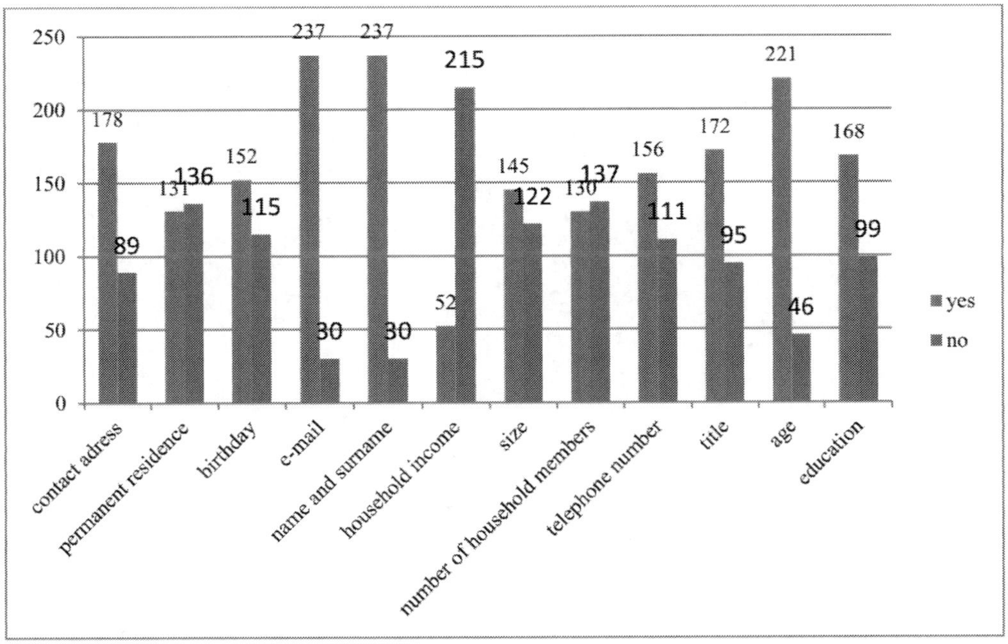

Fig. 6. Structure of respondents according to the provision of information

Source: Authors.

The Fig. 7 shows respondents' answers to the question "From which firm do you prefer the loyalty program?". The first 3 ladders were generally drugstore, others (ie, different companies in one piece) and more than 2 loyalty programs from other companies. The loyalty program from Tesco was best evaluated. Next followed Penny, DM, Billa, Coop, Globus, Shell, Alza, Cinestar, Orsay, Sportisimo, Hervis and Dr. Max.

The Tesco chain of business in the Czech Republic has launched the Tesco Clubcard loyalty program since 2010. Clubcard is a loyalty card that returns money for shopping in the Tesco business chain. For each spent CZK 10 (excluding a petrol station, alcohol, cigarettes) the credit card is credited with 1 point. The value of 1 point is 0.10 CZK. After collecting at least 300 points, you will receive money vouchers and discount vouchers for regular purchases in regular quarterly billing. Participation in the loyalty program is subject to registration. When you register directly at Tesco, you will be able to wait for an active card to be picked up, to which he can immediately collect points [7].

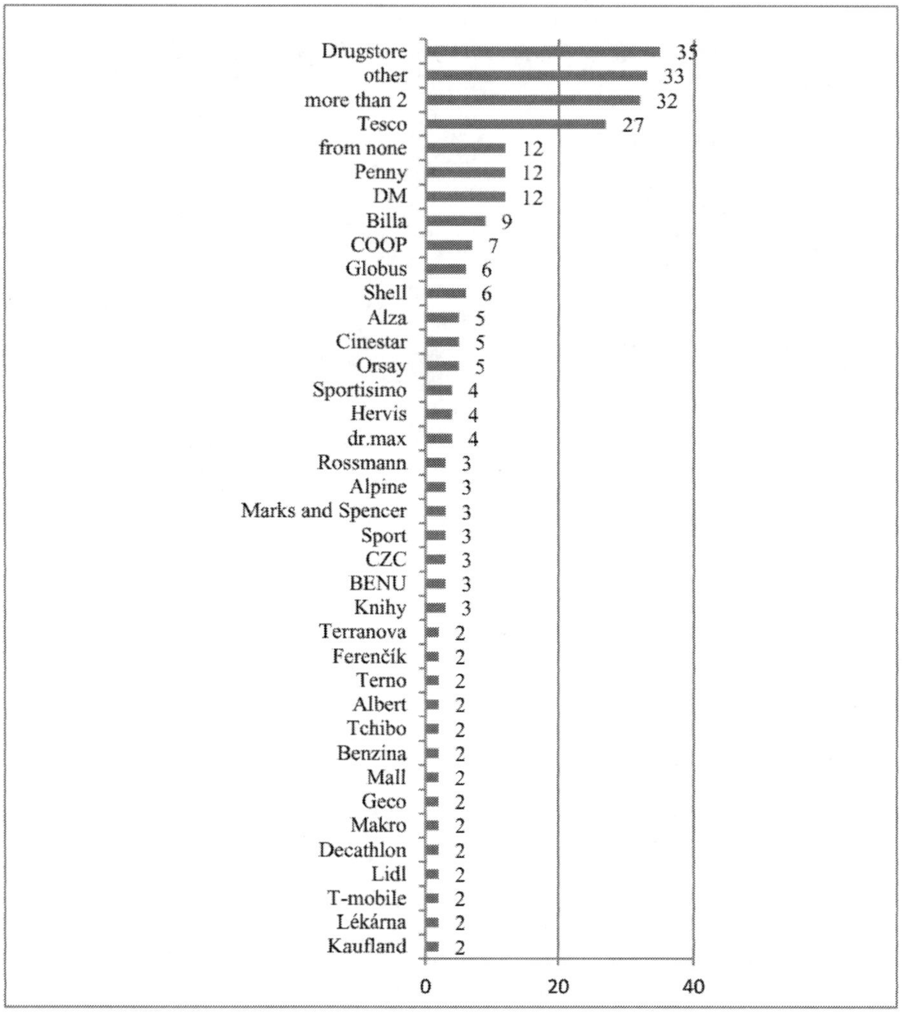

Fig. 7. Loyalty program structure by companies

Source: Authors.

4 Conclusion

The article deals with the various forms of benefits provided by businesses in the South Bohemian Region to their customers and the preferences they provide to these forms of loyalty program. In this research, a group of younger and middle-aged customer groups prevailed. In this region, an average customer uses more than six loyalty programs from over 40 different businesses. These companies deal with common consumer, food and pharmacy business rations.

Numerical characteristics of the actual and expected increment point to the fact that a significant difference between customer expected and actual increment of percentage discount of loyalty program. This is up to two times the difference.

Dispositions of providing personal information to businesses show that young people are not afraid to provide personal data but have a negative attitude towards family relationships such as family reunions and family members. It certainly has this new stance on the newly introduced EU regulation on the GDPR, which started to apply the General Data Protection Regulation (GDPR) in all EU member states.

A new trend is loyalty programs that make customers book or order ahead of the internet with an emphasis on the mobile world. This hides the mystery of an elegant solution for identification, while improving customer service.

For the customer, the benefits of the loyalty program must be really interesting and must be realistic. If you set up a complicated system from which shoppers will not even understand what they actually get in a customer club, do not expect to be successful. Gradual trend is also turning away from point schemes. People with delayed rewards are already tired and do not want to wait.

These results suggest that although non-financial benefits and lower customer preference than financial ones, they are underestimated in the loyalty program making process and are assumed to have their potential in the younger and middle category of customers. In the global economy, these elements of loyalty programs also have an irreplaceable position, and this trend is gradually being reflected in the South Bohemian region [11 and 12].

References

1. P. Kotler, L. Keller, *Marketing Management*, 12 ed. Grada Publishing, Prague, (2007)
2. D. Mark, Uncles, R. G., Dowling, K. Hammond, Customer loyalty and customer loyalty programs. *Journal of Consumer Marketing*, **20**(4), 294-316, (2003)
3. L. Yuping, The Long-Term Impact of Loyalty Programs on Consumer Purchase Behavior and Loyalty. *Journal of Marketing*, **71**(4), 19-35, (2007)
4. M. Karlíček, *Marketing communication: how to communicate in our market*, 2nd ed. Grada Publishing, Prague, (2016)
5. J. Stuchlý, *Statistické analýzy dat*. České Budejovice: Institute of Technology and Business in České Budějovice, 220, (2015)
6. Statistical Office, *Demographic data of the Czech Republic*. [online], Available at: https://www.czso.cz/csu/xc/casove_rady (2017)
7. *30 nejlepších věrnostních programů pro každodenní slevy* [online], Available at: https://www.mesec.cz/clanky/vernostnimi-programy-kompletni-prehled/ (2018)
8. L. Branská, Z. Pecinová, H. Lošťáková, Building Customer Loyalty as the Basis for the Stabilization of Relationships in the Supply Chain. *Proceedings of Conference. Carpathian Logistics Congress CLC*, Jeseník, Czech, 231-236, (2015)
9. V. Cempírek, V. Drahotský, P. Vrbová, The Concept of Prices from Antiquity to today in Comparison with Selected Current International trends. *Proceeding of the 14th International Scientific Conference of Doctoral Students and Young Scientists on Economic, Political and Legal Issues of International Relations*, Bratislava, Slovakia, 68-75, (2015)
10. M. Karlíček, M. *Marketing communication: how to communicate in our market*, 2nd ed. Grada Publishing, Prague, (2016)
11. T. Jambal, L. Ližbetinová, J. Stuchlý, Attitudes of South Bohemian customers to forms and set loyalty program conditions. *18th International Scientific Conference Globalization and Its Socio-Economics Consequences*, Žilina, Slovakia, (2018)
12. J. Stuchlý, T. Jambal, L. Ližbetinová, Loyalty programs and customer preferences. *32nd IBIMA Conference*, Seville, Spain, (2018)

SHS Web of Conferences **61**, 01008 (2019) https://doi.org/10.1051/shsconf/20196101008
IES2018

Prospects for development of Ukraine and EU in the field of renewable energy sources

Mykola Khomenko[1,*], *Kateryna* Pryakhina[2], and *Kostyantyn* Latyshev[3]

[1]Doctor of Economics, professor, Kremenchuk Mykhailo Ostrohradskyi National University, Ukraine
[2]PhD candidate, Kremenchuk Mykhailo Ostrohradskyi National University, Ukraine
[3]PhD, associate professor, Kremenchuk Mykhailo Ostrohradskyi National University, Ukraine

Abstract. In article assessment of tendencies and prospects of cooperation between Ukraine and the European Union in the field of renewable energy sources was carried out. Over the past decades, the theme of climate change, especially caused by human activity, has become one of the most acute and most exploited in scientific disputes. Today, the problem of finding new sources of energy is more than acute, with a special place occupy non-traditional ones. The reduction of Ukraine's energy dependence on traditional fuel resources is possible only through the development and use own alternative energy, which uses local resources – bioenergy as a fuel or does not require a fuel component at all – solar, wind and small hydropower. The European Union is one of the leaders in the implementation of alternative energy and renewable energy sources and initiates the creation of many organizations that finance alternative energy projects. Ukraine should ensure the effective convergence of its own legislation with the European Union, since it defined the European integration vector as a priority area of foreign policy development. At the same time, within the framework of the Union, the mechanism of environmental protection and progressive approaches to the implementation of energy policy, including in the field of renewable energy, can serve as the most constructive benchmarks for Ukraine's development in this area. Ukraine signed the Paris Agreement, and undertook certain commitments, namely: in accordance with the National Action Plan on alternative energy up to 2020, the objective was to reach 11% of RES in gross final energy consumption. Achieving this goal is possible only in the case of international cooperation with the EU in the field of alternative energy. However, today, the study of the prospects of cooperation between Ukraine and the EU in this field is not developing sufficiently and there is a certain vacuum in solving strategic problems in this context, which determines the relevance of the topic.

Key words: renewable energy sources (RES), Energy Strategy, European Union (EU), Total Primary Energy Supply (TPES), Energy Strategy of Ukraine (ESU).

* Corresponding author: katerinapryakhina@gmail.com

1 Introduction

The development of renewable energy sources is a very urgent issue for Ukraine, especially in the context of developing cooperation with the European Union. Alternative energy sources are renewable energy sources that include solar, wind, geothermal, wave and tidal energy, hydropower, biomass energy, organic waste gas, sewage treatment plant gas, biogas, and secondary energy resources, which include blast furnaces and coking gases, methane gas for the degassing of coal deposits, transformation of the waste energy potential of technological processes. Renewable energy sources are renewable non-expanding sources of energy, namely solar, wind, aerot energy thermal, geothermal, hydrothermal, energy of waves and tides, hydropower, biomass energy, gas from organic waste, gas of sewage treatment plants, biogas [1].

The main stages of the study are determined for assessment of tendencies and prospects of cooperation between Ukraine and the European Union in the field of renewable energy sources:

– the value of alternative energy in the economy of Ukraine and the European Union;

– research of historical preconditions of cooperation between Ukraine and the European Union in the field of alternative energy;

– analysis of the current state of alternative energy sources in Ukraine;

– definition of promising directions of development of cooperation between Ukraine and the EU in the field of RES.

2 Data and Methods

Dialectical and historical approaches used to study the historical preconditions for cooperation between Ukraine and the European Union in the field of alternative energy. The system approach was used to assess the current state of alternative energy sources in Ukraine and the countries of Europe and analysis of Ukraine's energy strategy. For this purpose, the statistical indicators are used. Major projects on cooperation between Ukraine and the EU in the field of RES are analyzed and synthesized on perspective directions of development of Ukraine-EU cooperation in the field of alternative energy. The object of the study is the processes of establishing cooperation between Ukraine and the EU in the of RES. The subject of the study is the assessment of RES in Ukraine and EU countries.

3 Results

3.1 The value of Renewable energy sources in the farms of Ukraine and the European Union

Information on the amount of energy from renewable sources from the EU countries is given in Table 1.

Table 1. The share of RES in EU farms (% of gross final energy consumption)

	2004	2012	2013	2014	2015	2020 (goal)
EU	8,5	14,4	15,2	16,1	16,7	20
Belgium	1,9	7,2	7,5	8,0	7,9	13
Bulgaria	9,4	16,0	19,0	18,0	18,2	16
Czech Republic	6,8	12,8	13,8	15,1	15,1	13
Denmark	14,9	25,7	27,4	29,3	30,8	30
Germany	5,8	12,1	12,4	13,8	14,6	18
Estonia	18,4	25,8	25,6	26,3	28,6	25
Ireland	2,4	7,2	7,7	8,7	9,2	16
Greece	6,9	13,5	15,0	15,3	15,4	18
Spain	8,3	14,3	15,3	16,1	16,2	20
France	9,4	13,4	14,1	14,7	15,2	23
Croatia	23,5	26,8	28,0	27,9	29,0	20
Italy	6,3	15,4	16,7	17,1	17,5	17
Cyprus	3,1	6,8	8,1	8,9	9,4	13
Latvia	32,8	35,7	37,1	38,7	37,6	40
Lithuania	17,2	21,4	22,7	23,6	25,8	23
Luxembourg	0,9	3,1	3,5	4,5	5,0	11
Hungary	4,4	15,5	16,2	14,6	14,5	13
Malta	0,1	2,8	3,7	4,7	5,0	10
Netherlands	2,1	4,7	4,8	5,5	5,8	14
Austria	22,6	31,4	32,3	32,8	33,0	34
Poland	6,9	10,9	11,4	11,5	11,8	15
Portugal	19,2	24,6	25,7	27,0	28,0	31
Romania	16,3	22,8	23,9	24,8	24,8	24
Slovenia	16,1	20,8	22,4	21,5	22,0	25
Slovakia	6,4	10,4	10,1	11,7	12,9	14
Finland	29,2	34,4	36,7	38,7	39,3	38
Sweden	38,7	51,1	52,0	52,5	53,9	49
UK	1,1	4,6	5,7	7,1	8,2	15
Iceland	58,9	72,5	71,7	70,5	70,2	64
Norway	58,1	65,6	66,7	69,4	69,4	67,5
Albania	28,1	35,2	33,2	32,0	34,9	38
Montenegro	–	41,6	43,7	44,1	43,1	33
Republic of Macedonia	15,7	18,1	18,5	19,6	19,9	28
Turkey	16,2	13,1	14,0	13,7	13,6	–

Source: [2].

Thus, at the end of 2015, the share of alternative energy was 16.7%, which is close to the planned 20% target, which the European Union is planning to achieve no later than 2020.

So, Sweden is the leader in this indicator, which receives more than half – 53.9% of the required energy from renewable sources. Immediately, we note that Norway and Iceland have much higher rates – 69.4% and 70.2% respectively, but these countries are not part of the European Union. Eleven of the 28 EU Member States have already reached their target. It should be noted that in 2014, the share of renewable energy in the EU was 16.1%, and in 2004, when Eurostat first published such statistics, this figure was only 8.5%. By 2030, the EU plans to reach at least 27% of renewable energy [3].

3.2 Stages of the development of cooperation in the field of RES in Ukraine and the EU

In tabl. 1.2 provides the main stages of establishing cooperation between Ukraine and the European Union in the field of RES. In the process of establishing cooperation, you can distinguish five main and most significant stages.

Table 2. Stages of the development of cooperation in the field of RES in Ukraine and the EU

Stage	Date of the beginning of the stage	Content
1	May 9, 1992	On May 9, 1992, at the international summit in Rio de Janeiro, most countries in the world, including Ukraine, signed an agreement on sustainable development, which ensures the reproduction of a living standard without degradation of the environment. The agreement was named the United Nations Framework Convention on Climate Change. It came into force on March 21, 1994 [4].
2	February 4, 2004	In 2004, the Verkhovna Rada of Ukraine adopted the Law "On Ratification of the Kyoto Protocol to the UN Framework Convention on Climate Change". The signing and ratification of the Kyoto Protocol allowed the determination of commitments, the establishment of specific goals and mechanisms for their achievement. The Kyoto Protocol came into force on February 16, 2005 [4].
3	December 7-18, 2009	In December 2009, the United Nations Climate Change Conference in Copenhagen ended with "taking into account" the extremely controversial compromise document "Copenhagen Accord", the future of which is unclear. Most States acknowledge that the Kyoto Protocol to the Framework Convention on Climate Change is ineffective and needs to be replaced by a document that would involve both developed and developing countries, which already account for most of the global greenhouse gas emissions [1].
4	December 8, 2012	Due to the expiration of the Kyoto Protocol, the Eighteenth United Nations Climate Change Conference was held in Doha, Qatar in 2012, which resulted in agreement to extend the Kyoto Protocol from 2013 to 2020 [3].
5	July 14, 2016	On April 22, 2016 Ukraine signed the Paris Agreement on combating climate change, and on July 14, 2016, the agreement was ratified by the Verkhovna Rada [5].
6	December 27, 2017	On December 27, President of Ukraine Petro Poroshenko signed the Law of Ukraine "On the Accession of Ukraine to the Statute of the International Agency for Renewable Energy Sources (IRENA)". Membership in IRENA will allow Ukraine to more actively develop the renewable energy sector to reduce energy dependence. It has the expertise of other IRENA member countries to do this.

Source: Own processing.

Thus, the international cooperation of Ukraine with the EU in the field of alternative energy has been going on for 25 years now. During this time, Ukraine has achieved some results in the field of alternative energy and use of RES. Therefore, it will be necessary to carry out an analysis of the state of affairs of Ukraine in the field of alternative energy, to allocate the most promising RES, which is provided by the state, as well as to analyze energy strategies of the leading European countries.

3.3 Analysis of the current state of alternative energy sources in Ukraine

Energy Strategy of Ukraine for the period up to 2035 "Safety, Energy Efficiency, Competitiveness" dated September 25, 2017 provides that by 2025 Ukraine's energy sector reform will be completed and priority targets for security and energy efficiency achieved, innovation renewal and integration with the energy sector of the EU will be ensured [6].

Despite global trends in the use of alternative energy, Ukraine is still lagging far behind developed countries. According to Eurostat by September 2017, only 6% of the energy produced in Ukraine was produced using renewable energy sources [7]. However, this is a significant progress, as in 2016, the RES contribution to the overall energy balance was only 1.05-1.16%.

In 2015, the Total Primary Energy Supply structure of Ukraine was characterized by a high proportion of natural gas of 28.9 %, (26.1 million tons of oil equivalent). The share of nuclear power was 25.5 % (23.0 million tons of oil equivalent); coal – 30.4 % (27.3 million tons of oil equivalent); crude oil and petroleum products – 11.6 % (10.5 million tons of oil equivalent); biomass (biomass, fuel and waste) – 2.3 % (2.1 million tons of oil equivalent); Hydroelectric power station – 0,5 % (0,5 million tons of oil equivalent); thermal energy (thermal energy of the environment and waste resources of technogenic origin) – 0.5 % (0.5 million tons of oil equivalent) and WEU and SES together – 0.1 % (0.1 million tons of oil equivalent). The total share of all RES was 3.6 million tons, or only 4 % (Table 3) [8].

Table 3. Total Primary Energy Supply structure of Ukraine, million tons of oil equivalent

Name of the source of TPES	2010	2015	2020 (forecast)	2025 (forecast)	2030 (forecast)	2035 (forecast)
Coal	38,3	27,3	18	14	13	12
Natural gas	55,2	26,1	24,3	27	28	29
Petroleum products	13,2	10,5	9,5	8	7,5	7
Atomicenergy	23,4	23,0	24	28	27	24
Biomass, biofuels and waste	1,5	2,1	4	6	8	11
Solar and wind power	0,0	0,1	1	2	5	10
HPP	1,1	0,5	1	1	1	1
Thermal energy	0,5	0,5	1	1	1,5	2
Total	132,3	90,1	82,3	87	91	96

Source: [8], own processing.

The nuclear power plants will remain the base power of Ukraine, and by 2035 their share in the overall energy balance will be more than 50%. An important task in this situation is diversification of nuclear fuel supplies. The role of coal generation in the energy balance by 2035 will remain significant – 25-30%, which will require supplies in excess of 20 million tons of coal per year. Ensuring own production at competitive prices will be an important component of Ukraine's energy security [6]. Regarding alternative energy, the Strategy envisages a steady expansion of the use of all types of renewable energy, which will become one of the tools for guaranteeing the energy security of the state (Table 4).

Table 4. The share of RES in the energy balance of Ukraine

Description of the key performance indicator	2015	2020	2025	2030	2035
Share of RES (including hydro generating capacity and thermal energy) in TPES,%	4%	8	12	17	25
The share of RES (including hydro generating capacity) in electricity generation,%	5	7	10	13	25
The share of local alternative sources of oil in local fuel and energy balances,% of total consumption		10	15	18	20

Source: [6], own processing.

In the short to medium term (by 2025), the ESU predicts an increase in the share of renewable energy to 12% of the TPES and at least 25% by 2035 (including all hydro generating capacities and thermal energy).

The main RES in Ukraine is wind and solar energy, as well as biomass. In addition, large hydroelectric plants make their contribution, but small hydroelectric power stations are almost not built. In 2016, the objects of alternative energy budo produced 1.7 billion kWh, and in 2017 to produce 1.9 billion kWh, which is 11.76% more [8].

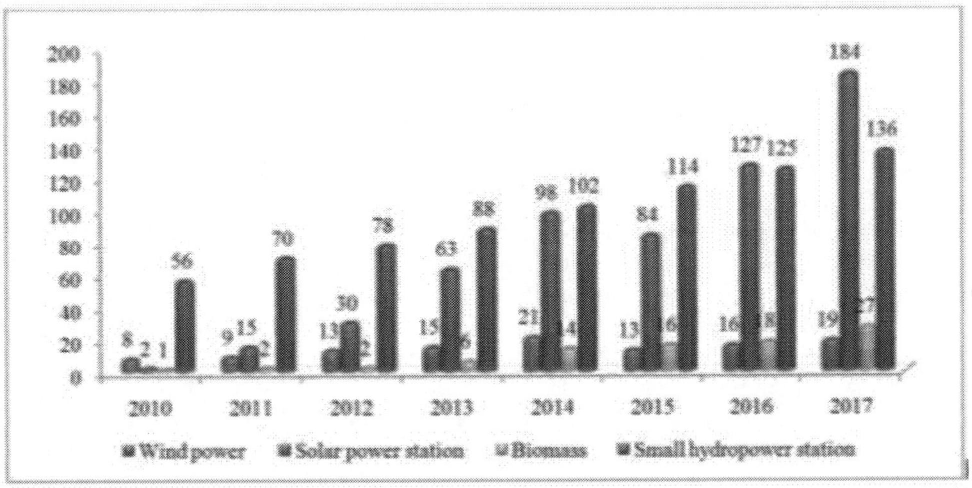

Fig. 1. Number of renewable energy generation by type in 2010-2017

Source: [8], own processing.

Acording to Fig. 1 in 2017 the most popular RES is Solar station (SPS) – 184 station, at the second place – Small hydropower station (SHS) – the number is 136. The least popular are wind power and biomass [13]. In the first half of 2017 in Ukraine 79 new RES facilities were built, including 67 SPS. Their total capacity was 182.7 MW. The power of the SPS has increased by 23%, or 132 MW, compared to the same data as of January 1, 2017 and amounted to 705 MW. According to the results of three quarters of 2017, the installed power of sources of alternative energy increased by 18% and amounted to 1320 MW. Also 201 MW of new RES capacities was introduced, which is 3.7 times more than in the same period of the previous year. Among these capacities, 83% were solar power plants.

Recently, solar power plants are becoming very popular, in particular small private SPS, whose owners sell excess energy production at a green tariff. To date, in Ukraine there are 2,223 households have their own solar station connected to the common electricity grid.

The segment of wind power plants has grown by 4.7%, or by 20.7 MW, from the beginning of the year and amounted to 458.7 MW of energy. The wind energy market is led by such companies as LLC "Ukrainian Wind Turbines of Ukraine", WindPower Ltd. and Windcraft Ukraine LLC.

An important renewable energy source in Ukraine is biomass and biogas. Over the last four years, the growth rate of production and use of biofuels has been around 42% per year. In 2017, the power of biomass and biogas power plants increased by 2 MW and amounted to 33 MW [9].

As of June 1, 2017, the total capacity of RES in Ukraine was 1461.7 MW. This includes wind and solar power plants, small hydro power plants, as well as power plants that operate on biofuels. So one of the most important tasks for today is to search for such energy sources that would reduce the use of traditional energy sources and reduce the dependence of these countries on imports. Currently, issues of energy independence are very relevant for Ukraine. The country's potential allows efficient use of RES, in particular solar and wind power, as well as biogas for the country's energy supply. The current Energy Strategy of Ukraine until 2035, as a whole, meets Ukraine's accrued obligations towards the EU on renewable energy sources and aims to increase the share of RES to 25% in 2035.

3.4 Promising directions of development of cooperation between Ukraine and the EU in the field of RES

Ukraine closely cooperates with the EU on the reform of corporate governance and the implementation of municipal projects. On May 17, 2017, Pavel Petrenko and Director of the European Bank for Reconstruction and Development, Frances Maliz, agreed on cooperation on energy, small and medium business support, and land reform. The EBRD provides Kyiv with significant advisory assistance on many issues. In addition, the EBRD has a dedicated USELF support program aimed at investing in REE. The program ranges from 500 million euros to 1 billion euros. The scheme of work of this program is quite simple and involves exclusively the co-financing of the project. That is, the EBRD agrees to finance 70% of the cost of the alternative energy facility, provided that 30% will be paid by the founders of the project or other sources of financing, while the own part of the costs must be invested in the project in the first place. Similarly, there are other banks and financial organizations in the EU that are ready to invest in alternative energy in Ukraine.

Significant potential for cooperation is represented by the International Agency for Renewable Sources IRENA. On October 5, 2017, the draft Law of Ukraine "On Ukraine's Accession to the Statute of the International Agency for Renewable Energy Sources (IRENA)" was submitted to the Verkhovna Rada for consideration, and on December 5 the law was passed. Membership in IRENA allows applications to the Abu Dhabi Development Fund to receive long-term loans at low interest rates. Admission to IRENA also provides additional guarantees to foreign investors regarding contributions to alternative energy projects in Ukraine.

In particular, the Finnish corporation "NEFCO" is interested in realization of "green" projects in Ukraine. Between this corporation and the State Energy Efficiency, negotiations are being held to create a joint venture that will produce materials and energy efficient equipment according to European technologies and standards.

On November 29, 2017, the Finnish Ministry of Foreign Affairs and NEFCO signed an agreement to create a fund for alternative energy projects in Ukraine [10], which interested Finnish businesses in investing in Ukrainian renewable energy projects.

Also, the Lviv region receives significant assistance in the implementation of environmental projects and clean energy projects. In 2017, the Eastern European Energy Efficiency and Environment Partnership and the John Nurminen Foundation provided grants to “Lvivvodocanal” to rebuild and renew the wastewater treatment and biogas production system.

On December 6, 2016, the Lviv Regional Administration signed a memorandum with the Turkish company Atlas GlobalEnergy LLC on the construction of a new wind power plant in the Skolevshchyna with a total capacity of 20 MW. Construction is scheduled to begin in 2018. Turkish investors are also planning to build a new wind farm within the Volovets and Svalyava districts of Lviv Oblast. The station will consist of 50 windmills and will have a total capacity of 120 MW.

Thus, it is relevant to identify the most promising directions for the development of cooperation between Ukraine and the EU in the field of alternative energy and implementation of RES. In addition to the EU and Ukraine, international and domestic actors are also domestic and foreign enterprises of all forms of ownership, consulting agencies, higher educational institutions and scientific research institutions. Today one of the most effective areas of cooperation is joint research in the field of alternative energy. To this direction, such forms of cooperation as realization of joint projects and initiatives, which leads to the creation of joint ventures, the allocation of international banks and funds of subsidies and loans [11, 12].

The second area of cooperation is the organization of cooperation at the expert level. It is this direction that is most widespread at the present stage, because it has a much lower level of risk. The forms of implementation of this direction are holding forums, meetings at the international level, consultations and recommendations of European experts in the field of legislative regulation of alternative energy, environmental taxation, implementation of environmental projects. Thus, for example, EBRD and IRENA are actively providing advisory services to Ukraine.

The third area of cooperation is to increase the level of environmental education of Ukraine’s population in the field of alternative energy, which takes place in the form of organizing and conducting internships at higher education institutions in alternative energy companies, exchanging students, conducting international scientific conferences in this field.

4 Discussion and Conclusion

The performed scientific research allows us to formulate the following conclusions:

1. Due to climate change, the gradual depletion of traditional energy resources and the trend towards increased energy consumption and the desire of countries to reduce their dependence on imported fuels the development of alternative energy and the transition of farms to renewable energy sources is relevant.

2. One of the first attempts to regulate emissions was the Framework Convention on Climate Change. As we see, the Kyoto Protocol to the UNFCCC, which ended its operation in 2012, has proved ineffective. The Paris agreement has replaced it. International cooperation with the EU in the field of alternative energy has been going on for 25 years now. During this time, Ukraine has achieved some results in the field of alternative energy and use of RES.

3. According to the United Nations Framework Convention on Climate Change, the EU member states commit to reaching 20% of alternative energy in farms until 2020. According to the adopted commitments and the National Energy Strategy, Ukraine should achieve 11 % renewable energy in the energy sector by 2020 and 25 % by 2035. The analysis led to the conclusion that the EU certainly achieved its goal in due time, but Ukraine plans look unrealistic because renewable energy in the country today are a very small proportion and require significant funding.

4. An assessment of the current state of alternative energy sources in Ukraine has shown that the most promising are solar and wind energy, as well as the use of biofuels.

5. The prospect of international cooperation between Ukraine and the EU in the field of alternative energy is the development and implementation of "green energy" projects. The tool for this purpose in Ukraine is financing of such international organizations as EBRD, IBRD, NEFCO and IRENA. Further, there are own state sources of funding, as well as funds from private investors. The main areas of international cooperation are conducting joint scientific developments in the field of alternative energy, organizing cooperation at expert level, and raising the level of environmental education of Ukraine's population in the field of alternative energy. International cooperation should be based on the principles of partnership, innovation, strategic, coordination and coherence of action.

Thus, Ukraine and the countries of the European Union have been closely cooperating in the field of alternative energy for over 25 years. Recently, cooperation has deepened significantly, it includes not only investment projects, but also advising on reforms in the energy sector, support for small and medium-sized businesses, as well as land reform. Due to the investment of international organizations, private foreign investors and the implementation of joint projects, the share of RES in Ukraine's total energy production has increased significantly.

References

1. T. Avdeeva, Prospects for International Negotiations on Climate Change: Following the Copenhagen UN Conference. *Biosphere*, 1, 73-82, (2010).
2. Eurostat Statistics Explined. *Energy from renewable sources* [online]. Available at: http://ec.europa.eu/eurostat/statistics-explained/index.php/Archive:Energy_from_renewable_sources (2018)
3. *Initial report "Emissions Trading Research"* [online]. Available at: http://climategroup.org.ua (2018)
4. A, Pochtovyuk, K. Pryakhina, Trade in Quotas as One of the Mechanisms of the Kyoto Protocol. *Problems of Modern Economy*, 3, 300-304, (2012)
5. I. Trofimova, *Paris Agreement will stimulate modernization of Ukrainian economy* [online]. Available at: http://zhytlo.in.ua/ua/napryamok/chista_energya/rina_trofmova_parizka_ugoda_dast_poshtovh_do_modernzac_ukransko_ekonomki.html (2018)
6. *Ukraine's Energy Strategy for the period up to 2035: Security, Energy Efficiency, Competitiveness*, 25.09.2017 [online]. Available at: http://mpe.kmu.gov.ua/minugol/control/uk/doccatalog/list?currDir=50358 (2017)
7. Power reform. *Ukraine is capable of 74% of electricity produced from renewable sources* [online]. Available at: http://reform.energy/news/ukraina-sposobna-74-elektroenergii-proizvodit-iz-vozobnovlyaemykh-istochnikov-evrostat-3251 (2018)
8. *The State Statistics Service of Ukraine* [online]. Available at: www.ukrstat.gov.ua/ (2018)
9. G. Geletuha et al., Bioenergy in Ukraine: current state and development prospects. *Part 2, Industrial Heat Engineering*, 37(3), 65-73, (2015)
10. *Ministry of Foreign Affairs of Finland and NEFCO Corporation have signed an agreement on the creation of a Trust Fund to finance Clean Energy Projects in Ukraine* [online]. Available at: https://fromua.news/article/55259725/mzs-finlyandii-ta-korporatsiya-NEFCO-pidpisali-ugodu-pro-stvorennya-tsiljov/ (2018)

https://doi.org/10.1051/shsconf/20196101008

11. A. Pochtovyuk, V. Semenikhina, Business relations between Ukraine and China: development options. *SHS Web of Conferences, Innovative Economic Symposium 2017*, 39, 01019, (2017)
12. J. Dušek, European Grouping of Territorial Cooperation as a Way of Cross-Border Regional Cooperation within the European Union. *Conference Proceedings „16th International Colloquium on Regional Sciences"*, Brno, Czech Republic, 329-336, (2013)
13. *Ukrainian Association of Renewable Energy* [online]. Available at: http://uare.com.ua/

Banking competition and its development in Russia

Yury Korobov[1,*]

[1]Saratov Socio-Economic Institute (branch of Plekhanov Russian Economic University), Department of Banking, Money & Credit, 410003 Saratov, Russia

Abstract. The article discusses theoretical issues of banking competition as well as its ifluences on the development of banking system (transformation of the essence of a bank; change of the target setting in banking; transfer of emphasis from operations to services; universalization of banking; transition from individual services to integrated banking products; change of priority from price to non-price methods of competition; transition from extensive to intensive type of network development; increasing the role of near-banks and nonbank banks). Factors influencing banking competition in Russia are shown: both restraining (reduction in the number of banks; concentration of assets at the largest banks; uneven territorial distribution of banks; local character of banking markets) and strengthening (financial globalization; digitalization of economy; forming of new banking culture).
Key words: banking system, banking market, banking competition, banking product, methods of competition, factors of competition, financial globalization, digitalization of economy, banking culture

1 Introduction

Competition is one of the most important characteristics of banking markets. Strengthening of banking competition in the leading countries occurred throughout the 20th century. Therefore, it is not surprising that interest in the problems of banking competition in our country (in particular, the author of these lines) arose simultaneously with the emergence of the first sprouts of the market economy in the early 1990s.

Nowadays we can see that banking sector undergoes profound transformations [1], and one of the main driving forces of these transformations is the development of competition. Therefore, it seems to be interesting to consider the essence of banking competition, as well as to give an overview of its development in Russia.

General problems of competition have been studied by many authors, beginning with the classics of economic science and theory of management, such as *Marx* [2], *Hilferding* [3], *Chamberlin* [4], *Robinson* [5], *Hayek* [6], *Porter* [7], and ending with contemporary authors such as *Yudanov* [8], *Rubin* [9],etc.

Theoretical issues of banking competition were discussed in works of such scientists as *Aust* [10], *Griesel* [11], *Miretsky* [12], *Samoylov* and *Bachalov* [13], *Petrov* [14], etc. Since

* Corresponding author: yurikor1965@gmail.com

1994 banking competition became the main subject of research by the author of these article [15, 16].

Applied aspects of banking competition are covered in many publications, among which are articles of Tabacco [17], Cornaggia, Mao, Tian & Wolfe [18], Fernández, González & Suárez [19], Jayakumar, Pradhan, Dash, Maradana & Gaurav [20], etc.

However, despite the fact that the issues of banking competition are of great theoretical and no less practical interest, their coverage in the economic literature remains rather modest. This determines the relevance of the topic of this article.

The purpose of the article is to discuss the specific features of banking competition and its influence on the development of banking system, as well as show factors influencing banking competition in Russia.

2 Theoretical background: specific features of banking competition

Banking competition is much more difficult than competition in any other sector of economy. Specific features of banking competition are determined by: 1) composition of participants; 2) structure of the sphere of competition; 3) specific features of the product; 4) forms of competition; 5) influence on other spheres of the economy and public life.

Participants of banking competition. The main participants of banking competition are commercial banks. However, banks are not the only competitors in the banking market. There are two other groups of competitors: near-banks and nonbank banks. The first group include financial institutions that are not formally banks, but offer a product similar to banking products (pawnshops, leasing companies, factoring companies, forfeiting companies, pension funds, insurance companies, investment companies, clearing houses, plastic card systems, trust companies, etc.). The second group includes non-financial enterprises whose main activities are not related to the financial sphere, but, nevertheless, are introduced into the banking market (mail, industrial enterprises, trade and service organizations, etc.). Introducing a particular banking product to the market, a commercial bank almost always faces competition not only from other commercial banks, but also from listed organizations. Thus, banking competition cannot be defined as a rivalry between banks, it is more correct to define it as a rivalry between all participants of the banking market. No other sphere of the economy has such a heterogeneous composition of participants.

Sphere of banking competition. The peculiarity of the banking market, which is the sphere of banking competition, is the complexity and diversity of its internal structure. The banking market consists in fact of a number of different markets, the main of which are the market of credit services, the market of deposit and investment services, the market of cash and settlement services, the market of currency services, the market of trust services and the market of consulting services. Each of the listed banking markets, in turn, consists of a number of banking industries, which should be understood as markets for individual banking services.

Unfortunately, the term "banking industry" in literature and in practice is often used to refer to the whole sphere of banking activity, as a synonym for "banking sector" or "banking". But if we assume that the industry is a group of enterprises that produce similar or interchangeable goods or services, the banking market on which a multitude of diverse products and services are sold cannot be recognized as a single industry. For example, there is nothing in common in attracting deposits from the population and issuing loans to multinational companies. It is difficult to find more dissimilar activities. According to *Porter*'s remark, it is the same as comparing the production of steel with the production of motion pictures [21]. Thus, banking is not a single industry, but rather a set of industries and branches, and each of them deserves carrying a specific analysis and developing a separate competitive strategy.

Banking product. A banking product should be understood as a separate banking service or a package (package) of interrelated services. The peculiarities of banking product are intangibility, complexity for perception, inseparability from the source, impermanence of quality, non-conservation, contractual nature, connection with money, length of service in time, secondary nature of satisfied needs.

From the bank's point of view, all these features are negative, since they lead to the formation of barriers (basically, psychological) between the bank and its clients. As a result, banking products a priori lose in attractiveness in comparison to material goods, and it brings difficulties for banks in promotion their products to their customers.

Forms of competition. In addition, due to the high level of risks in the banking sector there is not only individual, but also group competition. Various types of banking associations are formed, within which competition is completely or partially replaced by cooperation. Unions can be created both on a non-commercial and on a commercial basis (for example, for consortial lending).

Socio-economic impact. Any competition is a sort of natural selection, during which the strongest wins. Those who are not competitive, suffer losses or even become bankrupt. But while in many spheres of the economy (industry, commerce, agriculture, etc.), the bankruptcy of a competitor will be reflected mainly on its employees and creditors, the "circles on the water" caused by bankruptcy of a bank are much broader and more destructive. Bankruptcy of one bank can provoke anxiety of depositors of other banks and mass withdrawal of their deposits. This, in turn, may cause problems with liquidity in the banking community, which will affect counterparties and depositors of all banks, and lead to a system crisis. Thus, banking competition, being a microeconomic phenomenon, is capable of providing macroeconomic impact.

Since the results of banking competition are not indifferent to society due to their significant socio-economic impact, it cannot remain outside the sphere of state regulation. As a rule, in all countries large system-forming banks have "immunity" from bankruptcy and can rely on the state in critical situations. Also the state should make efforts to control the unfair competition, because it can undermine the clients' confidence in the banking system.

3 Influence of competition on the development of the Russian banking system

Currently, in Russia, as in many other countries, the banking system is changing, and the main driving force behind the changes is the strengthening of banking competition.

Specific manifestations of the **influence of competition on banking system** are: 1) transformation of the essence of a commercial bank; 2) change of the target setting in banking activities; 3) transfer of emphasis in banking activities from operations to services; 4) universalization of banking activities; 5) transition from provision of individual services to integrated banking products; 6) change of priority from price to non-price methods of banking competition; 7) transition from extensive to intensive type of banking branch network development; 8) increasing the role of near-banks and nonbank banks.

Transformation of the essence of a commercial bank. The concept of a bank as an institution changed to the concept of a bank as an enterprise. There is a formation of a new banking culture in all aspects – intra-corporate culture, social culture, culture of relations with customers, partners, competitors.

Change of the target setting in banking activities. For a long period of time, the main purpose of a commercial bank was to gain profit. However, already in the last third of the

20th century, as a result of the introduction of marketing, the satisfaction of customers' needs is shifted to the top of the pyramid of objectives of a commercial bank, pushing the profit gain to the background. This trend can further be seen in the development of social and ethical marketing concepts, as well as in spreading of Islamic banking model.

Transferring the focus in banking activities from operations to services. In current conditions of intensive banking competition, it is necessary for bank employees to consider their activities not only as technical implementation of operations, but primarily as providing services to consumers. This, in turn, changes the requirements for the qualification of a bank employee, increasing the importance of communication skills.

Universalization of banking activities. The diversification of banking activity that took place in the 20th century led to the prevailing of universal banks in most countries.

Transition from the provision of individual services to integrated banking products. This allows to simultaneously take into account the interests of customers, whose needs are often complex, and the interests of the bank, aimed at expanding their sales volume.

Change of priority from price to non-price methods of banking competition. Basically, any bank offers pretty much the same assortment of services as all other banks, therefore banking competition a priori should be price-based. However, due to the economic, legal and psychological limitations of price methods, banks are making significant efforts to differentiate their products, to ensure consumers' loyalty based on the image formation. Today, to ensure their competitiveness, banks use a complex set of price and non-price methods of competition (with a gradual shift of priorities in the direction of non-price competition).

Transition from extensive to intensive type of banking branch network development. While in the 20th century in many countries (including Russia) there was a significant increase in the number of commercial banks and an intensive expansion of their branch network, the main trends of the 21st century are increase in bank capital concentration and rationalization of banking branch network.

Increasing the role of near-banks and nonbank banks. Some of these organizations are comparable to banks in terms of financial strength (investment companies, insurance companies, etc.), and their number is growing.

4 Factors influencing banking competition in Russia

Many factors influence the development of banking competition in Russia. At the same time, some of them have a deterrent effect on bank competition, while others have a stimulating effect.

Factors which are restraining banking competition in Russia are: 1) important role of the state in functioning of banking system; 2) reduction in the number of banks; 3) concentration of assets at the largest banks; 4) uneven territorial distribution of banks; 5) local character of regional banking markets.

Important role of the state in the formation of banking system. In most Western countries, the "pioneers" of the banking market were non-state banks, the state began to actively influence the banking sector only in 19th century, with the emergence of central banks. Moreover, even some central banks have existed for a long time on the basis of private property (for example, the Bank of England was nationalized only in 1946).

In Russia, banking was originally developed as a sphere of activity of the state. The first banks, which appeared in the 1730s, were state-owned. The functions of the central bank were entrusted to the State Bank of the Russian Empire, established in 1860. Private commercial banks began to develop only from the 1860s. The nationalization of banks in the Soviet period led to the state monopoly of banking.

The revival of private banks occurred during the commercialization of the banking system in the 90s of the 20th century.

Reduction in the number of banks. In the first 5-6 years of operation of the market banking system in Russia, there was an exceptionally intensive growth in the number of commercial banks and other participants in the banking market. By 1995, there were more than 2,500 commercial banks. However, in the second half of the 1990s, the number of Russian banks began to decrease, which was due to the financial crisis and subsequent state policy aimed at restructuring of the banking system. As of 01.08.2018, only 476 functioning banks are remaining [22]. As a result, there is a significant gap in terms of the number of banks and their branches per capita between Russia and countries with developed economies.

Concentration of assets at the largest banks. In fact, the main players of the Russian market are only about 20 the largest banks. The share of the 20 largest banks as of 01.08.2018 accounted for 79.9% of all banking assets, including 59.8% - to the 5 largest banks [22].

Uneven territorial distribution of banks. Today, about 50% of all Russian banks are territorially located in the Moscow region. These banks control over 90% of banking assets [22].

Local character of regional banking markets. Historically most of the banks in regions of Russia used to have a local orientation and their branch network did not go beyond the borders of the home region. Therefore competition on every single regional banking market was almost only between local banks, and the country's banking market was split into many relatively isolated local markets. However, the development of information technologies and the widespread introduction of remote banking services are gradually erasing the boundaries between regional banking markets.

On the other side, there are **factors which are strengthening banking competition in Russia**: 1) financial globalization; 2) digitalization of economy; 3) forming of new banking culture. Consider these factors in more detail.

5 Impact of financial globalization

Intensity of banking competition in developed countries pushes their banks to go beyond national borders. My previous studies led to the conclusion that the process of globalization forces to reconsider the established idea of the essence of the banking system [1]. Today we should abandon the interpretation of banking system as the set of credit institutions within a specific country. The banking system of the 21st century is essentially an international one.

The most visible process of transformation of the national banking system into an international one has manifested itself in the countries of Western Europe, but some of its manifestations are also present in Russia. In 2000s, there has been a steady trend towards increased penetration of foreign banking capital into Russian banking market. Now 115 out of 476 Russian banks (24.2%) have a non-resident share in their authorized capital [22].

The influence of financial globalization on the banking system of Russia is ambiguous. On the one hand, the inflow of foreign bank capital contributes to the expansion of assortment and quality of banking services, introduction of international banking standards, creation of new jobs, increase of tax revenues to the state budget, and attraction of foreign investments. On the other hand, banking globalization can lead to an increase in the dependence of the country's economy on foreign capital. This brings up the need to increase the **competitiveness of national banking system**.

In my opinion, there are two main approaches to assessing the competitiveness of the national banking system: 1) assessment of the competitiveness of individual banks; 2) assessment of the competitiveness of banking system as a whole.

The competitiveness of individual banks is determined by their competitive advantages and affects the financial results of their activities.

Main factors of competitiveness of individual banks are:

- resources (capital, borrowed funds);
- the level of costs (makes impact on interest rates and fees);
- product differentiation (quality and assortment of services, bank's image).

The competitiveness of banking system as a whole can be defined as its resistance to external influences. It affects the stability of the monetary system of the country and the economy as a whole.

Main factors of competitiveness of banking system as a whole are:

- economic factors (availability of credit resources and qualified personnel; developed infrastructure; quantitative and qualitative level of demand for banking services; intensity of banking competition; competitiveness of the economy as a whole);
- political and legal factors (conditions of licensing of banks, supervision standards, legislative protectionism);
- socio-cultural factors (national psychology, level of knowledge, banking culture);
- technological factors (modern information technologies).

The advantages of the Russian banking system include a large resource potential, adequate regulation, and susceptibility to technology. But at the same time, there are a number of "bottlenecks": underdeveloped infrastructure, indiscriminate demand, sluggish competition.

Thus, in Russia there is a need to establish a system of measures to strengthen the competitiveness of the national banking system. These measures should include:

- improvement of the competitiveness of individual banks (capital increase, optimization of internal structure, improvement of the range of banking services and the level of customer service);
- development of banking infrastructure (adaptation of the regulatory framework to modern conditions, modernization of information and methodological subsystems banking infrastructure, increase of banking market transparency);
- raise of awareness and economic literacy of bank customers, primarily in retail banking;
- formation of marketing thinking among bank employees, development of communication abilities, etc.

6 Impact of digitalization

In the context of digitalization of the economy, the introduction of information technologies is becoming one of the most important factors affecting the development of all sectors of the economy. The influence of this factor on competition in banking sector is primarily manifested in impact on the *characteristics of banking services*.

As mentioned above, specific features of banking services are their intangibility, complexity for perception, inseparability from the source, impermanence of quality, non-conservation, contractual nature, connection with money, length of service in time, secondary nature of satisfied needs. From the point of view of a bank, these features play generally negative role, since most of them lead to the formation of barriers between a bank and their customers (mainly psychological). As a result, banking services a priori lose in attractiveness to material goods, and it is more difficult for banks to carry out promotion of their products than for enterprises in the sphere of material production. Let's see what is changing with the introduction of information technologies in banking, examining each specifuc feature of banking services.

Intangibility of banking services. Services have no material substance, they can not be perceived by the senses. In this regard, consumers show increased attention to external elements of service – equipment, forms of documents, etc. (figuratively speaking, "the wrapper of banking services"). Of course, with introduction of information technologies banking services do not become tangible. However, the transfer of market relations to the information environment, where all goods are virtual, deprives sellers of material goods of their advantages over sellers of services.

Complexity for perception. Unlike many other types of services, use of banking servicing requires a certain cultural and educational level of consumers. To facilitate the perception of banking services by the customers banks are providing them with personal advice and information materials explaining the the benefits of banking services. The development of information technologies, does not simplify the content of services, but significantly expands the possibilities of bank's explanatory activities. For example, reference materials placed on banking websites take full advantage of multimedia technologies, and the use of hypertext adds an element of interactivity.

Inseparability from the source. The service does not exist separately from its provider. Buyers of material products (for example, food) first of all they pay attention not on the source (in most cases they have no chance to visit the producer), but on the product itself. In contrast, bank customers normally could not use the service without visiting the bank, and could not be indifferent to personal qualities of bank employee (appearance, knowledge, communication skills, etc.) or to exterior and interior view of bank office (location, design, etc.). When a customer uses services through internet or mobile banking interface, the physical localization of the source ceases to play a big role, and the cost of creating a website is not comparable to the cost of construction of buildings and their design.

Impermanence of quality. Although many banking institutions offer clients a similar or even the same range of services, the absolute identity of banking products (such as in mass production in industry) is not achievable. Many banking services involve intensive communication between clients and employees, and employees may hsve different levels of technical and communication skills, different attitudes to work, etc. Moreover, the same employee may show different levels of service depending on the situation, mood, health, etc. Digitalization gives banks a chance to eliminate impermanence of quality in providing standardized services. The era of "mass production" of banking services began with the emergence of ATMs that provide services to all customers equally, regardless of the time of day, queue length, behavior or appearance of the customer. The same is true for internet banking, mobile banking, etc.

Non-conservation of banking services. Services can not be stored and conserved, they can not be "prepared for the future". At the same time, the volume of customer demand is constantly changing, it can be different on different days of the week and even at different times of the day. As in the previous case, a partial solution to this problem has been already found with the development of ATMs, and contemporary digitalization offers more advanced solutions.

Contractual nature of banking services. The provision of the absolute majority of banking services involves the conclusion of civil contracts between the bank and the customer. This creates certain difficulties for customers, and banks are facing the need for a thorough explanation of the content of banking services and the contractual terms of their provision to the customers, and ultimately – for gaining the trust of the customers. Thus, the customer service in a bank become partly similar to educational activity. This "distrust factor" does not disappear even in the new digital era. On the contrary, one of the main problems today is the lack of adequate legal regulation of many aspects of digital economy (cryptocurrency, etc.).

Connection of banking services with money. People pay special attention to everything related to money, and this also makes the bank's activities dependent on the clients' trust and requires efforts aimed at strengthening this trust. Although the essence of services does not change with the change of technology, the fact that in the information space money functions only in cashless form, to some extent reduces the internal psychological barrier of the customers (non-cash money is easier to spend).

Length of banking service in time. When providing most banking services, the client enters into a more or less long-term relationship with the bank. This, like two previous features, attaches particular importance to the trust-based nature of the bank-client relationship. In fact, for example, when opening an account, the client trusts the bank with his/her money and enters into a transaction, the course of which is initially uncontrollable for him/her. New information technologies provide the customer with the opportunity to receive more detailed information at the stage of choosing a bank for servicing.

Secondary nature of satisfied needs. Banking services don't satisfy the primary needs of customers (i.e. food, clothing, housing, etc). They are designed to satisfy financial needs (which are derived from primary needs). As a result, banking services lose in attractiveness to material products as well as to services which directly meet primary needs of consumers (i.e. health care, etc.). This complicates the task to promote banking services on the market and requires additional efforts from banks. Although information technology does not change the nature of needs, it provides banks with additional opportunities to satisfy these needs. In addition, transactions with material product carried out in digital space do not provide immediate satisfaction of needs, it always takes time to deliver product to the consumer.

Our analysis showed that the development of information technologies can significantly "shorten the distance" between producers and consumers of banking services. Russian banks almost do not lag behind Western banks in terms of digital innovation, so the development of information technologies is nowadays one of the main factors in strengthening banking competition in Russia.

7 Impact of new banking culture

As has been shown in my previous research, banking culture in the broadest sense can be defined as a set of principles of behavior of all subjects of banking activity – banks, banking customers and social institutions that ensure the functioning of the banking system. According to this, there are **three levels of banking culture**: corporate banking culture, consumer banking culture and social banking culture [1].

The elements of the *corporate banking culture* are the internal banking rules (appearance, rules of conduct in the workplace, correspondence, telephone conversations), a set of professional competencies of bank employees, the culture of communication with customers, corporate identity (branding), management culture (leadership style), banking strategy, culture of competition, principles of relations with the state and society.

Elements of the *consumer banking culture* are the financial literacy of bank customers and their willingness to use banking services.

Elements of the *social banking culture* are banking legislation, the development strategy of the banking system, regulation of bank competition (antitrust measures, combating unfair competition), banking supervision, banking culture of the mass-media. The high risks inherent to the banking system call for the regulation of its activity by the state, and a calm and balanced attitude from mass-media.

In modern Russia, changes are taking place at all three levels of banking culture. The corporate banking culture is changing under the influence of banking competition, Russian banks are becoming more customer-oriented and more receptive to innovation – both in terms of the development of new banking products and in terms of the introduction of new financial technologies, as well as techniques of promotion and sale.

The main change in the consumer banking culture in Russia is connected with the growth of financial literacy of bank customers. Bank customers become more and more demanding in choosing a bank for servicing, providing more incentives for banks to increase their level of competitiveness.

Changes in the social banking culture are manifested primarily in improving the regulation of the banking sector. One of the manifestations of this is the presence of the state strategy for the development of banking sector. In Russia strategies of development of the banking sector were developed in 2005, 2008 and 2011. Since Bank of Russia received in 2013 the status of a mega-regulator of all financial markets of Russia, new strategy is called *Main directions of development of financial market of Russian Federation*.

In general, the formation of a new banking culture has an important impact on the development of banking competition and the competitiveness of the Russian banking system.

8 Conclusion

Banking competition is the main driving force behind the changes occurring in contemporary banking system. Specific manifestations of the influence of competition on banking system are: transformation of the essence of a bank; change of the target setting in banking; transfer of emphasis from operations to services; universalization of banking; transition from individual services to integrated banking products; change of priority from price to non-price methods of competition; transition from extensive to intensive type of network development; increasing the role of near-banks and nonbank banks.

In turn, the development of banking competition in Russia is influenced by many factors, some of which have a deterrent effect on bank competition, while others have a stimulating effect. Factors restraining banking competition in Russia are: reduction in the number of banks; concentration of assets at the largest banks; uneven territorial distribution of banks; local character of banking markets. Factors strengthening banking competition in Russia are: financial globalization; digitalization of economy; forming of new banking culture.

The impact of financial globalization brings up the need to establish in Russia a system of measures to strengthen the competitiveness of the national banking system which should include: improvement of the competitiveness of individual banks; development of banking infrastructure; raise of awareness and economic literacy of bank customers, primarily in retail banking; formation of marketing thinking among bank employees.

The influence of digitalization on banking competition is primarily manifested in impact on the characteristics of banking services. Development of information technologies reduces barriers between producers and consumers of banking services and leads to increased banking competition in Russia.

New banking culture consists of three levels – corporate banking culture, consumer banking culture and social banking culture. Changes at all this levels make an impact on the development of banking competition and the competitiveness of the Russian banking system.

References

1. Y. Korobov, Global banking: transformation, innovation & competition. SHS Web of Conference – *Innovative Economic Symposium 2017*, **39**(01012), (2017)

2. K. Marx, *Das Kapital. Kritik der politischen Ökonomie, Band 3: Der Gesamtprozess der kapitalistischen Produktion,* (1894)
3. R. Hilferding, *Das Finanzkapital. Eine Studie zur jüngsten Entwicklung des Kapitalismus,* (1910)
4. E. Chamberlin, *The theory of monopolistic competition: A re-orientation of the theory of value,* (1933)
5. J. V. Robinson, *The economics of imperfect competition,* (1933)
6. F. Hayek, Der Wettbewerb als Entdeckungsverfahren. *Kieler Vorträge*, **56**, (1968)
7. M. E. Porter, *Competitive Strategy,* (1980)
8. A. Y. Yudanov, *Competition: Theory and Practice*, (1998)
9. Y. B. Rubin, *Competition: orderly interaction in professional business,* (2006)
10. E. Aust, *Der Wettbewerb in der Bankwirtschaft,* (1963)
11. H. Griesel, Funktionsweise und charakteristische Merkmale des Bankenwettbewerbs als Gegenstand wissenschaftlicher Bankbetriebslehre. *Bankbetriebliches Lesebuch,* (1978)
12. A. P. Miretsky, *Competitive position of the bank*, (1999)
13. G. O. Samoylov, A. G. Bachalov, *Banking competition*, (2002)
14. M. A. Petrov, *Banking competition and its features in Russia*, (2004)
15. Y. I. Korobov, *The theory of banking competition*, (1996)
16. Y. I. Korobov, *Practice of banking competition*, (1996)
17. G. A. Tabacco, A new way to assess banking competition. *Economics Letters*, **121**, (2013)
18. J. Cornaggia, Y. Mao, X. Tian, B. Wolfe, Does banking competition affect innovation? *Journal of Financial Economics*, **115**, (2015)
19. A. I. Fernández, F. González, N. Suárez, Banking stability, competition, and economic volatility. *Journal of Financial Stability*, **22** (2016)
20. M. Jayakumar, R. P. Pradhan, S. Dash, R. P. Maradana, K. Gaurav, Banking competition, banking stability, and economic growth: Are feedback effects at work? *Journal of Economics and Business*, **96**, (2018)
21. B. I. Deutsch, A conversation with Michael Porter. *Bank Marketing,* (1990)
22. *Statistical bulletin of the Bank of Russia*, **8** [online], Available at: https://www.cbr.ru/publ/BBS/Bbs1808r.pdf (2018)

Using Kohonen networks in the analysis of transport companies in the Czech Republic

Tomáš Krulický[1,*]

[1]University of Žilina, Faculty of Operation and Economics of Transport and Communications, Univerzitná 1, 01026 Žilina, Slovak Republic

Abstract. The transport sector has a significant impact on the performance of the Czech economy. Transport companies, of course, have their own specificities, whether they deal with ecology or the financial and economic situation. It is precisely the economic position of a transport company that needs to be analysed in order to identify the need for change, to predict the further development of such company. For analysis, a variety of methods is used, of which artificial neural networks are a very interesting and effective tool. The aim of this paper is to make a cluster analysis of transport companies operating in the Czech Republic based on this tool. The data of the financial statements of transport companies in the Czech Republic in 2016 are taken into account. Only some items from the financial statements are selected for analysis. The file is then subjected to a cluster analysis, specifically using the Kohonen networks – Statistica software. In accordance with the methodology of the contribution, the data is divided into three sets - training, testing and validation. Companies were divided into clusters in the 10x10 Kohonen Map. Some clusters are significant in terms of number of companies. These clusters are further analysed. Specific conclusions are made: A larger company generates, on average, a higher operating profit, larger companies achieve higher ROE and, in the case of a larger company, the financial leverage acts more positively.

Key words: Kohonen networks, transport companies, cluster analysis, neural networks, financial situation

1 Introduction

The importance of company analysis lies in defining needs and solutions that bring value to stakeholders. Company analysis is a disciplined approach to implementing and managing changes in organizations, whether they are beneficial businesses, governments or non-profit organizations [1]. Company analysis is used to identify and formulate the need for changes in organization functioning and the simplification of these changes. By making effective use of company analyses, it is possible to ensure that a company improves the way it is doing its business [2].

Company analysis can be measured in different ways, but Kohonen networks are considered to be a very attractive model. Kohonen networks are types of neural networks

* Corresponding author: krulicky@mail.vstecb.cz

that can be used to group sets of data into different groups [3]. The data is grouped so that the records within the group tend to be the same and the records in the different groups are different. Many experimental results show that Kohonen's networks are very effective for assessing companies [4].

A Kohonen network consists of an input layer that is completely interconnected with the output layer and learns by self-organization, without a teacher. This network has a very wide range of uses, as it is an alternative network that is applicable to most neural network calculations. However, it is mainly used for audio editing, speech processing, photos, videos, security applications, and allows the projection of highly dimensional data to lower dimensional data [5].

Kohonen Networks will be used in this paper for the analysis of transport companies in the Czech Republic. The transport sector has a significant impact on the performance of the Czech economy and is an indispensable part of the national economy. It is a field that requires strategic vision and timely and courageous, though often financially demanding, decisions [6]. The transport industry includes a wide range of means of transport. According to the type of transport routes, we distinguish road, air, rail, water, combined transport, etc. The transport industry includes all companies transporting persons or goods, by land, sea or air, from one place to another [7].

However, the transport sector has been one of the main factors contributing to environmental pollution in recent years. Much research works on energy consumption analysis. Ou et al. [8] for example, compared fuel types, including alternative fuel systems, which lead to greater fuel technology development. This development can, according to Hao et al. [9] help to reduce oil consumption in transport.

Evaluating the financial situation of companies is a complex and multidimensional process. First of all, the following aspects of business activity should be considered: profitability, financial liquidity, operational efficiency, debt and market value. Certainly there are many financial ratios that are used to measure these five aspects. However, it should be noted that different ratios can provide conflicting signals that could be observed during the recent economic crisis in the transport sector. This situation may cause problems in assessing the financial situation of the transport companies examined [10]. According to Pánek [11], it is very useful to use multidimensional statistics in studies on the financial situation of transport companies that allow for the determination of the synthetic rate that will replace the many sets of characteristics of the surveyed entity with one aggregate variable. Synthetic measure allows for the measurement of a multidimensional phenomenon, which is the financial condition of the transport company.

Compared to other companies, a transport company does not store the transport services offered in its inventory and its investments require considerable costs and a long construction time, which can affect the efficiency of the organization in the long term [12]. A comprehensive expenditure accounting system for transport companies should describe in detail all of the costs incurred, all services provided, how all of the funds of participing agencies were expended and evenly divide the costs between those participating agencies by allocating costs according to services received [13]. To this end, an accounting scheme has to be developed that can monitor all types of expenditure related to transport services [14]. The accounting approach recommended and used by successful business operations and transport systems is called full cost accounting. The use of full cost accounting means that all the costs of providing transport services are taken into account and that all different types of collection are recorded [15].

As far as freight and passenger transport in the Czech Republic is concerned, the number of goods transported grew steadily in freight transport from CZK 435,450 thousand tons to 549,086 thousand tonnes, from 2012 to 2015. However, 2016 saw a decline of almost 10,000 thousand tonnes. In passenger transport, however, the situation was different.

In 2012 the number of persons transported was 2,448,956 thousand persons, higher than in 2013-2015, by almost 60,000 thousand persons. However, in 2016, the number of persons transported grew compared to 2012 by 50,000 thousand persons. Transport performance in freight transport has steadily increased from 2012 to 2015 by 8 0000 million tkm. In 2016, however, the transport performance dropped from 76,000 million tkm to 68,000 million tkm. The passenger transport performance in all monitored years grew steadily, from 42 723 million pkm to 46 702 million pkm [16].

The aim of this paper is to perform a cluster analysis of transport companies.

2 Data and methods

For the purposes of this paper, a data set of 3989 companies operating in the transport sector in 2016 will be created. The set of companies was selected randomly from the Albertina database. The complete financial statements of these companies will be available (all financial information is in thousands of CZK). For the analysis, only some items will be used:

1. Total assets: This is information about the size of the company, namely the volume of assets (fixed, current, other).
2. Fixed assets (also long-term assets): The item describes the extent of the involvement of fixed assets and gives an idea of the level of technology used and the projected volume of outputs. It can be assumed that the higher the volume of fixed assets, the more the transport machinery (cars, aircraft, etc.) is available to the company. Of course, it is necessary to mention that this will not affect the means of production that the company has in terms of leasing. Such long-term assets are indirectly reflected in other variables - in outputs, and above all in the form of the economic result of an economic unit.
3. Current assets: represent the amount of funds, receivables and inventories in the company. It is in most cases the so-called working capital, ie those asset components that change their substance in less than one year and which are directly involved in the core business.
4. Own resources: reflect the level of business risk of business owners. This is, therefore, the amount of funds that owners can lose in case of bankruptcy.
5. Foreign resources: It is a capital that does not have a share in the company's business management. In a way, it reflects the external view of the potential success of the company.
6. Performance: characterizes the main activities of the business, ie production.
7. Value Added: This is the value added by a company to its core business of basic input material.
8. Operating result: Provides information on how the business is successful in its core business.
9. ROE: return on equity means an assessment of the equity of a company, that is, the capital provided to the business owner.
10. Profit before tax: describes the success or failure of a company in its entire business.

These are not generators of value but rather characteristics that may have, albeit indirect, influence on the performance of a company.

He file will then be subjected to cluster analysis, specifically using Kohonen networks. For cluster analysis, Dell's Statistica software will be used in version 12. Data Mining will be used as a specific neural network tool. Here we select neural networks without a teacher – Kohonen's networks. We will specify the data for analysis. In all cases, these are continuous predictors. The file will be divided into three parts:

1. Training Data Set: Represents 70% of companies in the set. The Kohonen network will be created on this dataset.
2. Testing Data Set: This is 15% of the original set of companies. Using this data set, we will verify the parameters of the generated Kohonen network.
3. Validation dataset: 15% of companies in the file will also be involved. Using this dataset, we will also test the Kohonen network to see whether it is or is not usable.

We will determine the topological length and the topological width of the Kohonen map at 10. The number of iterations of the calculation will be set at 10,000. However, I will recall that the level of error is decisive. Unless there is an additional iteration to improve the Kohonen network parameters, the training will be terminated before the 10,000th iteration is completed. If the network parameters are improved even at 10,000 iterations, we need to repeat the process and set the higher value of the required iterations to make sure the result is the best possible.

3 Results

In accordance with the methodology of the contribution, the data was divided into three sets - training, testing and validation. The descriptive characteristics of all three sets and the original data set are given in Table 1.

Table 1. Descriptive characteristics of the dataset

Samples	Total assets	Fixed assets	Current assets	Own resources	Foreign resources	Perform.	Value added	Operating result	ROE	Profit before tax
Minimum (Training)	-533	-65	-704	-255943	-19155	-9	-7167334	-2465847	-	-1089
Maximum (Training)	72746546	66982500	14350000	55525850	45481000	7042804	3627035	6170000	319,000	4326000
Average (Training)	124438	86387	37161	67443	56422	22438	17526	7466	0,153	2116
Standard deviation (Training)	1850267	1674838	343483	1161169	970913	240585	187133	141499	7,923	84268
Minimum (Testing)	-167	-1156	-167	-22814	-310	-1184	-121017	-133373	-33,000	-385000
Maximum (Testing)	73664000	64656000	9008000	33946000	39718000	1263852	5168222	1371000	26,476	1152000
Average (Testing)	227116	178974	46690	105445	121110	13907	26831	8724	0,037	1391
Standard deviation (Testing)	3165045	2760381	411755	1459102	1708704	69514	221384	69524	2,267	49714
Minimum (Valid.)	-113	-637	-457	-586014	-31	0	-129229	-142348	-25,513	-382
Maximum (Valid.)	79078485	72006551	6935054	63040526	15241527	18071252	11372726	1569062	23,444	111167
Average (Validation)	182192	139408	41069	124394	55292	44604	41926	8277	0,150	234
Standard deviation (Validation)	264989	170133	106565	176070	95440	170274	89501	27764	1,919	5667
Minimum (Overall)	-533	-1156	-704	-586014	-19155	-1184	-7167334	-2465847	-	-385000
Maximum (Overall)	79078485	72006551	14350000	63040526	45481000	18071252	11372726	6170000	319,000	4326000
Average (Overall)	148488	108215	39175	81678	65950	24482	22579	7776	0,135	1725
Standard deviation (Overall)	2339005	2098590	352679	1503140	1076596	351253	257576	125890	6,736	73110

Source: Author.

Ideally, the characteristics of the partial sets of data would be approximately the same. Since companies have been randomly assigned, it is not entirely so. However, the random distribution of data may not be an error in the result.

Based on the assignment, the Kohonen network was calculated. It is also referred to as SOFM 10-100 in the text. The companies were divided into individual clusters within the 10 x 10 topological grid. The frequency of individual clusters is evident from the three-dimensional graph in Figure 1. It is evident from the figure that the largest number of companies in the individual topological grid fields is in the position (4, 10) then in position (4, 9). The following ones are (5, 9) and (6, 10). Overall, we can say that the largest number of nearly 4000 companies monitored is in the sectors (4, 10), (6, 10) and (3, 7), (6, 7).

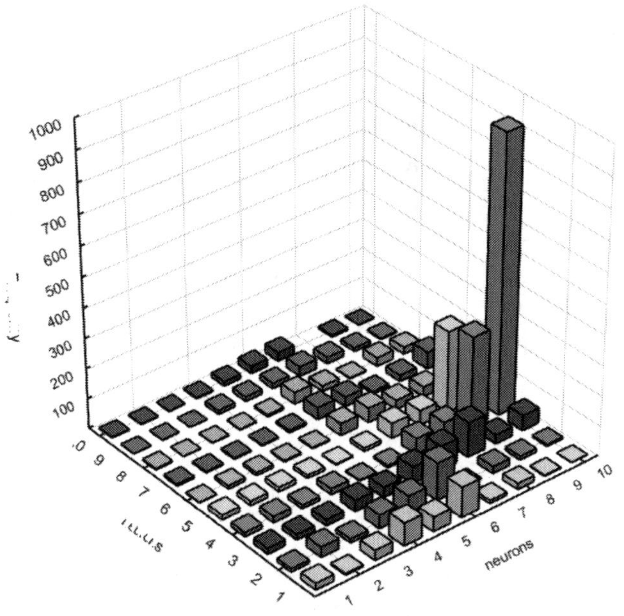

Fig. 1. Graph of company frequencies in the topological grid

Source: Author.

The results shown in Table 2 are more specific.

Table 2. SOFM 10-100 Network frequency Table

	Network: 1.SOFM 10-100 / Samples: Training, Testing, Validation									
	1	2	3	4	5	6	7	8	9	10
1	22	5	33	76	48	100	10	21	3	1
2	11	34	13	45	61	125	3	23	12	6
3	20	18	20	37	34	58	81	115	30	51
4	15	20	12	20	3	7	53	62	304	919
5	16	15	15	11	3	11	50	49	271	137
6	2	11	10	5	4	40	48	31	33	167
7	7	5	5	4	8	41	33	8	21	57
8	5	1	2	3	7	39	18	3	29	23
9	1	2	1	13	15	18	32	22	12	5
10	3	3	2	4	15	25	29	0	7	1

Source: Author.

Even from the table, it is advisable to focus on three clusters, namely (4, 10), (4, 9) and (5, 9). This is a very interesting set of data representing more than one third of all companies in the file. Dominance may be more evident from the Kohonen graph in Figure 2.

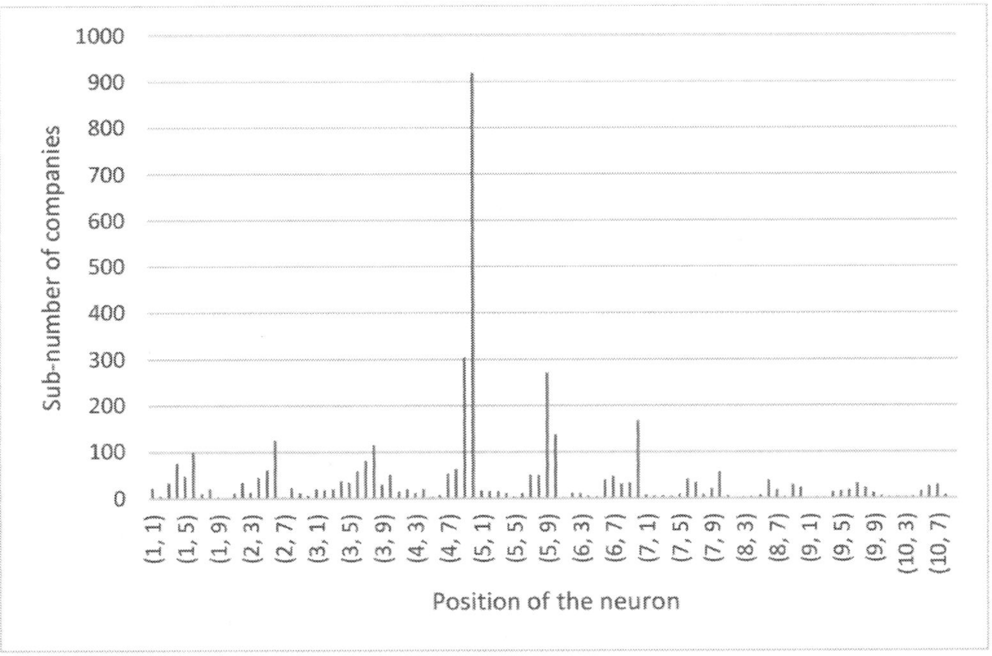

Fig. 2. Kohonen graph for selected cells of the SOFM 10-100 network

Source: Author.

While the data set is rather heterogeneous, cluster analysis suggests that there are characteristics in which the companies are similar.

Cluster (4, 10)

The cluster (4, 10) of the topological grid encompasses 919 companies. It is the largest set of data from the partial clusters. From the point of view of input parameters, we can derive the following average values:

1. Total assets: 1,686,400 thousand CZK
2. Total fixed assets: 810,882 thousand CZK,
3. Total current assets: CZK 823,749 thousand CZK,,
4. Total liabilities: 1,686,400 thousand CZK,
5. Total equity: 224,781 thousand CZK,
6. Foreign resources total: 1 471 796 thousand CZK,
7. Total performance: 38,515 thousand CZK,
8. Total value added: CZK 256,189 thousand CZK,
9. Operating result: 13 988 thousand CZK,
10. ROE: 10.5%.

From the average we can derive a typical business for this cluster of companies. It is a small company whose operating result is positive, yet close to zero. Despite this, its existence is not jeopardized. Business automation is relatively low. The company holds a relatively large volume of current assets (probably inventory and receivables).

When it comes to the homogeneous cluster of companies, we can verify the validity of basic logical links. The relationship between total assets and the level of operating income seems interesting (see Figure 3 below).

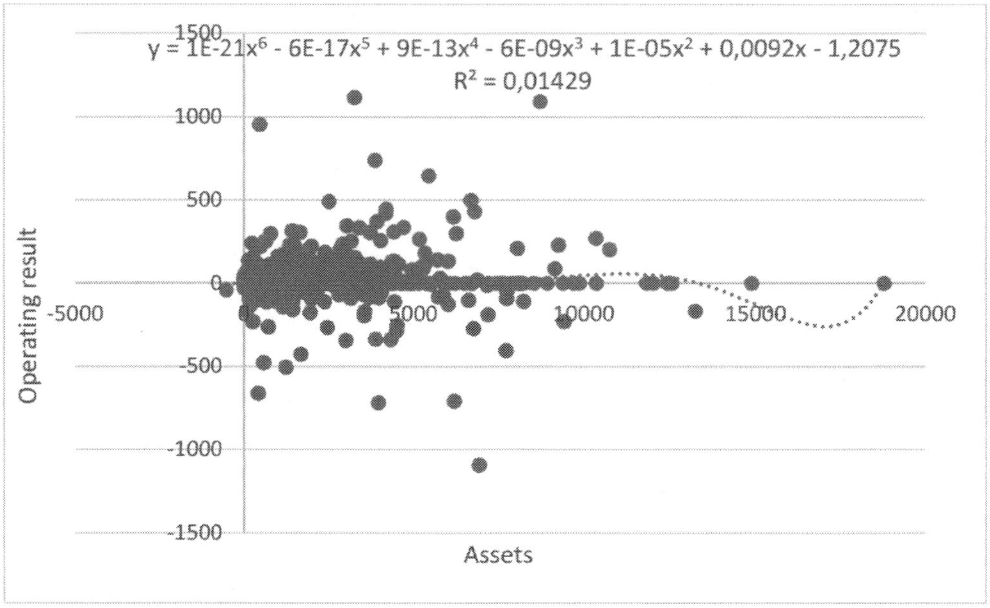

Fig. 3. Relation of total assets and operating profit of the cluster (4, 10)

Note: All values in graphs are in thousands of CZK.

Source: Author.

Total assets and operating result correlate to close to 0.12. Therefore, there can be no significant dependence of the two variables. Yet the data is interleaved by a regression curve, a 6th order polynomial. Based on this, it is possible to estimate the development of the operating result on the basis of the size of the company determined by the volume of total assets.

Figure 4 shows the relationship of foreign resources and operating income.

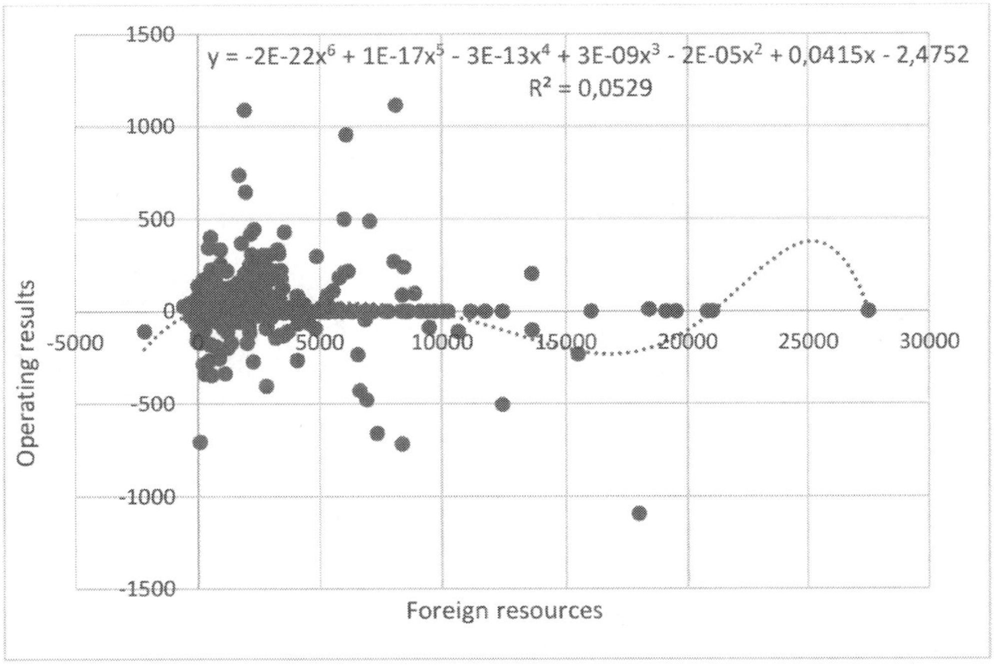

Fig. 4. Foreign resources and the operating result of the cluster (4, 10)

Source: Author.

The relationship can show us the effect of the financial leverage, assuming a relatively high correlation between the two variables. But this is only minimal. Yet, the graph points are interleaved ny the 6th order polynomial. It can be seen that the ideal debt ratio of companies in the cluster (4, 10) is up to 2 million CZK. The average debt will then be up to 100,000 CZK.

Cluster (4, 9)

The cluster (4, 9) of the topological grid occupies the imaginary second position in frequency. It has 304 companies. From the point of view of input parameters we can deduce a typical representative of this cluster:

1. Total assets: 2 698 333 thousand CZK,
2. Total fixed assets: 877,032 thousand CZK,
3. Total current assets: 1,741,898 thousand CZK,
4. Total liabilities: 2 698 333 thousand CZK,
5. Total equity: 1,037,298 thousand CZK,
6. Foreign resources total: 1,638,506 thousand CZK,
7. Total performance: 15 037 thousand CZK,
8. Total value added: 312 950 thousand CZK,
9. Operating result: 41,158 thousand CZK,
10. ROE: 17.9%

A typical company of this cluster is five times larger than the company cluster (4, 10) (meaning from the point of view of the total assets). This is a company with a lower degree of automation and a higher share of the circulation of total assets than in the case of cluster (4, 10). A larger difference between the two clusters can be identified in the case of ROE, which is 17.9% in the cluster (4, 9) vs. 10.5% of the cluster (4, 10). Even in this case, it is certainly interesting to explore the relationship of some variables.

Figure 5 shows the relationship between total assets and operating result.

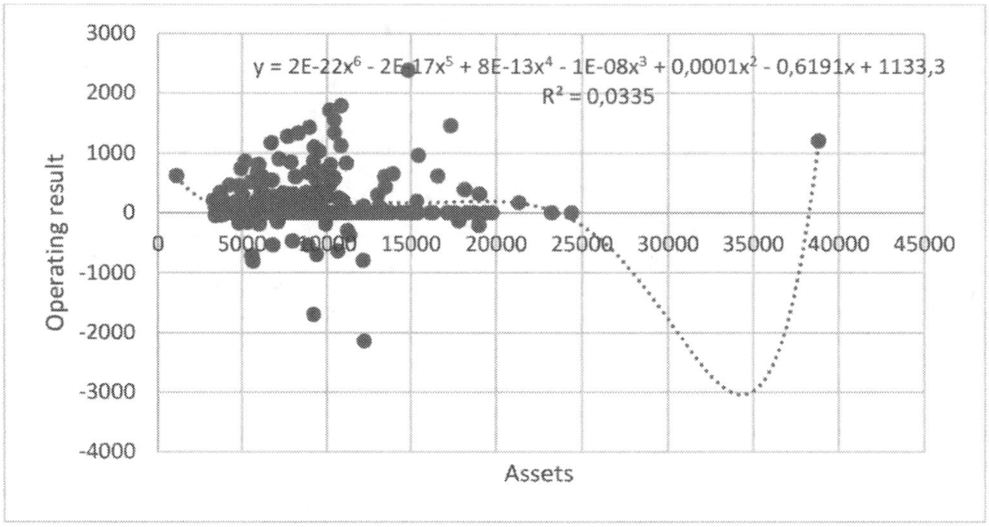

Fig. 5. Relationship of total assets and operating result of cluster (4, 9)

Source: Author.

The determination coefficient is 0.033. We can not talk about the dependencies of the two variables either. However, in this cluster we can see the fact that a higher volume of assets implies a higher operating result. The trendline has a 6th order polynomial again.

Figure 6 illustrates the relationship of foreign resources and operating result.

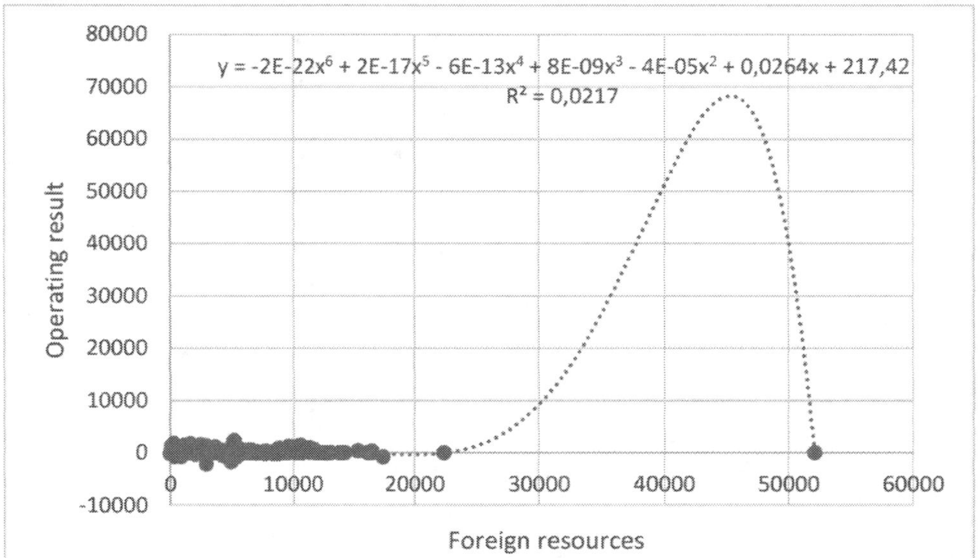

Fig. 6. Foreign resources and the operating result of the cluster (4, 9)

Source: Author.

Even in this case, the correlation coefficient is not high. However, even in this case, we can say that a higher debt ratio may mean a higher operating result.

Cluster (5, 9)

To illustrate the overall situation, we also present the characteristics of the third most significant cluster, (5, 9). The cluster contains 271 transport companies. If we take the total assets as a measure of the size of the company, we find that the companies of this cluster are approximately 1.5 times larger than those of the company cluster (4, 10), but 4 times smaller than the company cluster (4, 9):

1. Total assets: 567,190 thousand CZK,
2. Total fixed assets: 228,634 thousand CZK,
3. Total current assets: 328,622 thousand CZK
4. Total liabilities: 567,190 thousand CZK,
5. Total equity: 65,818 thousand CZK,
6. Foreign resources total: 496,975 thousand CZK,
7. Total performance: 48,996 thousand CZK,
8. Total value added: 218,318 thousand CZK,
9. Operating result: 27,713 thousand CZK,
10. ROE average: 14.5%

Even from the point of view of average profitability, clusters (5, 9) can be in terms of success categorized among companies of clusters (4, 10) and (4, 9).

4 Conclusion

The aim of the paper was to carry out a cluster analysis of transport companies.

The aim of the paper was fulfilled.

A cluster analysis was performed using neural networks without a teacher – Kohonen networks. Based on the cluster analysis, the companies were divided into clusters in the Kohonen map (10 x 10 clusters). Some clusters are significant in terms of number of companies. The most represented are clusters (4, 10), (4, 9) and (5, 9). These clusters concentrated more than a third of all companies operating in transport in 2016. These clusters have therefore been subjected to further analysis. In the case of the evaluated clusters, the typical representatives of these clusters were defined. They varied by their size, the volume of operating income generated and the return on equity. The companies of cluster (4, 10) are small businesses and generate lower operating profits compared to the remaining significant clusters. A typical company of cluster (4, 9) is five times larger than that of the cluster (4, 10). Such a company also achieved a significantly higher ROE, above 17%. The companies of cluster (5, 9) are larger than the companies of cluster (4, 10), but smaller than those of cluster (4, 9). Even in the case of ROE, they reach average values among the three analysed clusters. From the analysis we can make three very specific conclusions:

1. A larger company (or a company with a larger volume of assets) generates, on average, a higher operating result.
2. Larger companies achieve higher ROE rates.
3. In the case of larger company, the financial leverage is more positive, i.e. the more the growth of the volume of the foreign capital increases, the higher operating result and the profitability of the equity.

The scale of the work did not allow it, but it would be interesting to prepare a report on the state of the Czech industry in the range of all clusters of the Kohonen map. We could identify the most profitable companies for the entire industry, identify declining businesses, or others.

References

1. R. Andekina, R. Rakhmetova, Financial Analysis and Diagnostics of the Company. *Procedia Economics and Finance*, **5**, 50-57, (2013)
2. M. Slusarczyk Antosz, N. H. Morales Merchán, Análisis de las estrategias empresariales y de las TIC. *3C Empresa Investigación y pensamiento crítico*, **5**(1), 29-46, (2016)
3. M. Vochozka, J. Horák, Comparison of neural networks and regression time series in prediction of silver price development. *Proceedings of the 9th International Scientific Conference Company Diagnostics, Controlling, and Logistics*, 271-278, (2018)
4. X. Han, L. Wang, Stock Company Comprehensive Assessment Model Based on Kohonen Network. *2008 Second International Conference on Genetic and Evolutionary Computing*, 185-188, (2008)
5. M. Vochozka, Z, Rowland, V, Stehel, P, Šuleř, J, Vrbka, *Modelování nákladů podniku pomocí neuronových sítí*. České Budějovice: Institute of Technology and Business in České Budějovice, (2016)
6. J. Horák, M. Vochozka, V. Machová, Evaluation of transport and forwarding companies by means neural networks. *Proceedings of the International Scientific Conference for Ph.D. students and assistants – QUAERE*, 377-385, (2018)
7. S. M. Wagner, Innovation management in the German transportation industry. *Journal of Business Logistics*, **29**(2), 215-231, (2008)
8. X. Ou, X. Yan, X. Zhang, Z. Liu, Life-cycle analysis on energy consumption and GHG emission intensities of alternative vehicle fuels in China. *Applied Energy*, **90**(1), 218-224, (2012)
9. H. Hao, H. Wang, L. Song, X. Li, M. Ouyang, Energy consumption and GHG emissions of GTL fuel by LCA: Results from eight demonstration transit buses in Beijing. *Applied Energy*, **87**(10), 3212-3217, (2010)
10. J. Jedrzejczak-Gas, P. Kuzdowicz, Evaluation of the financial condition of the companies of transport and storage section in times of economic crisis - taxonomic analysis. *Management*, **20**(1), (2016)
11. T. Pánek, J. Zwierzchowski, *Statystyczne metody wielowymiarowej analizy porównawczej: teoria i zastosowania*. Warszawa: Oficyna Wydawnicza, Szkoła Główna Handlowa, (2013)
12. E. Quinet, R. W. Vickerman, *Principles of transport economics*. Northampton, MA: Edward Elgar, (2005)
13. O. Ponisciaková, *Some Aspects of Costs Management in the Transport Companies*. Lithuania, 763-766, (2016)
14. Transportation Accounting: How Understanding Transportation and Freight Costs Leads to Strategic Thinking and More Profitable Business. *Cerasis* [online]. 2014 [cit. 2018-07-16]. Available at: http://cerasis.com/2014/08/06/transportation-accounting/ (2014)
15. Transportation accounting fundamentals. *The national academies press: Sharing the Costs of Human Services Transportation* [online]. 2011 [cit. 2018-07-16]. Available at: https://www.nap.edu/read/14490/chapter/7 (2011)
16. Přeprava věcí a osob, přepravní výkony. *Český statistický úřad: Veřejná databáze – Czech Statistical Office* [online]. 2018 [cit. 2018-07-16]. Available at: https://vdb.czso.cz/vdbvo2/faces/cs/index.jsf?page=vystup-objekt&pvo=DOP05-

D&z=T&f=TABULKA&skupId=1613&katalog=31028&pvo=DOP05-D&c=v3~8__RP2016 (2018)

Real estate as an investment asset

Tomáš Krulický[1,*], *Jakub* Horák[1]

[1]University of Žilina, Faculty of Operation and Economics of Transport and Communications, Univerzitná 1, 01026 Žilina, Slovak Republic

Abstract. The present text focuses on perceiving real estate property as an investment asset that generates a certain amount of revenue to its owner, assuming expected risk and the expected level of liquidity. The first step is to determine the open market value of the selected property, which represent the expected expenses of the investment costs incurred (taking into account other acquisition costs), then we determine the open market rent value, which is the expected return on the selected property, then identify possible business risk associated with the commercial use of real estate and finally, the liquidity of the entire investment is estimated. In conclusion, methods for evaluating investments are applied to assess the realization of the investment – acquisition of the selected real estate for commercial purposes, the estimated return time and the percentage of the return on investment is calculated of the paper.
Key words: open market value, investment evaluation methods, investment triangle

1 Introduction

Together with the general increase in the standard of living of the population, the available funds of the population of the country also increase and, therefore, household savings, according to the Czech Statistical Office in the second quarter of 2018, the saving rate reached 9.8% of revenues, which was a year-on-year increase of 0.4 p. p. Growth in household savings also raises the need to keep these funds in the future until the need for such funds is expected [1]. The manners in which households can free their funds are many , depending on the approach to risk, the required liquidity, the required return, the confidence of the population in individual financial products, and especially on the level of awareness of the household. One of the options households invest in is buying real estate [2]. Investment real estate can be defined as real estate that is purchased with the intent of its commercial use and not solving own housing needs, which are already solved in another way (by owning other property, for example) [3]. The aim of the contribution is to determine the return on investment real estate and the payback period of such investment.

* Coresponding author: krulicky@mail.vstecb.cz

1.1 Investment according to economic theory

In terms of economic theory, an investment can be seen as sacrificing present value for future uncertainty, but with a reasonable amount of return reflecting, inter alia, the level of risk and liquidity [4]. In investment decision making, the investor usually decides according to the investment triangle shown in the figure below.

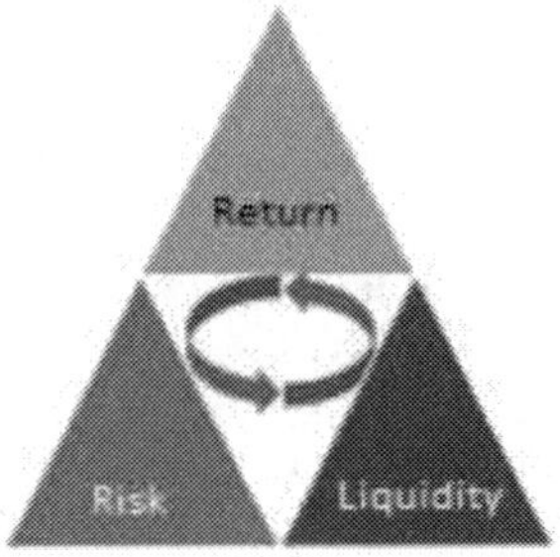

Fig. 1. Investment triangle

Source: Authors.

From the investment triangle, it is clear that the general investor is trying to decide on the investment according to three basic parameters, namely the expected return, the assumed risk and the liquidity of the investment. The relationship of all three variables is described by the behavior of the general investor – the general investor expects a higher return on higher risk, and vice versa [5]. Liquidity then expresses the ability to convert a given investment into a generally convertible asset (money) in a reasonable time at the usual price [6]. Each investor is specific and usually decides according to his / her relation to the assumed level of risk. It is clear that the three criteria do not represent a definitive list of criteria for investor investment decision making but serve as a demonstration of investment, and moreover, according to experience, these three characteristics are a typical feature of each investment [7]. Each investment is thus made with the vision of future return that may not only have the financial dimension but also the overall social dimension, the satisfaction of other needs and the wishes of the investor, etc. Correct investment should thus contribute to the future expansion of the investor's assets [8].

From the point of view of the development of the course during the lifespan of the investment activity, it is possible to distinguish the two basic phases of each investment:

- pre-investment phase,
- investment phase [9].

During the pre-investment phase, the investor collects and evaluates the available documents on the intended investment or investments, which are related to the first costs. It is not unusual for an investor to decide, after thorough analysis, to withdraw from the intended investment. In the construction sector, pre-project and project activities, geological surveys, costs of various permits, etc. are included in this stage. At this stage, the investment is abandoned either for objective reasons - for example, failure to grant approval for construction or for subjective reasons - investor strategy. It is clear that even if the investment is not realized, such costs are incurred by the investor, and it is not common for the investor to transfer the risk for an unrealized investment to third parties [10]. At this stage, households are looking for investment opportunities that they compare with each other and decide to realize some of the investments selected or a combination of them [11].

The investment phase is a realization phase, there is the actual investment and subsequent collection of return on investment. The investment phase is the longest of all phases, and the investor is under considerable risk. Unplanned completion of an investment at this stage occurs only exceptionally (depending on the riskiness of the investment) usually without the influence and will of the investor. Investors at this stage incur costs directly related to the operation of the investment [12].

1.2 Real estate as an investment

As mentioned above, investment real estate [13] is one of the options for the investment of household savings. Investment real estate is represented by real estate that can generate an expected return for their owner, most often real estate with a lower amount of floor area (approx. 60 m2) in larger residences intended for rental can be considered as investment real estate [14]. According to the authors, the demand for rental housing can be expected to be higher in this type of real estate, when they are presented as so-called start-ups for young or newly moved people from the countryside to cities who first choose the form of rent and then resort to buying their own housing.

Demand for the purchase of housing units between 2013 and 2018 enormously affected their price, housing price developments are monitored by the HB Index – see figure 2. The HB index is based on realistic estimates of market prices of individual real estate properties that were financed through a bank. The HB index is constructed using a hedonic model that tracks up to 30 different real estate property parameters in real estate resolution into three groups, namely flats, family houses and land.

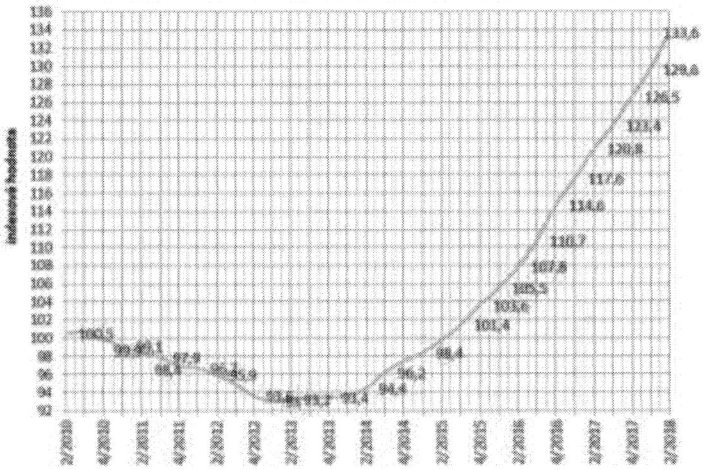

Fig. 2. HB INDEX – housing units

Source: [15].

In order to moderate the demand for the purchase of real estate, the Czech National Bank, with effect from 1 October 2018, tightened mortgage recommendation as one of the most popular tools for financing the purchase of real estate by Czech households, among the new recommendations are:

- The amount of income requirement – the amount of debt should not exceed the applicant's annual net income by 9 times.
- The applicant should spend a maximum of 45% of his monthly net income on debt repayment.

- The ratio of the loan amount to the value of the pledged property – 90% maximum.

The tightening of the recommended conditions for mortgage applicants would thus have to limit the number of applicants, ie buyers, so it is to be expected that this segment of households will be forced to stay in rental housing and will increase the demand for this type of housing [16].

The aim of the paper is to determine the rate of return on investment in the selected real estate.

2 Methodology

The paper deals with the determination of the rate of return on investment of the selected real estate and its inclusion in the investment portfolio of the investor. For the purpose of achieving the aim of the contribution, the initial price of the selected real estate representing the initial investment cost (capital expenditure) taking into account the initial costs will be determined in the first step, the open market rent representing the expected investment income will be determined, then the estimated costs associated with the investment, while taking into account the vacancy rate, will be calculated and lastly, methods of investment evaluation will be applied in the final step.

The open market value will be determined by a comparative method, capturing a strong interaction between the market situation and the price of the selected real estate. The comparative method will be based on bid prices of comparable real estate where, according to the author's experience, the housing market is flexible and it can be assumed that the final price will not be too different from the bid price (by max. 9 percent), this difference will be subject to application of the coefficient of price source reduction. For comparison, 6 comparable real estate properties will be considered, which must meet the qualitative criteria for comparison – the similarity of the location, the correspondence of the structural system of the building, the constructionally technical condition, the size and layout solution. For all comparable properties, a coefficient reduction of the source price will be applied, amounting to 5% of the price obtained, expressing the buyer's ability to negotiate the final realization price. The open market value of an investment real estate will be determined as the average of the modified comparable objects and will be rounded to hundreds of thousands.

Furthermore, the open market rent will also be determined using a direct comparison method, based on offers for rent of comparable real estate. The rental market of housing units in the city of České Budějovice is very dynamic, there is a high rate of fluctuation of individual advertisements, the final rental price set in the lease will therefore be very close to the bid price and the landlord does not have a high motivation to discount the required rent. Open market rent will be determined as the average of 6 comparable properties advertised for rent. Comparable objects should be as close as possible to the investment asset in all criteria – location, technical condition, layout, etc., and rounded to hundreds. The determination of the expected income will take into account the occupancy rate and the costs associated with the operation of the investment. Determination of the occupancy rate will be made using the data from the actual survey of the offer market for rent. The survey will be carried out by monitoring the rental market for apartments in the city of České Budějovice for a period of one calendar month, the conclusion will be an evaluation of how many new offers during the monitoring period on the market have been added, how many were withdrawn from advertising, which will mean the average time from the publication of the advertisement until the time of its withdrawal (rental).

The qualified estimate will determine the additional costs associated with the investment, such as the one-off costs (open market real estate value, real estate acquisition tax, legal service, building adjustments) related to the acquisition of the investment, as well as the regular costs incurred during the lifespan of the investment (property tax, insurance, maintenance costs). The annual income from the investment will be determined as the difference between the open market annual rent, the expected costs during the investment and the expected vacancy rate (loss of rent).

Finally, methods for investment evaluation will be applied, taking into account investment risks, such as determining the return on investment and the payback period. The rate of return on investment will be expressed as a percentage of the expected annual investment income and estimated cost of ownership. The payback period will be determined in years as the ratio of the acquisition costs to that of the investment income.

Finally, the real estate earnings evaluation will be done using the discounted cash flow method in a two-phase variant, with the first phase taking only one year (2019), the second phase then include the period of 2020-infinity. For calculating the second phase, the Gordon model will be used, and the rate of profitability of the investment will be used as the required return rate.

3 Results

The chapter Results presents the results of the process described in the methodology section.

3.1 Determination of the real estate open market value

The open market value is the price that could be reached locally and on time between voluntary independent sellers and buyers, after proper marketing without considering extraordinary effects.

For comparison, the number of comparable objects is very important – but their number is often limited by the available amount on the local market in a given segment. Thus, the number of comparable objects is objectively limited from the top because the selected objects for comparison must meet strict quality criteria. It is therefore clear that the comparative method can only be applied in cases where a market with similar assets is sufficiently developed. A suitable representative of the investment real estate is the apartment unit constructed as 2+1 with a total floor area of approx. 60m^2 located in a revitalized panel apartment building where the house and the apartment have undergone reconstruction, and the apartment is at least on the third floor above ground and is located in one of the apartment complex areas of České Budějovice.

In total, six comparable real estate properties were offered for sale at the Máj apartment complex area in the period from September 1, 2018 to September 30, 2018 were found. All comparable properties found are very close in qualitative terms and need not be corrected by correction coefficients, except for application of the reduction coefficient of the price source. This coefficient takes into account the difference between the estimated real estate sale price and the offer price. The reduction coefficient of the price source was determined for individual real estate properties after expert consideration, especially with regard to the negotiating possibility of the seller and the buyer with the final price. The selection of the coefficient value for each comparable property is given below.

Table 1. Determination of the open market value of investment real estate property

ID	Total price	Floor area	Layout	K1	Adjusted price
ID 1	CZK 2,290,000	62m^2	2+1	0.95	CZK 2,175,500
ID 2	CZK 1,850,000	64m^2	2+1	0.95	CZK 1,757,500
ID 3	CZK 1,950,000	64m^2	2+1	0.95	CZK 1,852,500
ID 4	CZK 2,270,000	63m^2	2+1	0.95	CZK 2,156,500
ID 5	CZK 2,180,000	65m^2	2+1	0.95	CZK 2,071,000
ID 6	CZK 1,840,000	67m^2	2+1	0.95	CZK 1,748,000
Average					CZK 1,960,167

Source: Authors.

The open market value of the average investment real estate was determined as the average of the adjusted price of comparable objects and amounts to CZK 1,960,167, after rounding to CZK 2,000,000.

3.2 Determination of the open market rent

The best use of the investment property is its rental. The property owner has basically two lease options – long-term (in the order of years) and short-term (days, weeks). The long-term lease is characterized by long-term cooperation and lease of a tenant, for the lessor, it means, depending on the choice of the tenant, lower lease management costs, unlike short-term rental, which for the lessor represents a higher administrative burden, but is offset by higher income. The following section will deal with determining the open market rent rates for long-term rentals.

For the purpose of determining the open market rent rates, a total of 6 comparable properties, which were advertised for lease in the period between 1 September 2018 and 30 September 2018, managed to meet these properties for their qualitative proximity to the price-setting factors, it was not necessary to apply correction coefficients of difference.

Table 2. Determination of the open market rent

ID	Total price	Floor area	Layout	Adjusted price
ID 1	CZK 8,000	64m^2	2+1	CZK 8,000
ID 2	CZK 7,800	50m^2	2+1	CZK 7,800
ID 3	CZK 8,300	50m^2	2+1	CZK 8,300
ID 4	CZK 7,800	61m^2	2+1	CZK 7,800
ID 5	CZK 7,500	50m^2	2+1	CZK 7,500
ID 6	CZK 7,500	64m^2	2+1	CZK 7,500
Average				CZK 7,817

Source: Authors.

The open market monthly rent of the average real estate property (Pn – monthly income) was determined as the average monthly rent of comparable objects and is CZK 7,817, after rounding 7,800 CZK, the annual rent (Annual income) can be determined from the formula 7,800 CZK * 12 = CZK 93,600. Open market rents do not include fees associated with owning an apartment (contribution to the fund for repairs, services, etc.).

3.3 Estimated costs associated with the investment

The estimated costs associated with the realization of the investment can be divided into two parts in terms of time spent, the first group being the costs incurred in connection with the acquisition of the investment and its immediate realization (acquisition costs); further costs are then spent during the main investment phase (operating costs). These costs do not express the value of the investor's time devoted to the acquisition and management of the investment.

Costs incurred for the acquisition and operation of the investment (one-off):

- Property acquisition (open market value) – CZK 2,000,000
- Acquisition of immovable property tax (4%) – CZK 80,000
- Legal service, counseling, etc. – CZK 10,000
- Building adjustments – CZK 10,000
- Total CZK 2,100,000

Total capital expenditure (K0 – acquisition cost) corresponds to the cost of acquisition of the investment and is estimated at CZK 2,100,000.

Costs incurred during property rental (recurring, annual):

- Property tax – CZK 500
- Property + household insurance – CZK 1,000
- Maintenance costs incl. reserves – CZK 8,500
- Total CZK 10,000

Regular annual costs (N_r – annual costs) associated with the operation of the property are estimated at CZK 10,000.

In determining the income from investment property it is also necessary to take into account the vacancy rate of the property. As a result of the change of the tenant, rents are lost when the apartment is vacant. From the survey conducted on offers for renting housing units in the whole city of České Budějovice during September 2018, most (approx. 75%) of advertised apartments were rented within 1 month of publication of the advertisement. Loss of rent, or vacancy rate (N_{VN} – Costs, Loss of rent) will be assumed at 10% of the total annual rent.

The total revenue (P) of the investment in the first year after taking into account the expected costs and the rent loss can be determined from the relationship:

$$P_r - N_r - N_{VN} = P$$

$$\text{CZK } 93{,}600 - \text{CZK } 10{,}000 - 10\% = \text{CZK } 75{,}240$$

3.4 Investment evaluation

The estimated incomes above are in absolute terms for the first year of rental, it can be assumed that the rational behavior of the lessor will make the income (rent) copy the development of inflation, otherwise a real decrease in rent will occur. Other compensation for landlords is the rise in property prices in the market.

Purchase of real estate is essentially an investment with an unlimited lifespan, when its lifetime is objectively limited by the life of the property itself from a structural point of view, but it can be very long, with adequate social demand (house maintenance costs).

Return on investment

The return of the selected investment (I) in percentage terms will be determined from the relationship:
$(P / K_0) * 100 = I$, therefore (CZK 75,240 / CZK 2,100,000) * 100 = 3.58%.

The return on the investment at unchanged input parameters is 3.58%.

Payback time

The payback period is the period for how long the expected revenue (P) equals the capital expense K0. For the purposes of the present article, a simplified calculation will be used as follows: 2,100,000/75,240 = 27.9.

The payback period of the investment 27.9 years from only rental income.

Earnings evaluation of the property

Property earnings evaluation methods can be applied to real estate evaluation when the property meets the requirements of going concern - unlimited lifespan, profit potential. The calculation will take into account year-on-year inflation, ie annual revenue and cost growth of the same amount (2%), while the rate of return does not include inflation.

Table 3. *Property earnings evaluation*

	2019	2020-x
Net annual income	CZK 75,240	CZK 76,745
Rate of return	3.58%	3.58%
Current value	CZK 72,638	CZK 2,142,000
Evaluation	CZK 2,214,638	--

Source: Authors.

From the point of view of the unlimited lifespan of investment property, the use of other methods for investment evaluation is of no great significance for the investment evaluation, because in the case of a sufficiently long investment period, the investment will overcome the economic benefits and the investment will be recommended for realization.

4 Conclusion

The purpose of this contribution was to determine the rate of return on investment in the selected real estate, in this case an apartment intended for commercial use (long-term lease), which is the most frequent investment property. In the first step, the capital expenditure was determined as the sum of the open market value of the selected property determined using the comparative method, and the estimated costs associated with the acquisition were added to this amount. The annual income from this property was determined as the open market rent income, taking into account the estimated operating costs and the time when the rent was not collected (loss of rent). The return on selected investment property was 3.58%, this profit did not include the expected inflation, and other implicit revenues in the form of a rise in property prices. If a comparison is made with other investment opportunities on the market, it is necessary to add 2% to the rate of return and the rate of inflation, the total return on the property is 5.58%.

The aim of the paper was fulfilled.

The contribution could be followed by further research into the return on investment in real estate, their mutual comparison and the identification of the most profitable type of investment real estate property.

References

1. Czech Statistical Office, *Tiskové sdělení: Míra úspor domácností se zvýšila* [online]. Available at: https://www.czso.cz/csu/czso/tiskove-sdeleni-mira-uspor-domacnosti-se-zvysila
2. C. Eves, Residential Property Investment: Assessing Home Ownership Long Term Viability. *International Conference on Construction and Real Estate Management*, China, 631-634, (2012)
3. J. Jílek, J. Svobodová, *Účetnictví podle mezinárodních standardů účetního výkaznictví*. Praha: Grada, (2011)
4. F. Gao, A. Xi He, P. He, A theory of intermediated investment with hyperbolic discounting investors. *Journal of Economic Theory*, **177**, 70-100, (2018)
5. E. Bikas, V. Saponaitė, M. Tvaronavičienė, Behavior of the Lithuanian investors at the period of economic growth. *Entrepreneurship and Sustainability Issues*, **6**(1), 44-59, (2018)
6. M. Bade, H. Hirth, Liquidity cost vs. real investment efficiency. *Journal of Financial Markets*, **28**, 70-90, (2016)
7. M. A. S. Bradbury, T. Hens, S. Zeisberger, How persistent are the effects of experience sampling on investor behavior? *Journal of Banking & Finance*, **98**, 61-79, (2019)
8. S. Sung, H. Cho, D. Ryu, The Behavior of an Institutional Investor with Arbitrage Opportunities and Liquidity Risk. *Emerging Markets Finance and Trade*, **55**(1), 1-12, (2018)
9. J. Fotr, I. Souček, *Podnikatelský záměr a investiční rozhodování*. Praha: Grada Publishing, (2005)
10. J. Janeková, J. Fabianová, D. Onofrejová, E. Puškaš, M. Buša, Implementation of deviation analysis method in the utilisation phase of the investment project: a case study. *Polish Journal of Management Studies*, **15**(1), 99-109, (2017)
11. C. Giordano, M. Marinucci, A. Silvestrini, The macro determinants of firms' and households' investment: Evidence from Italy. *Economic Modelling*, in press (2018)
12. Y. Chen, M. Chiang, C. Weng, Are investors always compensated for information risk? Evidence from Chinese reverse-merger firms. *Review of Quantitative Finance and Accounting*, 1-38, (2018)
13. F. Prizzon, A. Cullino, Investment Property in Rental: Profitability and Risk Analysis. New Metropolitan Perspectives. *International Symposium on New Metropolitan Perspectives*, 499-506, (2019)
14. N. Turner, M. Thomas, Property market indices and lease structures – the impact on investment return delivery in the UK and Germany: Part I. *Journal of Property Investment & Finance*, **19**(2), 175-194, (2001)
15. Mortgage Bank, *HB index: Nemovitosti dále zdražují i přes klesající poptávku* [online]. Available at: https://www.hypotecnibanka.cz/o-bance/pro-media/hb-index/nemovitosti-dale-zdrazuji-i-pres-klesajici-poptavk/ (2018)
16. Aktuálně.cz, *Hypotéky čeká přísné omezení: ČNB zavádí dvě nová opatření, vlastní bydlení bude méně dostupné* [online]. Available at: https://zpravy.aktualne.cz/finance/hypoteky-ceka-prisne-omezeni-cnb-zavadi-dva-nove-stropy-muze/r~9f6317e66e1511e892700cc47ab5f122/ (2018)

Estimation of the development of Czech Koruna to Chinese Yuan exchange rate using artificial neural networks

Veronika Machová[1,*], *Jan* Mareček[1]

[1]Institute of Technology and Business in České Budějovice, School of Expertness and Valuation, Okružní 517/10, 37001 České Budějovice, Czech Republic

Abstract. Through time series analysis, it is possible to obtain significant statistics and other necessary data characteristics. Prediction of time series allows predicting future values based on previously observed values. The exact prognosis of the time series is very important for a number of different areas, such as transport, energy, finance, economics, etc. It is within the topic of economy that the analysis and prediction of time series can also be used for exchange rates. The exchange rate itself can greatly affect the whole foreign trade. The aim of this article is therefore to analyze the exchange rate development of two currencies by analyzing time series through artificial neural networks. Experimental results show that neural networks are potentially usable and effective for exchange rate prediction.
Key words: artificial neural networks, exchange rate, time series, prediction, Czech crown, Yuan

1 Introduction

According to Sheikhan et al. [1], time series are characterized as sequences of spatially and de facto comparable observations that are time-based. Time series can also be defined as ordered sequences of values of variables at equally remote time intervals [2]. Time series analysis is a technique that involves the study of individuals or groups observed at the following moments. They are a series of data points in time order [3]. Through time series analysis, it is possible to obtain significant statistics and other necessary data characteristics. Prediction of time series allows predicting future values based on previously observed values [4]. The accurate prediction of time series is very important for a number of different areas, such as transport, energy, finance, economics, etc. [5]. The prognosis is considered to be the most important task of analyzing time series. Specifying a prediction problem is a basic step in the prognosis of time series. When predicting, it is necessary to have as accurate a picture as possible about the variables in the prediction, the availability of data, the nature of the data, and the prediction method and prognosis horizon [6].

The traditional view of the international economy lies in the fact that the exchange rate is a leading factor that can strongly affect foreign trade. Production fragmentation,

* Corresponding author: machova@mail.vstecb.cz

widespread around the world as a result of recent globalization, may weaken the role of the exchange rate in international trade. Given that the Czech economy is very open, quantitative knowledge of the impact of the exchange rate on exports is valuable information for all actors in the economy [7].

International currency can be defined as one generally recognized in the world and used for trade agreements, investments, and foreign exchange reserves [8]. This paper will analyze the time series of the Czech currency development compared to the Chinese currency. The Czech crown (with the international abbreviation CZK) is the official currency of the Czech Republic. CZK 1 is composed of 100 hellers. Although hellers, as coins, have not been used since 2008, they are still included in the commodity prices. The final price is then rounded to the nearest value [9]. According to Cimburek and Řežábek [10], CZK in the Czech Republic, according to Czech legislation, is a legal tender that is generally accepted. Cash is easily available in the Czech Republic not only thanks to the CNB's distribution network, but also thanks to a dense network of banks, their branches and ATMs. The koruna is a safe currency and, thanks to its security features, ranks among the most successful ones in comparison with other currencies. An important role is played by the primary objective of the Czech National Bank as defined by the Constitution of the Czech Republic, namely price stability. The Czech Republic has been able to record a low-inflation environment for several consecutive years. A low-inflationary and stable economic environment may partially motivate certain demographic groups to accumulate cash and retain a certain portion of cash assets while maintaining the maximum liquidity requirement of these assets. As with most central banks, the CNB focuses primarily on the stability of consumer prices. In practice, price stability is generally not understood to mean price stagnation, but rather a modest growth. Price growth, consistent with price stability, should include the upward statistical upturn in the measurement of price growth and should also provide sufficient room for minor changes in price relations that are consistently occurring in every economy with an effective pricing system [11].

China had a range of currencies, most of which were denominated in the "Yuan" unit. Yuan in Chinese literally means "round object" or "round coin" [12]. Chinese Yuan (abbreviated CNY) is the official currency in China. 1 yuan is divided into 10 jiao and 1 jiao is divided into 10 fen [13]. Unlike other currencies, Chinese Yuan is subject to strict currency controls by the People's Bank of China [14]. CNY has evolved from a trading currency to an investment currency and now has the potential to be a global reserve currency. The growth of CNY as an international currency could balance the system controlled by the US dollar and contribute to regional and international financial stability [15]. If the Chinese policy were to return to effective monetary stability, it would strengthen the stability of the regional currency. This would create more favorable conditions for developing towards monetary co-operation [16], [17]. CNY rates can be considered as financial time series, characterized by high uncertainty, non-linearity and with different behaviors over time [18]. In this paper, artificial neural networks will be used to analyze the time series of the two aforementioned currencies. In this respect, for example, Liu et al. [19] conducted exchange rate predictions that were conducted using RBF neural networks. Detailed designs of the RBF neural network model architecture, transfer functions of the hidden layer nodes, input vectors, and output vectors have been performed with many tests. Experimental results show that the performance of RBF neural networks for the CNY exchange rate prediction is acceptable and effective.

2 Data and methods

Data for analysis are available on the World Bank website [20] etc. For the analysis, information on the Koruna (also referred to as "CZK") and Yuan (also referred to as

"RMB") exchange rate will be used. The timeframe for which the data will be available begins on October 6, 2009 and ends on October 21, 2018. We always record the daily exchange rate of both currencies. This is in total 3,303 records of input data. Unit is CZK per 1 RMB. The descriptive characteristics of the datafile are given in Table 1.

Table 1. Characteristics of the datafile

Samples	Data statistics	
	Date (Input variable)	Chinese Yuan to Czech Koruna (Output (goal))
Minimum (Training)	40092.00	2.485800
Maximum (Training)	43394.00	4.163000
Průměr (Training)	41734.79	3.265645
Standard deviation (Training)	939.26	0.392383
Minimum (Testing)	40102.00	2.496100
Maximum (Testing)	43393.00	4.155700
Average (Testing)	41755.71	3.272882
Standard deviation (Testing)	957.97	0.394309
Minimum (Validation)	40111.00	2.498900
Maximum (Validation)	43388.00	4.152900
Average (Validation)	41768.68	3.246446
Standard deviation (Validation)	1438.25	0.499822
Minimum (Overall)	40092.00	2.485800
Maximum (Overall)	43394.00	4.163000
Average (Overall)	41743.00	3.263852
Standard deviation (Overall)	953.64	0.392668

Source: Authors.

For data processing, DELL's Statistica version 12 will be used. For calculating neural structures, we use the Data Mining tool, specifically neural networks (ANS – Automated Neural Networks). We will generate multilayer perceptron networks and neural networks of basic radial functions.Time will be the independent variable (or designated measuring – case). We will determine the CZK to RMB exchange rate as the dependent variable. We divide the time series into three sets – training, testing, and validation. The first group will be 70% of input data. Based on the training set of data, we generate neural structures. In the remaining two sets of data, we always leave 15% of the input information. Both groups will serve us to verify the reliability of the found neural structure, or the found model. The delay of the time series will be 1. We will generate 100,000 neural networks. We will preserve 5 of them with the best characteristics†. In the hidden layer, we will have at least two neurons, at most 50. In the case of a radial basic function, there will be at least 21 neurons in the

† We will be orientated using the smallest square method. We will terminate network generation if there is no improvement, ie to reduce the sum of squares. So we will preserve those neural structures whose sum of squares of residuals to the actual development of the CZK to RMB exchange rate will be as low as possible (ideally zero).

hidden layer, at most 30. For the multiple perceptron network we will consider these distribution functions in the hidden layer and in the output layer:

- Linear,
- Logistic,
- Atanh,
- Exponential,
- Sinus.

Other settings are left by default (ANS – automated neural network).

3 Results

Based on the established procedure, 100,000 neural networks were generated. Five networks have been preserved, showing the best parameters. Their overview is given in Table 2.

Table 2. Overview of preserved nerual networks

Network name	Training perf.	Testing perf.	Valid. perf.	Train. error	Test. error	Valid. error	Train. Algor.	Error function	Activ. of hidd. lyr	Output activ. funct.
RBF 1-30-1	0.983490	0.983020	0.984843	0.002516	0.002616	0.002319	RBFT	Sum of squares	Gauss	Identity
RBF 1-26-1	0.984841	0.985412	0.984883	0.002312	0.002255	0.002309	RBFT	Sum of squares	Gauss	Identity
RBF 1-25-1	0.986071	0.986443	0.985769	0.002126	0.002109	0.002179	RBFT	Sum of squares	Gauss	Identity
RBF 1-26-1	0.985491	0.985337	0.984503	0.002213	0.002262	0.002367	RBFT	Sum of squares	Gauss	Identity
RBF 1-30-1	0.984297	0.983784	0.984732	0.002394	0.002499	0.002339	RBFT	Sum of squares	Gauss	Identity

Source: Authors.

These are only the neural networks of the basic radial function. The input layer has only one variable - time. In the hidden layer, neural networks contain 24 to 30 neurons. In the output layer, we have logically a single neuron, and the only output variable is the CZK to RMB exchange rate. For all networks, the RBFT training algorithm was applied. All neural structures used the same function to activate the hidden neuron layer, namely the Gaussian curve. They also use the same function to activate the outer layer of the neurons, and this function is identity (see Table 2).

Training, testing and validation performance is also interesting. In general, we are looking for a network that has the same performance in all sets of data (we recall that the distribution of data into sets was random). The error should be as minimal as possible.

The performance of individual sets of data is expressed as a correlation coefficient. The values of the individual data sets according to specific neural networks are presented in Table 3.

Table 3. Correlation coefficients of individual data sets

	Correlation coefficients		
	Chinese Yuan to Czech Koruna (Training)	Chinese Yuan to Czech Koruna (Testing)	Chinese Yuan to Czech Koruna (Validation)
1.RBF 1-30-1	0.983490	0.983020	0.984843
2.RBF 1-26-1	0.984841	0.985412	0.984883
3.RBF 1-25-1	0.986071	0.986443	0.985769
4.RBF 1-26-1	0.985491	0.985337	0.984503
5.RBF 1-30-1	0.984297	0.983784	0.984732

Source: Authors.

The table shows that the performance of all preserved neural structures is approximately identical. Slight differences do not affect the performance of individual networks. The correlation coefficient of all training data sets is above 0.983. The value of the correlation coefficient of the testing data sets reaches the same values as the training set of data, ie always more than 0.983. The correlation coefficient of the validation data set for all neural networks is above 0.984. Also, let's recall that the error is always slightly above 0.002 for all sets of data. Differences in the error of aligned time series in the individual datasets are almost negligible.

In order to select the most suitable neural structure, we need to analyze the results obtained. Table 4 provides the basic statistical characteristics of each set of data for all neural structures.

Table 4. Statistics of individual sets of data according to preserved neural structures

Statistics	Output (goal): Chinese Yuan to Czech Koruna				
	1.RBF 1-30-1	2.RBF 1-26-1	3.RBF 1-25-1	4.RBF 1-26-1	5.RBF 1-30-1
Minimum prediction (Training)	2.58183	2.55340	2.52919	2.52734	2.62556
Maximum prediction (Training)	4.04950	4.09743	4.00225	4.00540	3.95151
Minimum prediction (Testing)	2.58355	2.55342	2.52917	2.52741	2.62557
Maximum prediction (Testing)	4.04944	4.09741	4.00223	4.00544	3.95152
Minimum prediction (Validation)	2.58184	2.55531	2.53062	2.52749	2.62600
Maximum prediction (Validation)	4.04951	4.09682	4.00226	4.00505	3.95129
Minimum residues (Training)	-0.22414	-0.21614	-0.30694	-0.24314	-0.28141
Maximum residues (Training)	0.37317	0.23107	0.22900	0.21521	0.29266
Minimum residues (Testing)	-0.21388	-0.18746	-0.28546	-0.22842	-0.26051
Maximum residues (Testing)	0.37307	0.23378	0.22341	0.20519	0.29323
Minimum residues (Validation)	-0.21094	-0.17494	-0.17479	-0.20650	-0.23232
Maximum residues (Validation)	0.26023	0.22773	0.18504	0.21784	0.21065
Minimum standard residues (Training)	-4.46833	-4.49505	-6.65757	-5.16815	-5.75120
Maximum standard residues (Training)	7.43936	4.80567	4.96689	4.57450	5.98108
Minimum standard residues (Testing)	-4.18178	-3.94719	-6.21611	-4.80292	-5.21090
Maximum standard residues (Testing)	7.29438	4.92251	4.86501	4.31438	5.86540
Minimum standard residues (Validation)	-4.38041	-3.64037	-3.74445	-4.24472	-4.80316
Maximum standard residues (Validation)	5.40392	4.73891	3.96396	4.47779	4.35511

Source: Authors.

Ideally, individual neural network statistics are cross-sectionally (minimum, maximum, residue, etc.) similar in all sets of one particular neural structure. In the case of stored neural networks, the differences in time series are minimal, both in absolute terms and in the case of residues. For this reason, we are not able to clearly identify which of the preserved neural networks has the most appropriate results. All neural structures appear to be usable in practice.

Figure 1 is a graph showing the actual development of the CZK to RMB exchange rates as well as the development of predictions of the generated and preserved networks (or time series).

The graph shows that all neural networks predict a slight change in the development of the CZK to RMB exchange rate at individual intervals. However, the similarity between the predictions of individual networks, but the similarity (or the degree of consistency) with the actual development of the exchange rate of both currencies is not important. In this respect, all stored neural networks seem very interesting at first glance. They respect the basic guidelines of the curve assessing the development of the CZK to RMB exchange rate and at the same time tend to perceive the extremes of this curve.

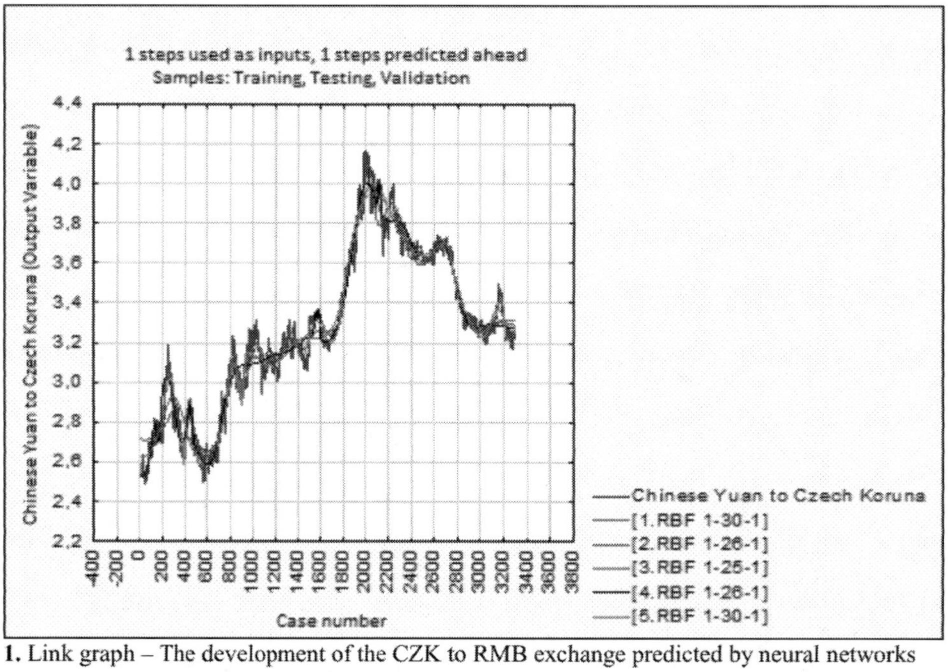

Fig. 1. Link graph – The development of the CZK to RMB exchange predicted by neural networks compared to the real exchange rate in the reference period
Source: Authors.

Given that the graph in Figure 1 includes 3,303 records of data on the CZK to RMB exchange rate, it may appear to be unclear. Therefore, it is appropriate to demonstrate the situation at the selected data interval. The graph in Figure 2 provides a comparison of the actual development of the CZK to RMB exchange rate at the interval of the last 100 days of the reference period, ie from July 14 to October 21, 2018.

 https://doi.org/10.1051/shsconf/20196101012

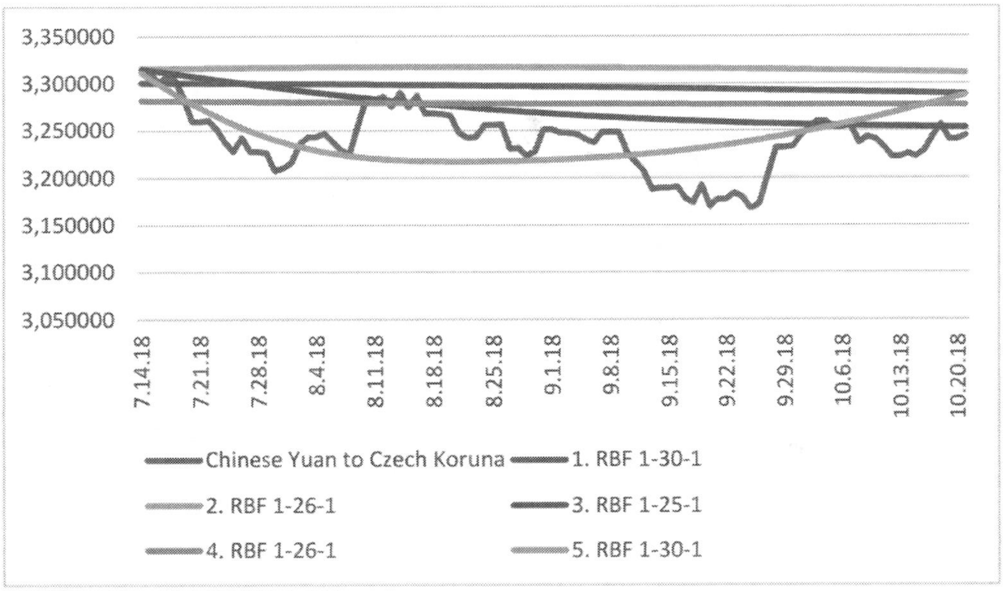

Fig. 2. Link graph – The development of the CZK to RMB exchange rate predicted by neural networks compared to the actual exchange rate in the period from 14 July to 21 October 2018
Source: Authors.

It is clear from the graph that none of the preserved neural networks can fully copy the development of the CZK to RMB exchange rate at the observed interval. The truth, however, is that the network 3.RBF 1-25-1 and the 5.RBF 1-30-1 network are the closest to reality. At the beginning of the monitored period, it almost coincides with the actual RMB value. We can identify a greater difference at the end of the reference period. The difference amounts to approximately 0.08 CZK per RMB. But even from this point of view, the least accurate network 2.RBF 1-26-1 differs from the actual rate by less than 0.011 CZK. Therefore, the examination of residues examination also appears interesting. The development of residues in the period from 14 July to 21 October 2018 is the subject of Figure 3.

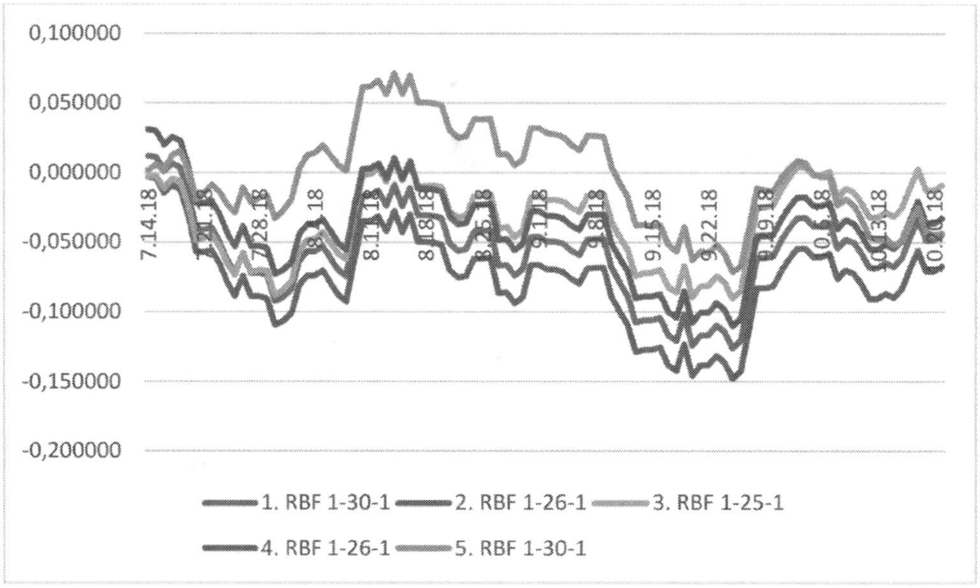

Fig. 3. Development of residues of aligned time series from 14 July to 21 October 2018
Source: Authors.

The graph shows that the sum of the residues approaching zero can be recorded in the observed period for all neural networks except the 5.RBF 1-30-1 network. In this case, residues produce relatively high positive values. To illustrate the situation, Table 5 offers the sum of the residuals of all aligned time series.

Table 5. Sum of the residues of the individual time series

	1. RBF 1-24-1	2. RBF 1-29-1	3. RBF 1-30-1	4. RBF 1-28-1	5. RBF 1-26-1
Sum of residues	0.150758	-1.025922566	-3.350398611	-1.785245346	-3.244516106

Source: Authors.

If the individual fluctuations of residues are to be eliminated throughout the whole investigation period, the absolute sum of the residues will ideally be zero. The closest to zero is the absolute sum of residues of the second neural network, which is almost -1,026. Conversely, the highest absolute values of the sum of residues are 3. RBF 1-30-1 and 5. RBF 1-26-1. Always above the value of 3. However, we must realize that in the case of 3303 measurements, the absolute value of the sum of residues at 3 is minimal. Therefore, we maintain that the most successful neural structures are the 3.RBF 1-25-1 and 5.RBF 1-30-1 networks.

4 Conclusion

The aim of the paper was to align the time series mapping the CZK to RMB exchange rate using artificial neural networks and to find such an artificial neural structure that would show a performance greater than 0.95.

In general, each prediction is given by a certain degree of probability with which it is to be fulfilled. As we predict the future development of any variable, we try to estimate the future development of this variable on the basis of previous years' data. Although we can include most of the factors influencing the target variable in the model, we always simplify reality, and we always work with a certain degree of probability that some of the predicted scenarios will be fulfilled.

In the case of our contribution, we aimed to predict the relationship between two currencies. We have not taken into account any other variables than time. We have not addressed growth, possibly a fall in gross domestic product, the volume of mutual transactions, the relationship of other countries to the currencies analyzed, the volume of mutual tourism and many others. We only dealt with the machine prediction of the future development of the two-currency exchange rate. Therefore, we performed the time series alignment using neural networks. All neural structures show above-average performance (higher than the desired 0.95) and present minimal error. Still, we analyzed five preserved neural networks. We cared about the time series that is more successful than the others. We concluded that the most successful networks are the network 3.RBF 1-25-1 and the network 5.RBF 1-30-1. Having the second highest sum of residues, in the case of basic statistics, keeps up with other preserved networks. At the same time, however, we must state that it is most similar the actual development of the CZK to RMB exchange rate. This network offers performance in a training, test, and validation set of data always above the 0.98 correlation coefficient. It also shows minimal error. So we can conclude that the aim of the paper has been met. We have generated a high-performance neural network with the ability to predict the exchange rate between CZK and RMB in days and weeks.

It is often a major mistake for neural networks to overcome them. This means that networks have excellent characteristics - minimal error and very high performance. Yet their results tend to be pointless. In this case, they were not. All preserved networks are usable with a certain degree of accuracy, and as previously noted, the 3.RBF 1-25-1 network and the 5.RBF 1-30-1 network give even more accurate results than other networks.

References

1. M. Sheikhan, N. Mohammadi, Time series prediction using PSO-optimized neural network and hybrid feature selection algorithm for IEEE load data: revue littéraire mensuelle. *Neural Computing and Applications*, **23**(3-4), 1185-1194, (2013)
2. S. De Baets, N. Harvey, Forecasting from time series subject to sporadic perturbations: Effectiveness of different types of forecasting support. *International Journal of Forecasting*, **34**(2), 163-180, (2018)
3. A. León-Álvarez, J. Betancur-Gómez, F. Jaimes-Barragán, H. Grisales-Romero, Ronda clínica y epidemiológica. Series de tiempo. *IATREIA*, **29**(3), (2016)
4. M. Vochozka, Formation of complex company evaluation method through neural networks based on the example of construction companies′ collection. *AD ALTA – Journal of Interdisciplinary Research*, **7**(2), 232-239, (2017)
5. F. Rodrigues, I. Markou, F. C. Pereira, Combining time-series and textual data for taxi demand prediction in event areas: A deep learning approach. *Information Fusion*, **49**, 120-129, (2019)
6. P. Rostan, A. Rostan, The versatility of spectrum analysis for forecasting financial time series. *Journal of Forecasting*, **37**(3), 327-339, (2108)
7. K. Jebran, A. Iqbal, Dynamics of volatility spillover between stock market and foreign exchange market: evidence from Asian Countries. *Financial Innovation*, **2**(1), (2016)
8. X. Chen, The Globalization of the Chinese Yuan (CNY) and Its Rising Role in the International Currency System. *China and WTO Review*, **2**(2), 303-320, (2016)

9. M. Mandel, V. Quang Tran, Empirická verifikace exportní funkce s akcentem na vliv kurzu české koruny k euru. *Politická ekonomie*, **65**(6), 649-668, (2017)
10. J. Cimburek, P. Řežábek, Hotovost v oběhu: světové trendy a situace v České republice. *Politická ekonomie*, **56**(6), 739-758, (2008)
11. Czech National Bank, *Úloha měnové politiky: Úloha měnové politiky ČNB podle zákona o ČNB* [online]. Available at: https://www.cnb.cz/cs/menova_politika/uloha.html (2018)
12. J. A. Frankel, S. J. Wei, L. Goldberg, Assessing China's exchange rate regime. *Economic Policy*, **22**(51), 576-627, (2007)
13. E. Ogawa, M. Sakane, *Chinese Yuan after Chinese Exchange Rate System Reform. China and World Economy,* **14**(6), 39-57, (2006)
14. C. W. H. Cheong, J. Sinnakkannu, S. Ramasamy, On the predictability of carry trade returns: The case of the Chinese Yuan. *Research in International Business and Finance*, **39**, 358-376, (2017)
15. S. Wang, X. Wei, Relationships between exchange rates, economic growth and FDI in China: An empirical study based on the TVP-VAR model. *Littera Scripta*, **10**(1), 166-179, (2017)
16. G. Ma, R. N Mccauley, The Implications of Renminbi Basket Management for Asian Currency Stability. *The Evolving Role of Asia in Global Finance*, 97-121, (2011)
17. Z. Zhang, K. Sato, Should Chinese Renminbi be Blamed for Its Trade Surplus? A Structural VAR Approach. *The World Economy*, **35**(5), 632-650, (2012)
18. Z. Cai, L. Chen, Y. Fang, A New Forecasting Model for USD/CNY Exchange Rate. *Studies in Nonlinear Dynamics and Econometrics,* **16**(3), (2012)
19. Z. Liu, Z. Zheng, X. Liu, G. Wang, Modelling and Prediction of the CNY Exchange Rate Using RBF Neural Network. *2009 International Conference on Business Intelligence and Financial Engineering*, 38-41, (2009)
20. World Bank. *The World Bank.* [online]. Available at: http://www.worldbank.org/ (2018)

Analysis of business companies based on artificial neural networks

Veronika Machová[1,*] and *Marek* Vochozka[1]

[1]Institute of Technology and Business in České Budějovice, School of Expertness and Valuation, Okružní 517/10, 37001 České Budějovice, Czech Republic

Abstract. Business companies have many kinds of products that they sell to other businesses, consumers, etc. They are a driving force of economies, especially in developing countries. The aim of this article is to analyse business companies in the Czech Republic using artificial neural networks and subsequently to estimate the development of this branch of the national economy. An analysis is performed to create a significant number of clusters of businesses. An analysis of the most significant clusters is also carried out. The result can be generalized and we can predict the number of companies that will be creditworthy or bankrupt in the following period. This makes it possible to estimate not only the overall growth or decline of business companies in the Czech Republic, but also to estimate the structure of the companies in terms of their size, turnover or volume of sales.
Key words: artificial neural networks, cluster analysis, business companies, analysis, prediction

1 Introduction

The primary strategic objective of every company should always be to increase its value. The value of a company is very important. There is a need not only to increase the wealth of individual shareholders but also to support the interests of other stakeholders [1]. Therefore, it is very important to analyse and predict the financial sustainability of the company. According to Grant [2], there are many different methods to this end. However, some methods are preferred, especially due to their ease of use, flexibility and versatility. However, there are also methods which are difficult to use and require skills that many analysts lack. Skalický and Puchýř [3] argue that one of the ways to analyse the potential risks arising from the company's financial data and how to oversee the financial health of a company is financial analysis. In advanced market economies, financial analysis has a long tradition. Among company assessment methods, we can also include benchmarking based on non-financial indicators. These indicators should predict the company's future potential and also evaluate its performance [4].

Economic Value Added (EVA) is another indicator that will allow us to uncover the success of a company on the market [5]. The importance of this indicator is increasing as it is used in practical management of a company as a motivation factor for managers and in analysing benchmarking [6]. It is also possible to evaluate the company, for example,

* Corresponding author: machova@mail.vstecb.cz

according to bankruptcy and creditworthiness models [7]. Companies can be evaluated by other methods, such as using value generators, to allow more accurate measurement of the efficiency of the processes and elements that generate the value. According to Valašková et al. [8], these approaches are referred to as value factors. Value generators can be defined as factors that contribute to value creation in a market evaluation of a company. Knowledge is necessary for involved parties to determine future and current internal values of shares [9]. Value factors allow to reflect the environment in which business entities operate and are able to measure the performance of the company and the value of its corporate strategy [10].

Another method for evaluating companies is by using neural networks, which can also be applied in business practice. The advantage of artificial neural networks is their ability to predict future periods. Sánchez and Melin [11] state that neural networks have a huge application in many different areas. Artificial neural networks have innumerable advantages over conventional methods. They allow you to analyse complicated patterns quickly and with high precision and, according to Santin [12], they are flexible in their own use. The disadvantage of these networks is the need for large sample data, since many test observations are needed to produce such data, however this is very complicated for users [13].

Kohonen networks are models of neural networks that can be used to configure data sets into different categories. The data is configured so that the data within one category tends to be the same and the data in different categories are different. A large number of experimental results shows that Kohonen's networks are very effective for business valuation [14]. According to Horák, Vochozka and Machová [15], the Kohonen network consists of an input layer that is connected with the output layer and is taught without a teacher, by self-organization. It can be said that the network is widely used because it is an alternative network that is usable for most neural network calculations. It is primarily used for managing photos, videos, audio, application security, speech processing, and high-dimensional data projection on lower-dimensional data [16].

In this paper, business companies will be analysed by Kohonen Networks. Businesses companies are companies that have many kinds of products that are sold for the needs of other businesses, consumers, etc. According to Dyhdalewicz [17], business companies keep a store, buy a huge amount of products, and buy and supply goods to customers. According to Doganay and Kocsoy [18], business companies are the driving force of economies, especially in developing countries. These business companies are able to connect sellers with buyers and earn revenue through sales commissions.

The aim of the paper is to perform a cluster analysis of businesses operating in trade.

2 Data and methods

For the purposes of this contribution, a set of data on 11,604 retail and wholesale businesses in 2016 will be created. These will be businesses with the main economic activity classified in Section G of economic classification of CZ NACE activities. The data set will be generated from the Albertina database. The complete financial statements of these companies will be available (all financial information is in CZK thousands). For the analysis, only some items will be used:

1. This is information about the size of the business, namely the volume of assets (fixed, current, other).
2. Fixed assets: the item describes the degree of involvement of fixed assets and gives an idea of the level of technology used in production. The higher the share of fixed assets, the higher the production automation can be expected.
3. Current assets: represent the amount of funds, receivables and inventory in the business. It is in most cases the so-called working capital, i.e. those asset

components that change their substance in less than one year and which are directly involved in the core business activity.

4. Equity: reflect the level of business risk for the business owners. This is the amount of funds the owners may lose in the event of bankruptcy.
5. External capital: This is capital that does not have influence on the company's business management. In a way, it reflects the external view of the potential success of the business.
6. Value Added: This is the value added by the company by its core business activity to basic input - material.
7. Operating income: Provides information on how the business is successful in its core business activity.
8. ROE: return on equity means an appreciation of the equity of a business, that is, the capital provided by the business owners.
9. Profit before tax: describes the success or failure of a company in its entire business activity.

These are not generators of value but rather characteristics that may have, albeit indirect, influence on the performance of a business.

The file will then be subjected to cluster analysis, specifically using Kohonen networks. For cluster analysis, Dell's Statistica software in version 12 will be used. Data Mining will be used as a specific Neural Network tool. Here we select neural networks without a teacher – Kohonen's networks. We specify the data for analysis. In all cases, these are continuous predictors. The file will be divided into three parts:

1. Training dataset: represents 70% of companies in the set. The Kohonen network will be created based on this dataset.
2. Testing dataset: this is 15% of the companies of the original set. Using this data set, we verify the parameters of the Kohonen network.
3. Validation dataset: this will also involve 15% of companies in the file. Using this dataset, we will also verify the Kohonen network to see whether it is or is not usable.

We determine the topological length and the topological width of the Kohonen map at 10. The number of iterations of the calculation will be set at 100,000. However, I shall recall that the level of error is decisive. Unless there is any iteration to improve the Kohonen network parameters, training will be terminated before the 100,000 iteration is completed. If the network parameters are improved even at 100,000 iterations, we need to repeat the process and set the higher value of the required iterations to make sure the result is the best possible.

3 Results

In accordance with the methodology of the contribution, the data was divided into three sets – training, testing and validation. The descriptive characteristics of all three sets and the original data set are given in Table 1.

Table 1. Descriptive characteristics of the data file

Samples	Total assets	Fixed assets	Current assets	Equity	External capital	Outputs	Value added	Operating income	ROE	Profit before tax
Minimum (Training)	-1564	-10744.0	-2071	-254305	-533106	-4956.0	-108795	-229908	-644.750	-35843.0
Maximum (Training)	1071267	804579.0	964246	911000	891328	574445.0	301525	234050	841.667	84791.0
Average (Training)	44408	11089.6	32830	20201	23898	3073.5	10335	3387	0.126	49.1
Standard deviation (Training)	78297	32174.8	59367	47471	49033	14845.0	21303	10589	13.297	1293.0
Minimum (Testing)	-8573	-65415.0	-15357	-475270	-8763	-2427.0	-42902	-227728	-871.500	-3998.0
Maximum (Testing)	1067016	430553.0	1039201	505227	901245	258720.0	388474	119588	52.233	44805.0
Average (Testing)	44690	11486.0	32662	18838	25515	2823.1	10146	3040	-0.623	90.2
Standard deviation (Testing)	82682	30419.8	64151	43633	57608	12059.3	21374	10846	22.279	1465.8
Minimum (Validation)	1	-10744.0	-5335	-298827	0	-6242.0	-293974	-84319	-216.000	-25100.0
Maximum (Validation)	1532330	996822.0	771988	1317026	653336	513452.0	407995	197614	18.272	26690.0
Average (Validation)	46190	12539.0	33121	21068	24756	3556.3	10066	3260	0.052	47.7
Standard deviation (Validation)	84300	34684.9	64020	43782	54274	14821.9	21327	9059	21.188	2621.3
Minimum (Overall)	-8573	-65415.0	-15357	-475270	-533106	-6242.0	-293974	-229908	-871.500	-35843.0
Maximum (Overall)	1532330	996822.0	1039201	1317026	901245	574445.0	407995	234050	841.667	84791.0
Average (Overall)	44718	11366.5	32848	20127	24269	3108.3	10267	3316	0.002	55.1
Standard deviation (Overall)	81285	34749.3	60024	48834	50748	15255.5	21768	10592	14.235	1295.5

Source: Authors.

Ideally, if the characteristics of the subsets of data were approximately the same. Since businesses have been randomly distributed, this is not the case. However, the random distribution of data may not be a defect in the result.

Based on the assignment, the Kohonen network was calculated. It is also referred to as SOFM 10-100 in the text. The companies were divided into individual clusters within the 10 x 10 topological grid. The frequency of individual clusters is evident from the three-dimensional graph in figure 1. It is evident from the figure that the largest number of enterprises in individual fields of the topological grid is in the position (1, 10) then in position (2, 10). The following ones are (1, 9), (2, 9) and (2, 8). Overall, we can say that the largest number of the 11,604 companies surveyed is in the sector (1, 7), (4, 7) and (1, 10), (4, 10).

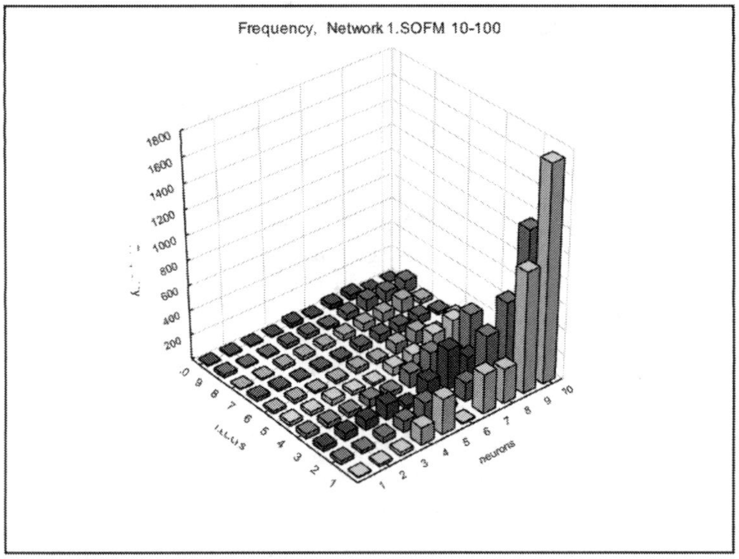

Fig. 1. Graph of business frequencies in topological grid
Source: Authors.

The results are shown more specifically in Table 2.

Table 2. SOFM 10-100 Network Frequency Table

	1	2	3	4	5	6	7	8	9	10
1	8	19	29	137	281	10	316	278	969	1726
2	29	24	45	88	150	26	151	446	636	1138
3	37	60	78	121	31	148	320	181	97	283
4	38	42	40	64	40	116	201	137	354	60
5	30	30	46	31	20	66	88	178	214	1
6	21	30	16	25	43	28	83	107	15	79
7	28	23	10	22	22	44	65	61	63	25
8	10	10	27	29	34	48	63	75	108	34
9	26	12	16	34	37	22	55	93	87	101
10	9	11	11	6	35	11	37	44	4	3

Source: Authors.

It also follows from the table that it is advisable to focus on four clusters, namely (1, 10), (2, 10), (1, 9) and (2, 9), possibly the fifth cluster (2, 8). This is a very interesting set of data representing more than one third of all companies in the file. Dominance is more evident from the Kohonen graph in Figure 2.

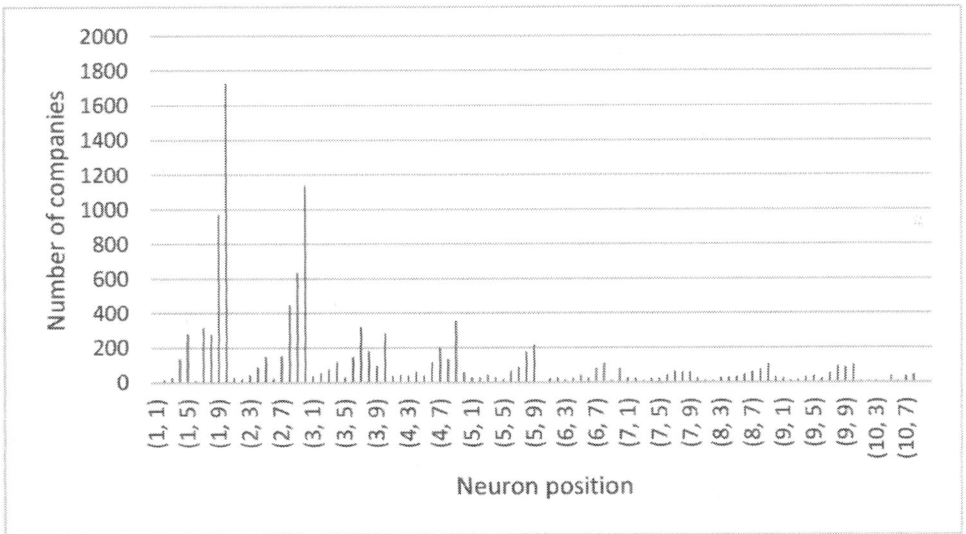

Fig 2. Kohonen graph for selected cells of the SOFM 10-100 network
Source: Authors.

While the data set is rather heterogeneous, cluster analysis suggests that there are characteristics in which businesses are similar.

Cluster (1, 10)

The cluster (1, 10) of the topological grid is occupied by 1,726 companies. It is the largest set of data from partial clusters. From the point of view of input parameters, we can derive the following average values:

1. Total assets: 5,740.1 mil. CZK.
2. Fixed assets: 948.2 mil. CZK.

3. Current assets: 4,705.4 mil. CZK.
4. Equity: 1,174.9 mil. CZK.
5. External capital: 4,520.5 mil. CZK.
6. Outputs: 206.3 mil. CZK.
7. Value added: 1,109.3 tis. CZK.
8. Operating income: 175.2 mil. CZK.
9. ROE: 0.19 %.
10. Profit before tax: 2.3 mil. CZK.

From the average values, we can derive a typical company for this cluster of businesses. It is a small company whose operating income is minimal, 101 thousand CZK per year. It generates a pre-tax profit on average of only 1.3 thousand CZK. Nevertheless, it has assets of CZK 3.3 million. However, the assets are financed for the most part by external capital. In particular, such a company has more than CZK 2.6 million of external capital. Value added is also not high. It is slightly over 680,000 CZK. Nevertheless, the existence of the business is not jeopardized. Based on the nature of its business activity, the company holds a relatively large volume of current assets (probably goods) in excess of CZK 2.7 million.

When it comes to a homogeneous cluster of businesses, we can verify the validity of basic logical links. The relationship between total assets and the amount of operating income appears interesting (see Figure 3 below).

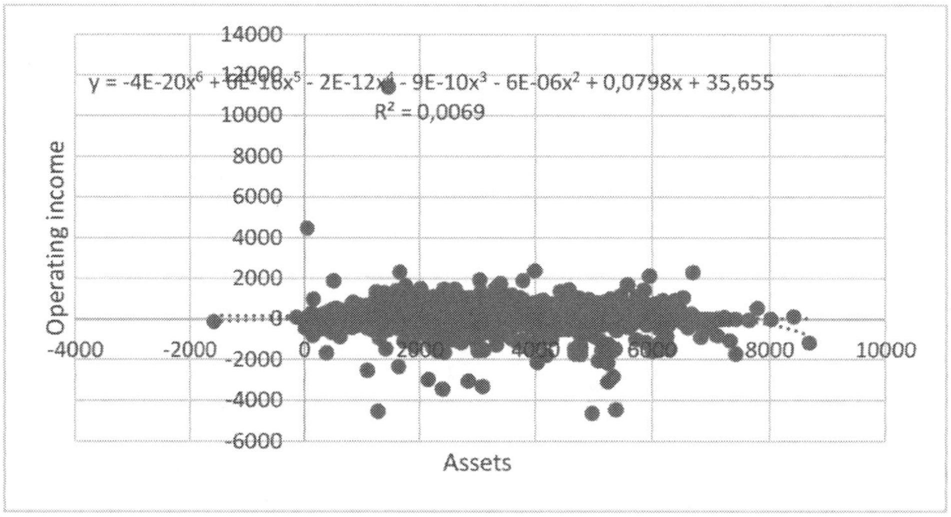

Fig. 3. The relationship between the total assets and the operating income of the cluster (1, 10)
Source: Authors.

Total assets and operating income are correlated to more than 0.08. Therefore, there can be no dependence of both variables. Yet the data is intersected by a regression curve, a 6th order polynomial. Based on this, it is possible to estimate the development of the operating profit on the basis of the size of the company determined by the volume of total assets.

Figure 4 provides a relationship between fixed assets and operating income.

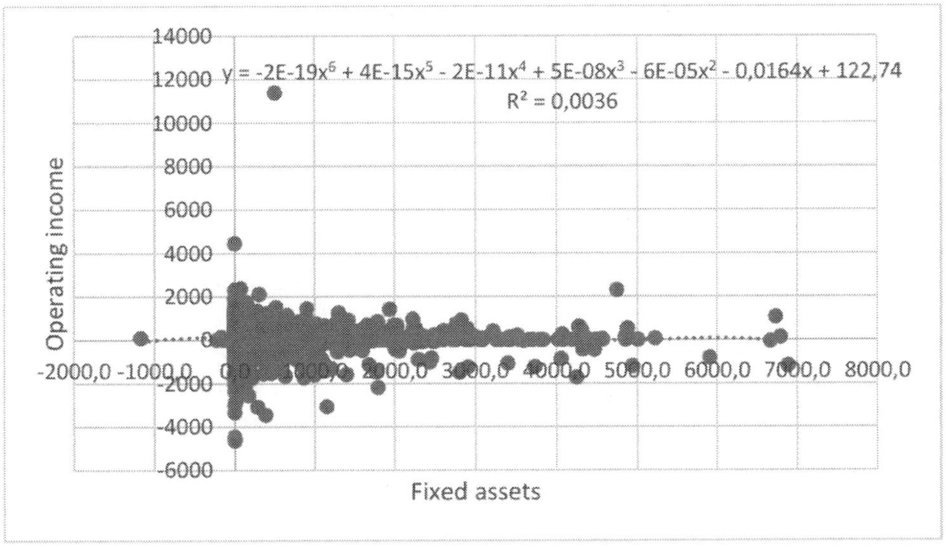

Fig. 4. The relationship of fixed assets and operating profit of cluster (1, 10)
Source: Authors.

The figure shows that there is virtually no dependency between the data. The correlation coefficient is 0.06. By default, the volume of fixed assets represents the degree of automation in the enterprise. In the case of trade business companies, it rather provides information about the ownership of the factors of production – namely, one – the real estate in which the business is located, if it is needed to carry out the business activities. A number of business transactions can go hand in hand with real estate – such as the operation of e-shops. Figure 4 illustrates that trade business enterprises often do not need long-term assets for their business, possibly only in the minimum amount (for example, in the form of a car.

Figure 5 thus presents the relationship between external capital and operating income.

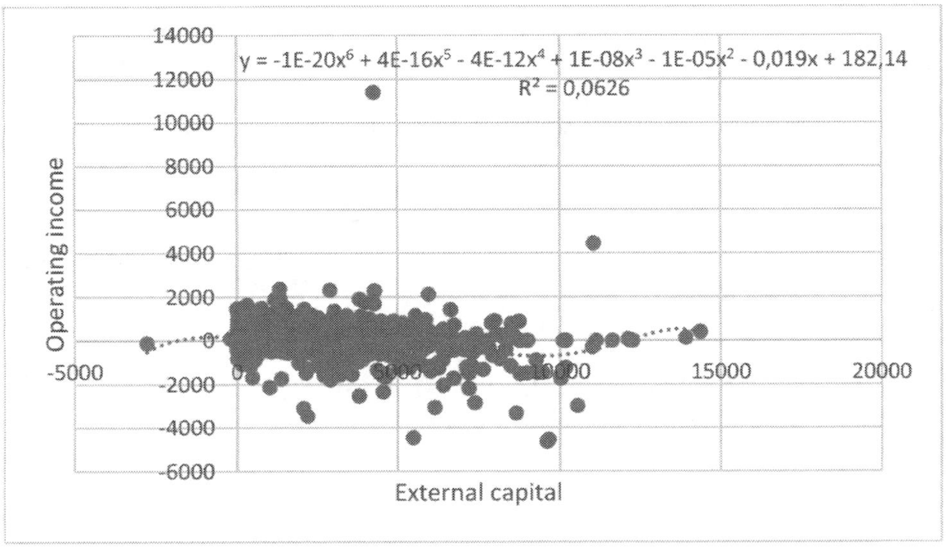

Fig. 5. The relationship between external capital and the operating income of cluster (1, 10)
Source: Authors.

The relationship can show us the effect of a financial leverage, but only assuming that there will be a relatively high correlation between the two variables. However, it's only 0.25. Yet, the graph points are intersected by a 6th order polynomial. However, we can see from the graph in the figure that businesses are in unhealthy debt. Additional external capital does not increase the return on equity. Both variables are indifferent and point to a poor capital structure and perhaps even unnecessary assets in the enterprise. Companies in this cluster should reduce the involvement of external debt.

Cluster (2, 10)

The cluster (2, 10) of the topological grid occupies the imaginary second position in frequency. There are 1,138 companies. From the point of view of the total input parameters we can deduce a typical representative of this cluster:

11. Total assets: 6 668.9 mil. CZK.
12. Fixed assets: 1 364.9 mil. CZK.
13. Current assets: 5 212 mil. CZK.
14. Equity: 2 534.3 mil. CZK.
15. External capital: 4 083 mil. CZK.
16. Output: 344.2 mil. CZK.
17. Value added: 3 667.7 mil. CZK.
18. Operating income: 976 mil. CZK.
19. ROE: 0.81 %.
20. Profit before tax: 20.4 mil. CZK.

A typical company of this cluster is more than 1.75 times larger than the average company of cluster (1, 10), meaning from the point of view of the volume of total assets. It is an enterprise that employs more than 5.8 million CZK worth of assets. The company owns fixed assets in the amount of almost CZK 1.2 million. At the same time, it manages current assets in excess of CZK 4.5 million. Its assets are funded by more than CZK 2.2 million from its own resources, with the remaining amount, i.e. almost CZK 3.6 million, coming from external capital. It achieves an operating income over 850 thousand CZK, which represents almost 18 thousand CZK profit before tax. In addition to the realized profit, a certain shift in the company of this cluster compared to the cluster (1, 10) is a lower indebtedness rate. The companies of cluster (1, 10) had a debt ratio of almost 80%. On the other hand, businesses of cluster (2, 10) have a debt ratio slightly above 62%. Yes, the high level of indebtedness is due to the nature of the business activities. Nevertheless, we can say that companies of cluster (2, 10) are closer to the optimum (even with a larger sum of money than companies of cluster (1, 10)). Even in this case, it is certainly interesting to explore the relationship of some variables.

Figure 6 shows the relationship between total assets and operating income.

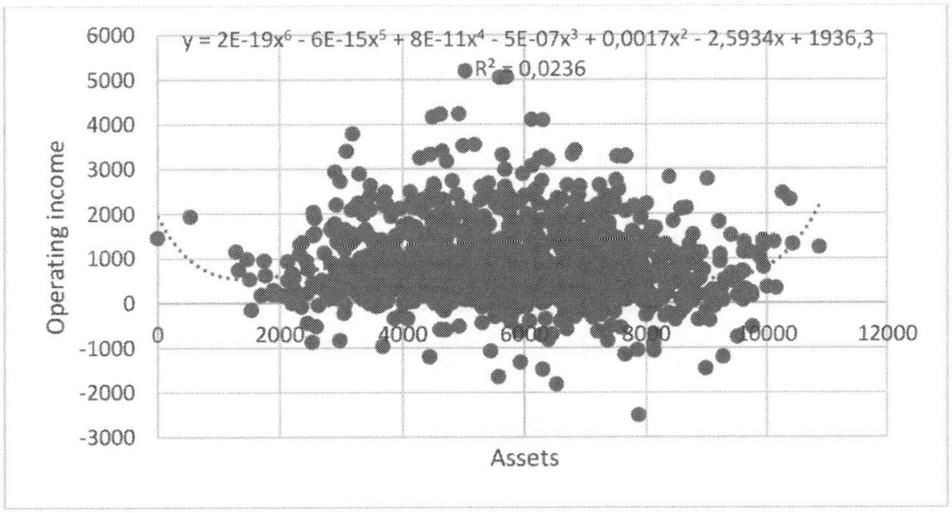

Fig. 6. The relationship between the total assets and the operating profit of cluster (2, 10)
Source: Authors.

The correlation coefficient of the two variables is 0.15. Also, we can´t talk about a dependency between the two variables. However, in this cluster, we can observe the trend in the fact that a higher volume of assets implies a higher operating income. The trend-line has a 6th order polynomial again.

Figure 7 illustrates the relationship between fixed assets and operating income.

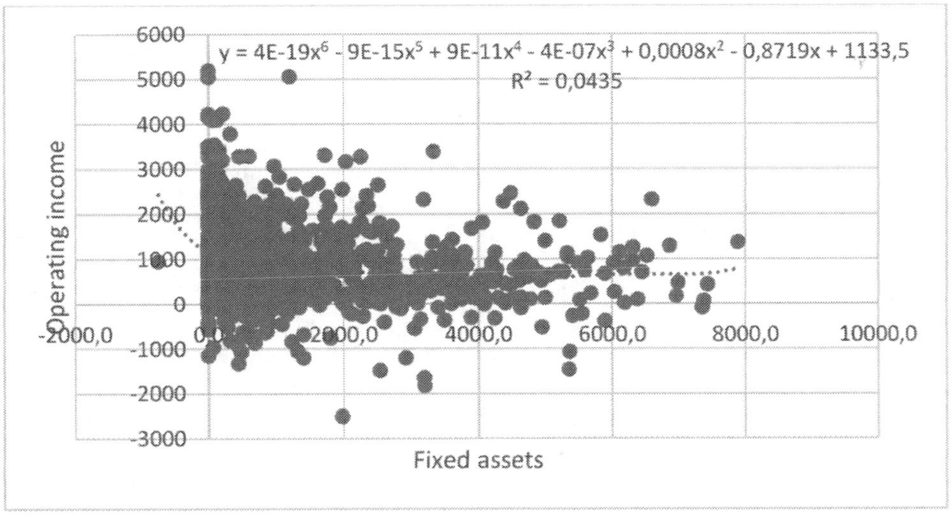

Fig. 7. Relationship between fixed assets and the operating income of cluster (2, 10)
Source: Authors.

For the relationship between fixed assets and operating income, the correlation coefficient is 0.21. While not a statistically significant dependency, we can already say that the higher volume of fixed assets has a positive effect on the generation of operating income.

Figure 8 provides a relationship of external capital and operating income.

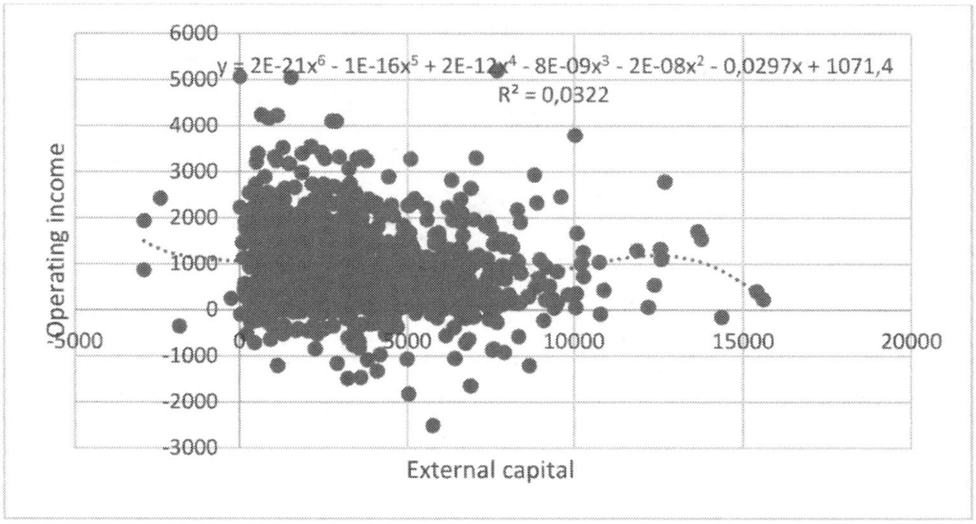

Fig. 8. The relationship of external capital and the operating income of cluster (2, 10)
Source: Authors.

Even in this case, the correlation coefficient is not high, namely 0.18. However, even in this case, we can infer that a higher debt ratio does not mean a higher operating income. As in the case of cluster (1, 10), it is advisable to reduce the volume of external capital involved.

4 Conclusion

The aim of the paper was to perform a cluster analysis of businesses operating in trade. The aim of the paper was fulfilled.

Cluster analysis was performed using neural networks without a teacher – Kohonen networks. Based on cluster analysis, companies were divided into clusters in the Kohonen map (10 x 10 clusters). Some clusters are significant in terms of number of businesses. The most represented are clusters (1, 10) and (2, 10). These clusters have therefore been subjected to further analysis. In the case of both clusters, the typical representatives of these clusters were defined. These varied by their size and volume of operating income generated. The companies of both clusters were on the borderline of profit margins in 2016. The average operating income for one company of cluster (1, 10) was approximately 1.3 thousand CZK, in the case of cluster (2, 10) it was 18 thousand. CZK. The companies of the two clusters differ far more in the amount of the operating profit. Companies of cluster (1, 10) make a profit of 101 thousand CZK. Companies of cluster (2, 10) reach 850 thousand CZK in profit. Businesses of both clusters are highly indebted. This can be partly related to the field in which they operate and partly to the non-rational decision about the involvement of external debt. In the case of debt, companies of cluster (2, 10) are better off, with external capital accounting for 61% of all capital. On the other hand, the indebtedness of companies in (1, 10) is almost 80%. According to the volume of fixed assets, it is possible to conclude, in particular, for companies of cluster (1, 10) that they do not operate in their own premises, but in rented premises or on the Internet. From the analysis we can make two very specific conclusions:

1. A larger company (or a company with more assets) generates, on average, a higher operating income.

2. In the case of a larger company, the financial leverage is more positive, i.e. with the growth of the volume of the foreign capital, the operating income and the profitability of the equity increase.

References

1. R. R. Gamayuni, The Effect Of Intangible Asset, Financial Performance and Financial Policies On The Firm Value. *International Journal of Scientific & Technology*, **4**(1), 202-212, (2015)
2. D. Grant, Business analysis techniques in business reengineering. *Business Process Management Journal*, **22**(1), 75-88, (2016)
3. M. Skalický, B. Puchýř, Use of financial analysis for elimination of risk factors in the construction company. *Political Sciences, Law, Finance, Economics and Tourism Conference Proceedings,* Albena, Bulgaria, 669-676, (2016)
4. M. Vochozka, Z. Rowland, J. Vrbka, Financial analysis of an average transport company in the Czech Republic. *Nase More*, **63**(3), 227-236, (2016)
5. I. Ismail, The ability of EVA (Economic Value Added) attributes in predicting company performance. *African Journal of Business Management,* **5**(12), 4993-5000, (2011)
6. J. Vrbka, Modeling the future development of top company indicator EVA Equity using indicator breakdown and sensitivity analysis on example of a specific company. *Proceedings of the 5th International Conference Innovation Management, Entrepreneurship and Sustainability*, Prague, Czech Republic, 1095-1106, (2017)
7. D. Alaminos, A. Del Castillo, M. A. Fernández, G. Ponti, A Global Model for Bankruptcy Prediction. *PLOS ONE*, **11**(11), (2016)
8. K. Valášková, T. Klieštik, L. Švábová, P. Adamko, Financial risk measurement and prediction modelling for sustainable development of business entities using regression analysis. *Sustainability,* **10**(7), 2144, (2018)
9. E. Vítková, J. Chovancová, J. VESELÝ, Value Driver and Its Impact on Operational Profit in Construction Company. *Procedia Computer Science*, **121**, 364-369, (2017)
10. M. Strnadová, M. Karas, M. Režňáková, Value drivers of Processing Industry Enterprises in the Czech Republic in the Years 2007 -2011.*Trends Economics and Management,* **7**(13), 91-99, (2013)
11. D. Sánchez, P. Melin, Modular Neural Networks for Time Series Prediction Using Type-1 Fuzzy Logic Integration. *Design of Intelligent Systems Based on Fuzzy Logic, Neural Networks and Nature-Inspired Optimization,* **601**, 141-145, (2015)
12. D. Santin, On the approximation of production functions: a comparison of artificial neural networks frontiers and efficiency techniques. *Applied Economics Letters*, **15**(8), 597-600, (2008)
13. M. S. Hossain, Z. C. Ong, Z. Ismail, S. Noroozi, S. Y. Khoo, Artificial neural networks for vibration based inverse parametric identifications: A review. *Applied Soft Computing*, **52**, 203-219, (2017)
14. X. Han, L. Wang, Stock Company Comprehensive Assessment Model Based on Kohonen Network. *2008 Second International Conference on Genetic and Evolutionary Computing*, Jingzhou, China, 185-188, (2008)
15. J. Horák, M. Vochozka, V. Machová, Evaluation of transport and forwarding companies by means neural networks. *Proceedings of the International Scientific*

Conference for Ph.D. students and assistants – QUAERE, Hradec Králové, Czech Republic, 377-385, (2018)

16. P. Šuleř, Using Kohonen´s neural networks to identify the bankruptcy of enterprises: Case study based on construction companies in South Bohemian region. *Proceedings of the 5th International Conference Innovation Management, Entrepreneurship and Sustainability,* Prague, Czech Republic, 985-995, (2017)
17. A. Dyhdalewicz, Innovation of Trading Companies in Relation to the Content of Annual Reports: Research Results. *Economisc and Social Development (ESD 2017)*, 629-638, (2017)
18. M. Doganay, M. Kocsoy, A field study concerning the effects of economic crisis on international trading companies and the measures taken. *African Journal of Business Management*, **55**(22), 9477-9489, (2011)

Formation of a marketing support complex for an export-oriented enterprise in conditions of risk and uncertainty

Olga Maslak[1,*], *Petro* Sokurenko[2], *Natalya* Grishko[3] and *Mariya* Maslak[4]

[1]Doctor of Economics, professor, Kremenchuk Mykhailo Ostrohradskyi National University, Ukraine
[2]PhD, associate professor, Affiliate «Kremenchuk institute of Higher Educational Establishment «Alfred Nobel University», Ukraine
[3]PhD, associate professor, Kremenchuk Mykhailo Ostrohradskyi National University, Ukraine
[4]PhD, associate professor, National Technical University «Kharkiv Polytechnic Institute», Ukraine

Abstract. The work covers the theoretical and methodological foundations of marketing for an export-oriented enterprise. At the present stage of development of the economy, the use of marketing tools in activities is a powerful means of increasing the enterprises` competitiveness, strengthening adaptive characteristics when external conditions and market conditions change. Consumption volumes are the driving force of the production activity of the enterprise, affecting the nomenclature, range and scale of production. The introduction of the marketing concept, which focuses on the production of products that are in the market demand, stimulates the production of products needed by the consumer. Theoretical and methodological issues of marketing aimed at restricting the demand for irrational demand products (opposing marketing) attract the attention of scientists and practitioners. At the same time, the choice of the topic was due to insufficient research of this scientific problem as a whole and its relevance for the economy of Ukraine, especially in the context of the development of cooperation with the European Union. A detailed study of the tools of opposing marketing allows export-oriented enterprises to ensure the profitability of commercial and marketing activities, while the harmful effects of the consumption of goods of irrational demand determine the feasibility of research and analysis of the conditions and their consumption motivation. The demand for these products is undesirable; therefore, countermeasure marketing at the state level is controlled by the relevant organizations.
In this study, attention is paid to the analytical and regulatory aspects of determining the current state and trends of marketing support for the tobacco industry enterprises activities. The features of a complex of marketing support formation of their activities in the conditions of risk and uncertainty are determined.
Key words: marketing, marketing support, management, model, marketing technologies, complex, irrational demand

* Corresponding author: 2nata.grishko@gmail.com

1 Introduction

Marketing should be taken as a study of customer needs, characteristics of consumer demand, work with target segments to generate potential demand. Marketing support of the tobacco industry has certain features, based on the properties of products related to products of irrational demand (harmful and socially dangerous goods). Tobacco products as goods are characterized by specialty of implementation, which is used in the development of the marketing complex, namely, social and ethical marketing in the tobacco market.

The main stages of the study were formed to assess the trends and opportunities for international, particularly:

- analytical and regulatory aspects of current state determination and trends of marketing support for the activities of export-oriented enterprises;

- determination of the formation specifics of a marketing complex for irrational demand goods producers , in particular: the establishment of legislative restrictions on tobacco advertisements in EU countries; justification of the model creation of a marketing information system in the enterprise.

2 Data and Methods

The aim of the study is to develop a methodological approach and practical recommendations for finding and implementing ways to improve the marketing mechanism of an export-oriented enterprise under conditions of risk and uncertainty of the market of goods with irrational demand.

The solution of the set tasks was carried out using the following methods: *theoretical generalization* – to study and systematize theoretical approaches for the definition and justification of the problem studied; *analysis and synthesis* – to determine the object and subject of research; *graphic; mathematical statistics* – when establishing the dependence of the integral performance index of the marketing complex; *method of multi-criteria optimization* – when modeling the costs of implementing the marketing complex in a particular market.

Information base selected materials of official publications of the State Statistics Committee, domestic and foreign scientific literature data, primary reporting enterprises documentation data, annual financial and statistical reporting.

3 Results

3.1 Analytical and regulatory aspects of determining the current state and marketing supportive trends for the tobacco industry enterprises' activities

Sales promotion is a set of techniques to increase sales. Features of products or services determine the use of incentives and the choice of a specific channel to promote products, the use of logistics supply intermediaries, a combination of several distribution channels, which forms a specific distribution system.

Thus, effective marketing provides the groundwork for a profitable enterprise; while the marketing complex forms an individualized system, which affects the performance of the enterprise through the product (product), price (price), place (distribution) and promotion.

According to the legislative acts of Ukraine - tobacco goods - products fully or partially made from tobacco leaf as a raw material, prepared in such a way as to be used for smoking, sucking, chewing or smelling: filter or non-filter cigarettes, cigarettes, cigars, cigarillos, as well as pipe, snuff, sucking, chewing tobacco, tobacco and other products made of tobacco or its substitutes for smoking, smelling, sucking or chewing [1].

The development of conversion marketing activities that are used by manufacturers of irrational demand should be preceded by a detailed study of the production and sale of tobacco products. Products of irrational demand are also targeted marketing. The ban on smoking in public places and establishments as well as the trend of rising prices for tobacco products and alcoholic beverages, have affected the decrease in sales volumes and, consequently, their production volumes in Europe and the USA. In Ukraine, the decline in production was observed during 2016-2017, while the production volumes in the processing industry had a positive trend. The wave of growth in the production of tobacco products was characterized by the period from 2009 to 2013, the next wave was observed during 2014-2016. The production of tobacco products in Ukraine in 2007-2017 data is shown in Fig. 1.

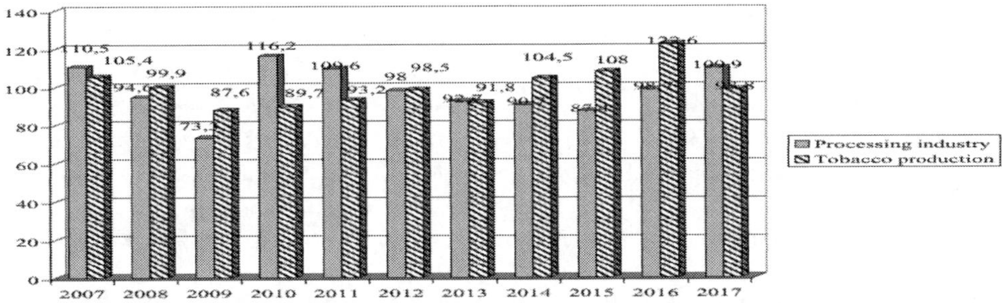

Fig. 1. Indices of tobacco production in Ukraine (percent to previous year)
Source: Formed by authors based on [2].

Since the 1990s, tobacco production has been characterized by trends of creation joint ventures with international and world-renowned corporations. The creation of joint ventures covered 95% of the production of tobacco factories in Ukraine. All this is primarily due to the fact that Ukraine was among the top 20 countries in terms of tobacco consumption. The prices of tobacco products corresponded to the waves of consumer price indices for goods during 2002-2016, while the general trend was growing. Producer price and consumer price indices for individual product groups are shown in Fig. 2.

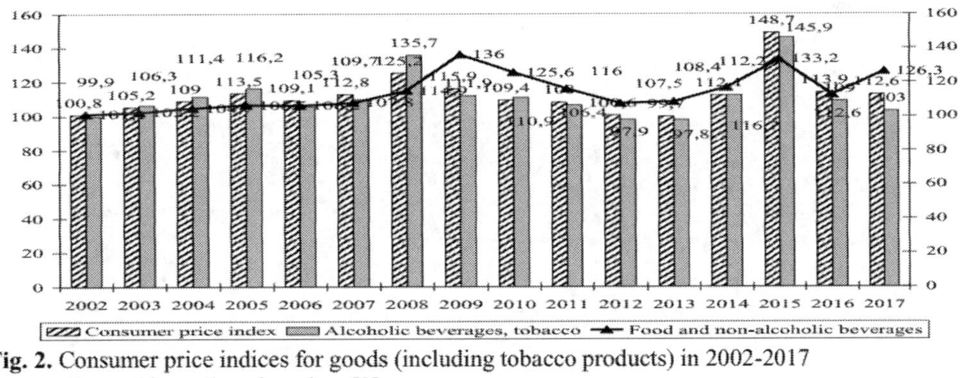

Fig. 2. Consumer price indices for goods (including tobacco products) in 2002-2017
Source: Formed by authors based on [2].

During the 2009-2012 there was a gradual increase in the growth rate of the physical volume of tobacco products (retail turnover of enterprises in Ukraine) from 79.1% to

143.5%, respectively. 2012 is a turning point. During 2012-2016 there is a decrease in the pace of retail turnover of enterprises in Ukraine. The growth rate of production of tobacco products - in 2016, production volume amounted to 122.6% of the volume of 2015; and consumption in the retail commodity networks of Ukraine - 95.1% of the previous period, that is, domestic consumption declined amid production growth, focusing on the export-oriented nature of sales.

In 2017, tobacco products produced 98.8% of 2016, nevertheless, at the end of the research period, prices for alcoholic beverages and tobacco products increased by 26.3%. When developing a toolkit for conversion and opposing marketing, it should be borne in mind that the most part of tobacco products is sold through retail chains, having certain specific features. This market has an oligopolistic character next to the underdevelopment of the domestic raw material base of fermentation and tobacco enterprises.

Tobacco products manufacturers work on imported raw materials. Among the main producers, which hold the main market share, the following should be highlighted: CNTC (China) – 33.7%, Altria (USA) – 17.6%, British American Tobacco (USA) – 15.1%, "Imperial Tobacco (UK) – 3.6% and other companies – 30% [3].

Enterprises producing tobacco products – Philip Morris Ukraine, PJSC "JT International Ukraine" sell some of the products for export, which requires them to take into account the requirements of the legislation for implementation in EU markets, EU legal norms

- restriction in advertising products, in particular irrational demand;
- restriction in advertising non-prescription medical drugs;
- Requirements for quantitative parameters of commercials
- the use of personal data for individual advertising.

Fig. 3 shows some legal restrictions on certain types of goods in EU countries (convention adopted by 18 countries) [4].

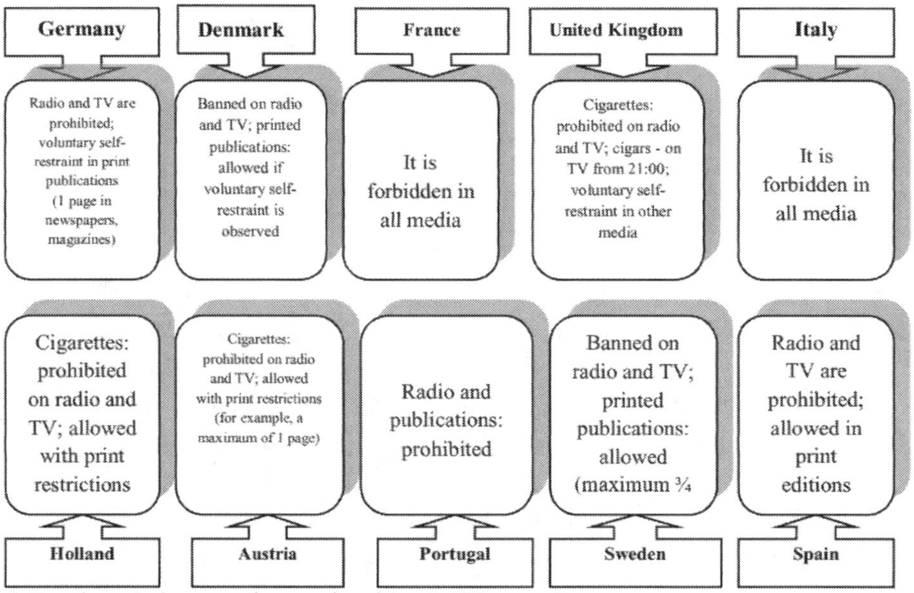

Fig. 3. Legal restrictions on tobacco advertising in EU countries
Source: Formed by authors based on [4].

When entering the external market, in particular the EU market, it is necessary to take into account the requirements of the convention of the European (international) television, which formulate provisions for ethical standards when promoting and advertising products.

The content of advertising products must also pass through the advertiser's self-control as required by the domestic market.

Thus, the socio-ethical marketing of tobacco production enterprises should take into account the opposing marketing by the state and society, based on social responsibility, which will ensure the profitability of the activity while maintaining a positive reputation in the market. In summary, the marketing complex for irrational demand goods producers should be formed, taking into account legislative restrictions on the markets for sales in the sphere of pricing regulation, sales and promotion features.

3.2 Prerequisites for the formation of a marketing information system model in an enterprise

An analysis of the current state and characteristics of the development of the tobacco industry in Ukraine showed that enterprises were reoriented to sell products on foreign markets, with the majority of enterprises being reorganized into joint ventures with a foreign capital share. So, PJSC "JT International Ukraine" is a member of Japan Tobacco group of companies, a leading international manufacturer of tobacco products, produces more than 90 brands of cigarettes, and three of them are among the five most popular in the world. JTI is represented in 120 countries and has over 25,000 employees. The company's headquarters is located in Geneva, Switzerland. During the research period, there is a positive trend in sales volumes, in particular for export. But, by the end of 2017, sales volumes decreased by 14.44%, which is associated with a decrease in sales, primarily in the domestic market. In 2014 - 2017 the share of sales for exports gradually increased from 42%; 52%; 52.18% to 56.6%. Sales of products have a seasonal nature, with the main sales markets being the regions of the CIS and the Caucasus, Georgia, the main sales channels are indirect distributor channels. The restriction on introducing innovative types of products is due to a ban on advertising tobacco products, which increases the riskiness of new products entering the market. Legislative restrictions on advertising and selling tobacco products are leveled by the entry of enterprises into markets with milder terms of regulation. Promotion of products for export takes place through contractors, in the domestic market - through the customer service department. The raw material base is the countries of Asia and the EU.

The implementation of marketing functions in order to improve the efficiency of the enterprise is carried out within the framework of marketing support of PJSC "JT International Ukraine". The implementation of management functions combines the following procedures: production planning for market needs; development of new types of products in the framework of the expansion of the range; introduction of process innovations, entry into new markets, the formation of effective well-established logistics supply chains, order fulfillment for profit. The customer service department of PJSC "JT International Ukraine" is responsible for the formation of a communication network and establishing connections with customers in the domestic market. In order to strengthen the marketing component in ensuring the effectiveness of the enterprise, it is advisable to consider a group of indicators that will help more effectively reflect its level when calculating the integral indicator [5].

The results of diagnostics of the marketing complex of PJSC "JT International Ukraine" are illustrated in fig. 4.

The negative dynamics of the integral indicator of marketing support of the enterprise at the end of 2017 is primarily due to shortcomings in the strategic planning of the company's marketing activities. It is also necessary to consider indicators that have a quantitative dimension. The first group of indicators in accordance with the complex of marketing refers to the element "Product". This group may include indicators reflecting various aspects of this element: width, depth, harmony and saturation of the range; coefficient of rationality of the assortment structure; the stage of the life cycle on which each assortment of goods is located; profitability of products and individual commodity groups; the number of new types of products that have replenished the commodity nomenclature; quality of products in terms of incoming complaints from consumers, etc. Among the most significant marketing tools used by the enterprise is a combination of subsystems "goods", "price policy", "sales channel", "distribution system".

Fig. 4. The level of marketing support of PJSC "JT International Ukraine" for 2014-2017
Source: Formed by authors based on [6].

The peculiarities of promotion of such goods as tobacco products, which is a product of irrational demand, should be taken into account. Next to the marketing activities that the company implements, there is a system of counteracting marketing by the society and the state. That is, PJSC "JT International Ukraine", together with a focus on maximizing profits, which should not lose its reputation in the market and is obliged to maintain social responsibility and social and ethical standards in promoting products. Thus, the state, implementing measures to regulate the activities of enterprises producing alcoholic and tobacco-containing products, limits the amount of irrational demand. Along with the implementation of conversion and countering marketing activities, the company's marketing services must provide high sensitivity to trends and rates of change in the market, assess the adequacy of the internal environment of the enterprise to new tasks and exercise powerful information influence in moving the company in a strategically chosen direction. Effective indicators of the integral model of assessing the performance of PJSC "JT International Ukraine" by the functional components of the elements of the marketing complex are shown in Table 1. The calculation of the integral indicator was based on the standardization procedure (standardization) of the selected indicators [6] for comparability of the evaluation base.

Determination of the level and dynamics of the integrated performance index of PJSC "JT International Ukraine" in terms of the functional components of the marketing complex, depending on the evaluation criteria during 2014-2017, allows to note 2016 as the best in terms of ensuring effectiveness. This doubtful state is observed during 2014 and 2017. To strengthen the marketing component in the performance of the enterprise, a group of indicators that help to more effectively reflect its level when calculating the integral indicator is considered. The results show that PJSC "JT International Ukraine" is in a state of considerable dependence on the means of regulating behavior on the market: further legislative measures to ban smoking, advertising and distribution of tobacco products.

The effectiveness of the sales activity of the production enterprise depends on the ability to predict the influence of relevant factors on the level of the integral performance index of the marketing complex, possibly using information support systems to solve problems of effective information interaction with market actors, expanding the information base for making management decisions.

Table 1. Effective indicators of the integrated model of the assessment of PJSC "JT International Ukraine" by the functional components of the elements of the marketing complex during 2014-2017

Indicators	2014	2015	2016	2017	minimum	maximum	scope of variation	normalized indicators			
								2014	**2015**	**2016**	**2017**
The element of the marketing complex "Product"											
The coefficient of rationality of assortment structure, coefficient	0.87	0.95	0.98	0.96	0.87	0.98	0.11	0.000	0.727	1.000	0.82
The coefficient of product innovations implementation, coefficient	0.45	0.54	0.58	0.45	0.45	0.58	0.13	1.818	0.692	1.000	0.00
Update range index, coefficients	0.24	0.25	0.26	0.24	0.24	0.26	0.02	-5.727	0.500	1.000	0.00
The coefficient of complaints, coefficient	0.02	0.01	0.03	0.03	0.01	0.03	0.02	0.091	1.000	0.000	0.00
The integral indicator for the average value of the trait								**-2.364**	**0.730**	**0.750**	**0.205**
The element of the marketing complex "Price"											
Costs for 1 hryvnia of sales, UAH	0.56	0.589	0.647	0.858	0.565	0.647	0.082	2.664	0.082	0.720	0.00
The adequate value (adequacy of the selling price compared to other products), coefficient	0.97	0.98	0.99	0.99	0.97	0.99	0.02	0.909	0.500	1.000	1.00
The coefficient of rationality of providing discounts on prices, coefficient	0.88	0.89	0.92	0.89	0.88	0.92	0.04	0.091	0.250	1.000	0.25
Net profit / loss ratio of product sales, coefficient	0.18	0.1797	0.1754	-0.009	0.1754	0.1885	0.0131	-6.195	0.955	0.934	0.00
The representation in the network of sales ratio, coefficient	0.66	0.78	0.83	0.78	0.66	0.83	0.17	-1.909	0.706	1.000	0.71
The integral indicator for the average value of the trait								**1.221**	**0.277**	**0.907**	**0.417**
The element of the marketing complex "Distribution"											
The coefficient of supply reliability, coefficient	0.97	0.98	0.98	0.97	0.97	0.98	0.01	0.909	1.000	1.000	0.00
The recoupment ratio of sales costs, coefficient	0.86	0.88	0.92	0.88	0.86	0.92	0.06	-0.091	0.333	1.000	0.33
The share of logistics costs in the structure of total costs, coeffic.	0.04	0.08	0.11	0.11	0.04	0.11	0.07	7.545	0.571	1.000	1.00
The integral indicator for the average value of the trait								**2.966**	**0.713**	**0.987**	**0.408**
The element of the marketing complex "Promotion"											
The growth rate of net income from sales of products, coefficient	1.31	1.33	1.55	0.85	1.31	1.55	0.24	4.000	0.686	1.000	0.00
Readiness for delivery (probability of shortage)	0.78	0.79	0.79	0.78	0.78	0.79	0.01	-0.818	1.000	1.000	0.00
Delivery quality, coefficient	0.96	0.97	0.98	0.96	0.96	0.98	0.02	0.818	0.500	1.000	0.00
Order flexibility, coefficient	1.02	1.04	1.06	1.02	1.02	1.06	0.04	1.364	0.500	1.000	0.00
The integral indicator of the average value of the trait								**1.341**	**0.671**	**1.000**	**0.000**
The integral performance index of the marketing complex								**0.492**	**0.598**	**0.911**	**0.257**

Source: Formed by authors based on [6].

Thus, the establishment of the above dependence allows us to form an analytical conclusion about the mechanism of ensuring the effectiveness of marketing activities of PJSC "JT International Ukraine". Consequently, an increase in the value of an enterprise among consumers, an increase in the efficiency of marketing processes (implementation and adaptation) will help adjust the management of an enterprise so that it acquires economic growth. PJSC "JT International Ukraine" has the goal of achieving sales performance, the important factors of which are precisely the volume of sales and the competitiveness of the enterprise.

4 Discussion

The implementation of information technology into the marketing management system is associated with the need to facilitate and accelerate the perception of the array of primary empirical information about the activities of the enterprise, to increase its effectiveness.

The implementation of a marketing information system in an enterprise involves the implementation of the following steps:

1. Determining the dependence of the integral performance index of the marketing complex on relevant factors;
2. Formation of an optimization model for managing the performance of the marketing activity of the enterprise and developing the algorithm for its solution in the conditions of the existing enterprise of PJSC "JT International Ukraine" using the PPP MS Excel;
3. Modeling of the costs for the marketing complex realization in a specific market for PJSC "JT International Ukraine" [7].

Stage 1. To assess the future results of marketing activities, in particular, in individual markets, which characterize the characteristics of sales activities of PJSC "JT International Ukraine", it is possible to use a multifactorial regression equation, in which the indicator reflecting the results of the marketing complex plays the role of a dependent variable, comprehensively characterizes its effectiveness as a target function.

The level of the integral performance index of the marketing complex is influenced by both material and non-material factors. To the most significant material factors we assign in the executed research the share of sales expenses in the total costs of the enterprise.

The specification in the economic-mathematical model of indicate values gives grounds for evaluating the directions of using attracted funds and the probability of obtaining the expected income from the money invested in the development of the enterprise.

1. dependent variable (value of the objective function) - the level of the integral performance index of the marketing complex, coefficient (Y),

2. the most significant factors that affect it differently: export / domestic sales ratio, coefficient (x_1), update range index, coefficient (x_2), the share of distribution costs in the total costs of the enterprise, coefficient (x_3), costs 1 UAH, sold products, UAH (x_4), the turnover ratio of funds in the calculations (x_5), the integral coefficient of marketing support activities (x_6).

Stage 2. The proposed methodology was applied for the enterprise of PJSC "JT International Ukraine" (Table 2).

The calculated value of the criterion is F_{calc} =564,974, and the critical one for P=0.99, k_1=6, k_2=4 is 6.09. Since $F_{calc}>F_{tabl}$, with reliability P=0.95 we can assume that the adopted mathematical model is adequate to the experimental data and can be used for economic analysis and forecast [8].

To identify the relationship between indicators of a multifactorial regression model of indicators of influence on the level of the integral performance index of the marketing complex of PJSC "JT International Ukraine", estimates of the model parameters, coefficients a_0, a_1, a_2, a_3, a_4, a_5, a_6 were found. The required sums and the sample mean "y" and "x" were calculated using the appropriate master functions. Thus, we can conclude that the impact of the share of sales expenses in the total expenses of the enterprise and the ratio of marketing support to the integral performance index of the marketing complex of PJSC "JT International Ukraine" is positive and significant. An increase in the index of the renewal of the range does not lead to an increase in the integral index of the effectiveness of the marketing complex, which characterizes the stability of habits and certain conservatism of consumers [9].

Table 2. Multi-factor regression model of indicators of influence on the level of the integrated performance index of the marketing complex of PJSC "JT International Ukraine"

Index	**Value**						
Evaluation of multiple linear regression parameters	a_6 = 0.971	a_5 = 0.470	a_4 = -2.242	a_3 = 11.522	a_2 = -3.135	a_1 = 3.743	a_0 = 0.453
Standard deviations of parameter estimates	0.203	0.089	0.760	6.641	5.464	1.335	1.939

The coefficient of determination R^2=0.999	Standard deviation of the indicator S = 0.027	**$Y=0.453+3.743X_1-3.135X_2+11.522X_3-2.242X_4+0.470X_5+0.971X_6$**
The calculated value of F - statistics F_{calc}=564,974		
he sum of the squares of the deviation f the calculated values from its averag value =2.860	esidual sum of square =0.001	

Source: Formed by authors based on [6].

So, the established dependencies allow us to set a range of relevant factors of influence on the performance of the enterprise marketing activity and are the basis for modeling the marketing complex costs on a specific market for PJSC "JT International Ukraine".

Stage 3. Modeling the costs for the realization of marketing complex on a specific market for PJSC "JT International Ukraine". We enter the source data (limiting factors) necessary for building the model in a table created using Excel spreadsheets and the "Search for Solution" add-in. Choose the amount needed to conduct research in a particular market, the total cost of material costs for marketing, the value of which does not go beyond the established limits. We set the conditions necessary to solve the problem. We receive the table which is presented in Fig. 5.

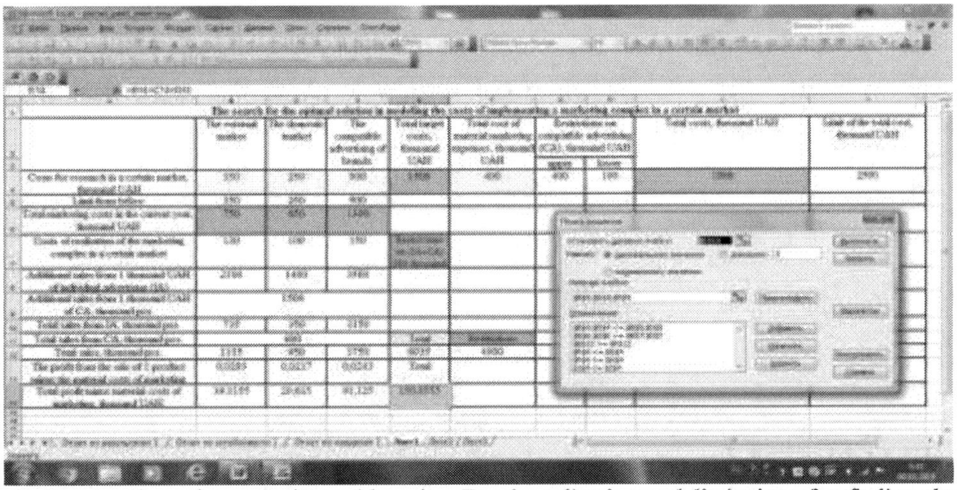

Fig. 5. Construction of source data table using Excel application and limitations for finding the optimal solution in modeling the costs of implementing a marketing complex in a specific market for PJSC "JT International Ukraine"
Source: Formed by authors based on [6].

After fulfillment of modeling the expenditure volume for realization of marketing complex in a particular market by building an optimization model in Excel "Search for Solution", and after analyzing the results, we can draw the following conclusions: an increase of the costs for the realization of marketing complex is effective for the enterprise, as the expenditure amount of 1,900 thousand UAH will bring it additional gross income.

5 Conclusion

Consequently, increasing the awareness of the activities and products of the enterprise among consumers, the formation of a positive reputation will help in adjusting the management of the enterprise so that it strengthens its market position. PJSC "JT International Ukraine" has the goal of achieving increased sales profitability, the most important factors of which are precisely the growth rates of sales volumes and increasing the competitiveness of the enterprise. Evaluation of the cost-effectiveness of the marketing complex realization c is a task, which solution depends on the correctness of the selected criteria and evaluation methods. At the same time, methods and means of influence are chosen; an expenditure budget is drawn up for the implementation of the marketing complex into a certain market. The company is facing an alternative: either to reject a number of goals as unattainable due to the insufficiency of the allocated funds, or, as agreed, to increase the estimated costs. For a reasonable decision about the tactics and strategy of marketing activities of the enterprise, it is important to evaluate the budget and reserve fund of the campaign. This can be done using the model that is proposed for implementation.

It should be emphasized, that the marketing complex for producers of goods of irrational demand must be formed with due regard for account legal restrictions on the markets for implementation in the sphere of pricing regulation, sales and promotion specificities.

References

1. S. Anholt, Competitive Identity: The New Brand Management for Nations. *Cities and Regions*, 732, (2007)
2. *The State Statistics Service of Ukraine* [online], Available at: www.ukrstat.gov.ua/ (2018)
3. *Eurostat Statistics Explined* [online], Available at: http://ec.europa.eu/eurostat/statistics-explained/index.php/ (2018)
4. The Global Comhetitiveness Report, *World Economic Forum* [online], Available at: www.weforum.org (2016)
5. O. Maslak, N. Grishko, O. Hlazunova, K. Vorobiova, Approaches to the management of the costs of innovation activity of mining enterprises: aspects of economic security. *Scientific Bulletin of National Mining University*, **5**, 137-145, (2017)
6. Annual financial statements of enterprises, *Stock market infrastructure development agency of Ukraine (SMIDA)* [online], Available at: www.smida.gov.ua/ (2018)
7. E. Dmitruk, Possibilities of model of the enterprise's reputation influence on its economic safety. *Project management and production development*, **4**(32), 152-159, (2009)
8. E. Boulianne, Empirical analysis of the reabilite and validity of balanced scorecard measures and dimensions. *Edvances in Management Accounting*, **15**, 127-142, (2006)
9. Z. Iaremko, G. Tkach, Concerning the national branding in Ukraine in the context of promoting its interests in the world. *Young*, **40**(12.1), 1035-104,0 (2016)

Innovative Business Model of the Cluster as an Ecosystem

Natalya Masyuk[1,*], *Marina* Bushueva[2] and *Zinaida* Bragina[3]

[1]Vladivostok State University of Economics and Service, Vladivostok, Russia,
[2]Russian University of Economics by GV Plekhanov (Ivanovo Branch), Ivanovo, Russia,
[3]International Academy of Business and New Technologies (MUBINT), Yaroslavl, Russia

Abstract. Business modeling is an innovative form of representation of inter-firm interactions of enterprises and organizations, united by formal and informal economic relations. The business model is a graphic illustration of the structure of these relations, helping to understand their logic and complexity in the context of globalization. Networking and clustering are the responses to the global challenges of the external environment. In these conditions, new variants of an interaction of economic agents arise in conditions of uncertainty and risk. The most common form of intrafirm interaction in the conditions of one territory is clustering, which, as practice shows, have network structures. In our opinion, the purpose of the regional cluster as a network is to implement the key competencies of the territory and improve the quality of the economic situation in the region by improving the business climate of the environment. There are many efforts of scientists and business practitioners to design and make configurations for illustration of the cluster's relations and interactions. The main goal of this article is to show a new way of graphic design of clusters from the position of ecological approach instead of the UML. The article describes the configuration of the cluster business model in its interpretation as an innovative network ecosystem. Theoretical views about the cluster were extended in the direction of the configuration of its business model and the definition of its main structural elements. Particular attention is paid to the partnership interaction of the main groups of economic agents in the cluster, which include: key business structures, key educational organizations, key research organizations, and the state.
Key words: cluster, network, innovative ecosystem, business model, configuration, business climate, key competencies, partnership, regional development

1 Introduction

Until recently, clusters were considered autonomously as a kind of independent economic phenomenon. However, studies of economists of the last decade make some adjustments to this point of view. Many researchers come to the opinion of the network character of regional

* Corresponding author: masyukn@gmail.com

clusters [1]. The concept of "ecosystem" is borrowed by economists from biology – along with the concept of "ecology". In an economic context, both terms are commonly used in combination, within the framework of the ecosystem approach, and are considered as concepts that describe the evolution of the nature of interactions of economic agents, models of their innovative activity and their relationship with the environment of functioning [2]. Authorship of the term "business ecosystem" belongs to James F. Moore [3, 4]. He starts with a simple question: "How is it, that the company that created the whole business community – how IBM created personal computer loses control and profitability in the same business?" [4]. The author is coming to the idea of the coevolution of natural and social systems, considering an example of joint development two types at different levels of food (trophic) chain – predator-prey at work anthropologist G. Bateson "Mind and Nature" [5]. Based on the above, we propose to consider the cluster as an innovative ecosystem.

The idea of attracting and borrowing analogies from wildlife into the economy is not new. A. Marshall (referring to the works of G. Spencer) in his work "Principles of Economic Science" noted that "... The mecca of an economist is rather economic biology than economic dynamics. But biological concepts are more complex than theories of mechanics ... ". And "... Economic science, therefore, deals mainly with human beings who are forced to move along the path of progress ... Fragmentary static hypotheses are used as temporary auxiliary tools for developing dynamic – or rather biological – concepts, but at the center of economic science, even when it comes to its fundamentals, there must be a living action and movement ... " [6]. Similar ideas with the views of J. F. Moore can also be found in the work of M. Rothschild "Bionomics. Economy as Ecosystem", where the author tries on biological concepts for real economic phenomena, already using the concept of "ecosystem" and highlighting its approach to the hotel direction of research, called "bionomics" [7].

Another group of the authors described six research streams in innovation ecosystem: industry platform × innovation ecosystem; innovation ecosystem strategy, strategic management, value creation and business model; innovation management; managing partners; the innovation ecosystem lifecycle; innovation ecosystem and new venture creation [8]. Ecosystem concept also adhere to Gomes et al. [9]. The ecosystem approach considers innovation systems of all levels (national, regional, cluster, etc.) as living social organisms, subject to continuous variability under the influence of new motivations of participants and new circumstances. In this perspective, the innovation ecosystem looks not only as a dynamic set of organizations and institutions, but also as a mobile set of their multidimensional internal connections [10]. According to one of the interpretations, it includes economic agents, their mutual relations, as well as an innovative environment consisting of ideas, technologies, rules of the game, social interactions and culture [2]. The ecosystem approach proposes a number of important policy principles for economic policy that diverge from the settings of the static system model. First, if static systems can be regulated exclusively by the method from above, by the state's influence on organizations and institutions, then the ecosystem has its own, market mechanisms of self-development, i.e. it is controlled by the bottom method, which creates preconditions for the continuity of innovation processes, eliminating excessive government intervention. Secondly, the ecosystem approach focuses not so much on the system participants themselves, but on the nature and dynamics of their interactions (with each other and with potential participants), emphasizing that the collaboration provides for the creation and diffusion of knowledge flows, transforming these flows into innovations and further spread of innovation throughout the economy [10, 11].

Although ecosystems, unlike systems, are not constrained by spatial boundaries, the collaboration of network participants occurs in specific territories and is associated with the factor of localization of innovation processes. The importance of this factor was realized before others in Finland and Sweden, where already in the early 1990s. using the bottom method, regional innovation systems based on network interactions began to be created. In

Finland, the State Innovation Research Council and the TEKES agency were involved in the development and implementation of the ecosystem approach, and in Sweden the Swedish Governmental Agency for Innovation Systems VINNOVA. In the following years, Scandinavian ideas spread widely across all OECD countries, which adapted them to their national and regional specifics. Today, not only in Scandinavia, but also in many other economies of the world (USA, Canada, UK, South Korea, China, Australia, etc.) regional innovation ecosystems are being formed that are designed to achieve world-class innovation effects [10]. They are innovative hubs, or network innovation communities, which allow territories to continuously update goods produced, flexibly responding to technological and market changes.

Adapting to the new claster-network paradigm, regional innovative economic systems are moving to a cluster structure with a predominance of network inter-firm and inter-organizational interactions. It can be stated that most of the regions that gravitate toward clustering enter the era of cluster-network interaction. One of the first authors who has demonstrated interest into the network structures was Manuel Castells [12] in the context of the formation of a post-industrial society. In his opinion, the main feature of the information society is the network logic of information use. At the same time, hierarchical systems gave way to a network method of coordinating links consisting of network information flows, network structures and network interactions [13]. The authors Bragina, Irodov, & Maslova [14] argue reasonably that network interactions ultimately inevitably lead to the formation of a networked economy. Of interest is the work of Romanova & Lavrikova, in which the thesis was expressed that «... network is a promising form of functioning of the regional economy, and clusters are tools for creating this form» [15].

The development of business modeling allows us to take a fresh look at the phenomenon of regional clusters. The tendencies of networking and clustering are traced in the development of regional economic systems not only on the territory of the Asia-Pacific Region, but also practically in all territories of developed modern states [16, 17, 18]. Business modeling is a modern innovative tool for representing regional economic systems developing in the face of global changes in the external environment [19]. The scientific problem is that, on the one hand, science knows a lot of attempts to graphically depict the spatially organized territorial systems to which the cluster belongs, on the other – none of them explains the patterns of their development and functioning. The use of analogues with biological systems of wildlife sharply changes the idea of clusters [20]. Capra believes that any socio-economic systems have many similarities with biological systems, since both are "living", mobile, constantly changing systems with many internal connections. Agreeing with this opinion, based on this perspective, we will begin to build a cluster business model as a network ecosystem.

2 Method and Methodology

If we consider the integration of economic entities into a cluster in an evolutionary context, then it can be noted that the models of innovation activity of actors participating in the cluster change over time, because each of them, through various devices, exchanges information with many other partners and simultaneously belongs to another one or several networks.

Therefore, this study takes the view that the cluster for the above reasons can be represented as a network ecosystem – a kind of «living» organism that, in order to ensure the ultimate effect of its existence and destination, goes through all stages from creation and development to stagnation. The term «ecosystem» for describing the cluster is also used in the context of the description of the evolution of the interactions of all its basic elements [17].

Before proceeding to the description of the proposed business model of the cluster, we note that the founder of business modeling in its current form is rightly considered to be the

Osterwalder and his followers [21]. Chesbrough also made a great contribution to the development of business modeling [22].

Turning to the living biosystems, we can see that they all have a complete cycle, passing through all the stages of growth and development from roots to flowers. According to the authors, the cluster as a socio-economic system should also be presented as a living" organism, which, in order to ensure a guaranteed final effect, must pass through all stages of growth and development. Investigations of key elements of regional systems made it possible to depict the cluster as a biological system or a "business ecosystem" explaining all the structural relationships in the cluster.

Until recently, clusters were considered autonomously as a kind of independent economic phenomenon. However, studies of economists of the last decade make some adjustments to this point of view. Many researchers come to the opinion about the network character of regional clusters. Adapting to the new paradigm, regional innovative economic systems are moving to a cluster structure with a predominance of network inter-firm and inter-organizational interactions. It can be stated that most of the regions that gravitate toward clustering enter the era of cluster-network interaction. One of the manifestations of networking is the emergence of associative holdings, based on the virtual interaction of actors. Such economic categories as a mutual trust of partners, a balance of mutual interests, etc. come to the forefront. As early as the end of the twentieth century, Reiss M. predicted a great future for network organizations. Some researchers emphasize that entrepreneurial networks and clusters among a variety of network structures, noting the similarities in terms of their management [23].

Before proceeding to the consideration of the business model of the cluster, we want to define the concept of "ecosystem", which is borrowed from biology. In the economic context, this term is usually used and is viewed as a concept that describes the evolution of the character of interactions of economic agents, models of their innovation activity and their relationship with the environment of functioning. According to Smorodinskaya N.V., at the meso- level, a typical ecosystem of network links that determines the model of the organization of the modern economy from the point of view of the recruitment of its productive sectors, are the trans-sectoral clusters that have their own specialization and territorial localization [24].

3 Results

Now let's look at the configuration of the cluster business model. Attempts to depict the cluster graphically at different times were undertaken by different scientists. For a long time, none of the graphic images was consistent with the proposed philosophy of clusters. Turning to the "living" biosystems, you can see that they all have a complete cycle, passing through all the stages of growth and development from roots to flowers. According to the authors, the cluster as a socio-economic system should also be presented as a "living" organism, which, in order to ensure a guaranteed final effect, must pass through all stages of growth and development [25].

In our opinion, the treatment of a cluster as an ecosystem deserves special attention and allows one to resort to an unconventional form of representing a business model of a cluster in the form of, for example, a flower. The image of a cluster in the form of a flower with leaves and roots not only reflects its structure, but also reflects the logic of the interrelations of its structural elements, as well as the principles of interaction of all subjects within the cluster. The main of these principles is interdependence. "All members of the ecological community are interconnected through a vast and complex network of relationships, the web of life. They acquire their vital properties and, in fact, their very existence through relationships with other objects. Interdependence – the mutual dependence of all life processes from each other – lies in the nature of any ecological relationships. The behavior of each living member of the ecosystem depends on the behavior of many others. The success of the entire community depends on the success of its individual members, just as the success of each member depends on the success of society as a whole. «Understanding ecological interdependence means understanding relationships. Therefore, it was this configuration of the business model of the cluster that we accepted as the basis for this research [26].

Analyzing the proposed configuration of the business model of the cluster (Fig. 1), it should be noted that the partnership between the four leading groups of economic agents-the institutional sectors of the cluster-key business structures, key educational organizations, key research centers and the state, placed in petals of a flower, by analogy with the «triple helix» of Etzkowitz-Leydesdorff [27]. The core of the flower – the business core – is an innovative cluster-forming products and services. The territorial basis of the cluster as an ecosystem is the link to its territory (following the classical definition of M. Porter) with its industrial orientation, infrastructure, geographical location, campuses, platforms for cooperation, etc. In the root of the root system, on which growth and development depends, are placed the ideas and talents of people, human capital. Here also are all possible sources of funding: venture funds, business angels, direct investment funds, seed investors, etc. The most important assets of the cluster are Clients, which can be attracted at the early stages of innovative solutions, as well as specialized suppliers of raw materials.

Considering a cluster as an ecosystem, one cannot help noticing that all the elements that make up its structure interact not only with each other, but also are involved in horizontal (network) interactions. And, for example, any university can become a participant in an educational alliance based on network relations; business structures are united in entrepreneurial networks; scientific research organizations form their networked scientific communities; banks, insurance companies and various credit institutions are also participants in various networks. However, this does not prevent them from fulfilling their direct role in the cluster. The network nature of the cluster gets its most complete embodiment; it begins to function most fully in this environment. Being in a cluster, economic agents in certain cases can make mutual concessions (local trade-offs) and financial losses to maintain their positions in the long term. But for this, business partners must absolutely trust each other. All of the above allows us to state the competence of the cluster representation in the form of a network innovation ecosystem. At the same time, in our opinion, the innovation of the cluster should be ensured not by the enterprises included in it, but by the cluster-forming types of products and services that make up the business core of the cluster.

4 Discussions

Evolutionary processes occurring in nature and society affect also socio-economic systems, including clusters. Business entities participating in the cluster are connected by horizontal network interactions with a multitude of non-cluster agents, so clusters can be considered network ecosystems, and innovation is given to these systems by innovative cluster-forming products and services. The configuration of the business model of the cluster in the understanding of its essence as an ecosystem can be the appearance of a plant, in our case, a four-leaf flower. Regional clusters tend to the territory on which they are located, and "absorb" all its advantages and disadvantages. The purpose of the regional cluster as a network, in our opinion, is to realize the key competencies and potential of the territory and improve the quality of the economic situation in the region by improving the business environment.

Further studies of the regional cluster business model as an innovation network ecosystem should be conducted in the direction of studying economic mechanisms for replenishing the missing resources and competencies of the territory not only through direct partners in the cluster, but also through network entities located in the second network ring of the geographical limits of the regional cluster. The logic of reasoning about the possibility of using examples of living nature to characterize inanimate objects has led us to the fact that, since a cluster can be regarded as an innovative network ecosystem, the configuration of a plant (tree or flower) can be used for its graphic representation, in our case it is a flower.

5 Conclusion

In conclusion, it can be noted that the parameters of the cluster as an innovation ecosystem vary from ecosystem to ecosystem, but in any of the metrics are based on measuring and measuring performance functioning in several aspects: 1) participants – financial analysis and organizational characteristics, roles and functions in a cluster, business models economic entities, their strategic and tactical behavior, capabilities and potential development companies, input and output material flows, production capacity and output, sales volumes and sales, the history of the development of successful firms; 2) structure – "types" of companies and their dynamics, information value channels and values business platforms, ways interactions and forms of cooperation, institutional aspects of economic practice; 3) competitiveness in comparison with other ecosystems, including by product, services, innovations, technologies, personnel, brands; 4) business activity in terms of interaction, network formation, transactionality, trade turnover between partners; 5) strategic vision in terms of opportunities, risk, and development. If you summarize the parameters of the cluster as innovation ecosystem and metrics to measure effectiveness then can be traced four basic principles of construction and organization this kind of systems: complexity, self-organization, coevolution, and adaptation. Generally, these principles are properties of any complex socio-economic systems and are the result interdisciplinary synthesis of the following scientific directions: systems theory, synergetic and technology. It may be noted that the theory of clusters also appeals to them. It means, that between ecosystems and clusters more common than differences. The formation and functioning of clusters represent the stage of development of the innovation ecosystem, and the cluster itself is a transitional form ecosystem in terms of evolution economic space. As they accumulate critical mass and capital cluster becoming a full-fledged business ecosystem.

In the work theoretical concepts about a configuration of business models of clusters in a direction of their image as some element of animate nature, in our case – a flower-four leaf are expanded. In this case, a meaningful explanation is given to all the key elements included in the business model. The main elements (blocks) of the business model of the cluster are "key business structures", "key educational organizations", "key research centers", "customers", "key suppliers", "territory" and "key sources of financing ". The increment of scientific knowledge on the subject matter in question is the proposed configuration of the cluster business model in the understanding of its essence as an ecosystem depicted as one of the elements of living nature, namely, the four-leaf flower. Further research, in our opinion, should be carried out in the direction of developing a configuration of entrepreneurial networks, which are similar in many respects to clusters, but also have their differences.

References

1. A. Hamdouch, Innovation Clusters and Networks: A Critical Review of the Recent Literature. *19th EAEPE Conference,* (2007)
2. B. Mercan, D. Goktas, Components of Innovation Ecosystems: A Cross Country Study. *International Research Journal of Finance and Economics*, **76**, 102-112, (2011)

3. J. F. Moore, Predators and prey: a new ecology of competition. *Harvard Business Review*, **71**, 76-86, (1993)
4. J. F. Moore, *The death of competition: leadership and strategy in the age of business ecosystems*. John Wiley & Sons, (1996)
5. G. Bateson, *Mind and Nature*. New York: E. P. Dutton, (1979)
6. A. Marshall, *Principles of Economics*. London: Macmillan and Co., (1993)
7. M. Rothschild, *Bionomics. Economy as Ecosystem*. New York: Owl Books, (2017)
8. W. M. Cohen, J. Fjeld, The three legs of a stool: Comment on Richard Nelson. *Research Policy*, **45(**9), 1708-1712, (2016)
9. L. A. de V. Gomes, A. L. F. Facin, M. S. Salerno, R. K. Ikenami, Unpacking the innovation ecosystem construct: Evolution, gaps and trends. *Technological Forecasting and Social Change*, **136**, 30-48, (2016)
10. A. Bramwell et al., *Growing Innovation Ecosystems: University-Industry Knowledge Transfer and Regional Economic Development in Canada*. Final Report to the Social Sciences and Humanities Research Council of Canada, (2012)
11. A. Townsend et al, *Future Knowledge Ecosystems. The Next Twenty Years of Technology-Led Economic Development*. Creative Commons Attribution-Noncommercial-Share IFTF Report Number SR-1236, (2010)
12. M. Castells, *The Rise of the Network Society*. 2nd ed. Wiley-Blackwell, (1996)
13. L. G. Mattsson, Management of Strategic Change in a «Markets-As-Networks» Perspective. *The Management of Strategic Change*, 234-256, (1987)
14. Z. V. Bragina, M. I. Irodov, A. V. Maslova, *From network interactions to a networked economy*, (2016)
15. O. A. Romanova, Y. Lavrikova, Cluster development of the region's economy: theoretical possibilities and practical experience. *The Economy of the Regio*, **4**, 40-51, (2007)
16. Y. Startsev, A. Davankov, Networks and clusters - the formation of modern approaches to regional development. *Bulletin of the Chelyabinsk State University: Economy*, **35**, 110-116, (2011)
17. N. Smorodinskaya, Changing the Paradigm of World Development and the Formation of a Network Economy. *Economic Sociology*, **13** (4), 95-115, (2012)
18. N. Masyuk, A. Karantseva, M. Bushueva, The Far East of Russia: synergy of clusters and territories of advanced socio-economic development. *Scientific review*, **13**, 182-187, (2015)
19. M. Zagoruiko, N. Masyuk, O. Perfilieva, The strategy of choosing the business model of the organization and the methodology for assessing the degree of internationalization of the company. *Scientific Review*, **15**, 419-423, (2015)
20. F. Capra, *The web of life: A new scientific understanding of living systems* (2003)
21. A. Osterwalder, Y. Pigneur, C. Tucci, Clarifying business models: origins, present and future of concept. *Communications of the Association for Information Systems*, **16**, 1-25, (2005)
22. H. Chesbrough, R. Rosenbloom, The role of the business model in capturing value from innovation: evidence from Xerox Corporation's technology spin-off companies. *Industrial and Corporate Change*, **11**(3), 529-555, (2002)
23. M. Raiss, Borders of «boundless» enterprises: the prospects of network organizations. *Problems of theory and practice of management*, **1**, 92-97, (1997)

24. N. Smorodinskaya, Network innovative ecosystems and their role in the dynamization of economic growth. *Innovations*, **7** (189), 27-33, (2014)
25. M. Bushueva, N. Masyuk, Z. V. Bragina, Conceptual Bases for the Construction of the Business - Model of a Regional Cluster as an Innovative Ecosystem. *Azimuth of Scientific Research: Economics and Administration*, **2** (19), 39-42, (2017)
26. N. N. Masyuk, M. A. Bushueva, Z. V. Bragina, A. B. Petrukhin, Presentation of the business model of a textile cluster as an innovative network ecosystem. *Izvestiya Vuzov*, **1**, 10-17, (2017)
27. H. Etzkowitz, L. Leydesdorff, The Endless Transition: A «Triple Helix» of University-Industry-Government Relations. *SSRN,* **36**(8), 203-208, (1998)

Relationship of activities in reverse logistics according to sell point in Czech Republic

František Milichovský[1,*]

[1]Doctor of Economics and management, Assistant Professor, Brno university of Technology, Faculty of Business and Management, Department of Management, Kolejní 2906/4, CZ-61200 Brno, Czech Republic

Abstract. The paper is focused on findings, if final customers in sell points reflect activities of reverse logistics. Main result of research provides relationship between sell point and reverse activity. The research was aimed at random chosen group of 811 respondents in the Czech Republic. Real participants have been 293, what is effectiveness in 36.13%. The primary research provides possible approaches for companies in sell points within reverse logistics activities to final consumers. To process the results of the questionnaire survey were used basic types of descriptive statistics on the selected data set. The data were processed by using the statistical program IBM SPSS Statistics 24, which was subsequently analysed the dependency between the two nominal variables by means of contingency tables and Pearson's chi-squared test. Limitation for this research is because of the chosen sample and targeting only on Czech Republic.
Key words: Reverse logistics, marketing communication, green communication

1 Introduction

On the grounds of the market globalization of all markets larger competitive environments have been created and higher pressure is put onto offered values – especially quality, price and production quantity [1]. Therefore, companies must focus on monitoring of effective realization of all activities (especially in marketing, business and reverse logistics). Based on these reasons the companies must focus on monitoring of effective realization of marketing activities.

Effectiveness of company has become one of the important part of corporate sustainability, on which each company focus on. Therefore, fundamental of each corporate strategy is success, which is supported by marketing and business. Strategy of every company deals with many cultural differences that are essential for understanding of different culture.

These conditions create necessity of implementation of supply chain management with individual activities, which would be win-win for both of seller and customer. In supply

* Corresponding author: milichovsky@fbm.vutbr.cz

chain management there is possible to consider reverse logistics as key part for global sustainability and building competitive advantages in the markets [2], [3], [4], [5], [6].

2 Theoretical background

Logistics itself has become more significant in present day because there is necessary to link three corporate fields together. These fields are marketing activities, production management and corporate financing, where logistics provides required flows – such material and product, information and financial. All logistics flows help to allocate every sources, which are important for effective processes [7].

Logistics has become process by which is distributed product from producer to final customer. From point of view of used material, this process started within raw material. That could be called as typical logistics chain (or supply chain) Main problem in logistics chains is, what to do with waste in production process, used packaging of new products and old products [8].

According to Ferrer & Whybark [9] there is necessary to use garbage as the way of gaining material sources instead of their stocking or burning (impact on nature environment). Waste problems could be solved by reverse logistics (RL) which requires adequate skills, capabilities, market awareness, and staff with experiences [10]. Reverse logistics (RL) has become important part of marketing activities from logistic point of view. It is usually located in the end of whole marketing loop where is after-sale phase defined. RL itself is difficult to describe because of applying individual activities in all phases. RL is process of moving products from place of typical usage for recapturing value or any other disposal [11], [12].

There is difficult to define exact list of RL activities. Abdulah & Yaakub [13] mention main groups of activities which are considered in RL. These groups are: (A) product returns, (B) disposition, (C) green manufacturing, (D) recycle, (E) refurbishing, (F) landfill, (G) salvage, (H) remanufacturing. Similar groups of RL's activities are described by Nikolaidis [14]. However, in comparison with Abdulah & Yaakub [13] he defines only seven groups in different way: (A) landfilling, (B) recycling, (C) cannibalization, (D) remanufacturing, (E) refurbishing, (F) repairing, (G) direct reuse or resale. Appropriate connection of logistics and marketing should be crucial for company.

Final consumers begin to look forward products with correct production such bio-production or green labelling [15]. According to requirements of green supply-chain management, Hervani et al. [16] define as the most significant RL activities (1) reuse, (2) remanufacturing, (3) recycling. Green corporate activities (e.g. production, marketing) are very important from point of view of sustainability [4]. Green marketing as part of RL is focused on bio-products, low operational costs and products of biodegradable waste [17].

By using information of technology development in marketing campaigns, sellers want to increase own income (especially for electronic products). These products have high influence of innovation cycle [18], [19]. Customers still require new products to replace old ones, and they are more interesting in alternative of landfilling. Almost all products need more natural re-sources and energy to make new ones. Electronic devices have become distinguished because of the low life cycle and non-friendly wastes [20].

RL itself has become part of sustainable industrial marketing because of the necessity of destroying of packaging and the rest of used materials. Main problem could be how these materials would be collected from the customers to applying RL activities. Therefore, there are compare both economic and ecological aspects of possible processing [21], [22]. RL activities and their realization need knowledge management and adequate experiences in all phases [23], [24], [25].

Usually, all activities in reverse logistics are applied in corporate campaigns which provides to customers relevant emotion for purchase. This emotion is depending on various conditions such sex, age, kind of product, region and their needs. Specific emotions for purchase is hope of high quality products or fear of unknown [26].

Schmitt [27] and Alkilani, Ling, Abzakh [28] propose the concept of strategic experiential modules that aims to help the marketers to create alternative types of customer experiences. The experiential modules managed by the experiential marketing include: sensory experiences (sense); affective experiences (feel); creative cognitive experiences (think); physical experiences, behaviours and lifestyles (act) and social-identity experiences that result from relating to a reference group or culture (relate). Schmitt [27] and Smilansky [29] add that the purpose of experiential marketing is to create holistically integrated experiences that possess, at the same time, sense, feel, think, act and relate qualities. Lee, Hsiao, Yang [30] mention that experiential marketing can be related to the customer satisfaction.

As way of building competitiveness, companies must focus on long-time period from marketing perspective. Complex manufacturing process need to be innovated in various ways – improving or changing technology of production, changing used material and others. They need to do production strategy in cooperation marketing strategy [31], [32]. Present trend for consumer products is Eco-friendly categorization, what means, that the products are dissoluble in nature by common ways. Perception of marketing campaigns includes many differences, especially from demographic point of view. The reason for this difference could come out mainly from demographic variables [33]: (1) female consumers are more interested in both expressive and functional aspects than males; (2) people with higher income pay more attention to functionalities, ease of use, and quality); (3) people with higher education put less significance to symbolic aspects in purchasing. One of the most important perspectives is gender on which many marketing campaigns focus on [34], [35], [36], [37]. All demographic factors impact corporate marketing activities, mainly in connection with launching new products, for which have become defined basic market segments with possible high income.

3 Applied Methodology

The main aim of this paper is to identify level of influence of reverse logistics' activities in connection with marketing communication in Czech business environment. Main hypothesis is that there is dependence between in sell-places and application of reverse logistics (reverse logistics has becomes very important part for decision making process of purchasing). Main hypothesis is that there is closed relations in their connection. Presented paper is based on primary research by questionnaire survey, conducted in 2016, focused on activities of reverse logistics in connection within purchase place. To process the results of the questionnaire survey there were used both of descriptive statistics and correspond analysis. The conclusions provide characteristics of the limitations of the research and its potential further direction. Sample population was created by 811 consumers, which were chosen in random way. There were returned questionnaires from 293 respondents (effectiveness was 36.13%). To process the results of the questionnaire survey were used basic types of descriptive statistics on the selected data set. The data were processed by using the statistical program IBM SPSS Statistics 24, which was subsequently analyzed the dependency between the two nominal variables by means of contingency tables and Pearson's chi-squared test.

4. Results

Based on the analysis of descriptive statistical characteristics of the sample (Mean, Median, and Values of Variability), conclusions will be presented merely as an explorative result limited by the resultant reliability. The conclusion provides characteristics of the limitations of research and its potential further direction. According to descriptive statistics of gained responses it is obvious, that customers are interesting in returns ($\bar{x} = 7{,}31$), changes ($\bar{x} = 7{,}58$) and repairs ($\bar{x} = 7{,}64$) of purchased products. They care about self-profit, but they do not care where the product is bought (see Table 1). For purpose of the paper there is necessary to design, what kind of sell-points have impact on activities in reverse logistics. Due the aim of the paper, an analysis of dependency of variables was carried out only between performance in marketing field and reverse logistics activities. In order to establish the dependency test, statistical testing using Pearson chi-square test was used.

Table 1. Descriptive statistics of variables

	Mean	Median	Std. deviation	Variance
Return of purchased products	7.31	8.00	2.877	8.276
Change of purchased products	7.58	8.00	2.316	5.365
Return of old products	4.81	5.00	2.885	8.321
Repair of purchased product	7.64	8.00	2.490	6.202
Reverse taking of packaging	2.92	2.00	2.550	6.501
General retail	1.67	2.00	.911	.830
Specialized retail	1.72	2.00	.864	.746
Electronic shop	2.02	2.00	1.092	0.934
Market	0.88	1.00	1.192	0.873
Mail-order store	1.26	1.00	1.031	1.063

Source: Author.

In further analysis, there was applied analysis of dependency between individual variable groups (sell-points, reverse logistics' activities). In order to establish the dependency test, statistical testing using chi-square test was used.

The results of the dependency test are provided in Table 2 which examines the dependency between reverse logistics' activities and sell-points. The results of the dependency examination in individual variable categories are depicted in the following results of Pearson's chi-square test. According to these values there were find out fifteen dependencies from twenty-five relations.

Table 2. Results of Pearson's test between individual variables

		General retail	Specialized retail	Electronic shop	Market	Mail-order store
Return of purchased products	Pearson χ^2	0.148	0.147	0.495	**0.000**	**0.037**
	Contingency coef.	0.519	0.518	0.484	**0.678**	**0.546**
Change of purchased products	Pearson χ^2	**0.040**	0.574	0.097	**0.020**	**0.018**
	Contingency coef.	**0.529**	0.447	0.518	**0.544**	**0.541**
Return of old products	Pearson χ^2	**0.000**	**0.025**	0.238	**0.004**	**0.000**
	Contingency coef.	**0.614**	**0.552**	0.513	**0.581**	**0.624**
Repair of purchased product	Pearson χ^2	**0.006**	0.338	**0.012**	**0.026**	**0.028**
	Contingency coef.	**0.559**	0.475	**0.558**	**0.523**	**0.535**
Reverse taking of packaging	Pearson χ^2	**0.007**	**0.000**	0.243	0.214	0.063
	Contingency coef.	**0.575**	**0.627**	0.515	0.514	0.539

Source: Author.

By maintaining percentage reliability of the Pearson's chi-square test, there was compared significance of the test with theoretical value 0.05, which represents 5% reliability level (if is accepted alternate hypothesis, than there is 5% probability of error). According to this condition, there were observed fifteen dependencies, which brings the conclusion that an alternative hypothesis applies. Other relations have not confirmed the dependency (individual significance is over 0.05). Subsequently, the degree of such dependency was examined. To that end, the intensity of dependency determined by means of contingency coefficient as per formula (1) was used.

$$C_p = \sqrt{\frac{\chi_p^2}{\chi_p^2 + n}} \tag{1}$$

where C_p is contingency coefficient; n represents number of cases; χ^2 is Pearson's chi square test

The intensity of dependency ranges between ⟨0; 1⟩, what means that as observed figure as is closed to 1, then is considered as dependence. In the level of 95 % confidence, the gained values of individual pairs are (according to Table 2):

- Return of purchased products is depended on Market (χ^2 = 0.000, C_p = 0.678) and Mail-order store (χ^2 = 0.037, C_p = 0.546);
- Change of purchased products is depended on General retail (χ^2 = 0.040, C_p = 0.529), Market (χ^2 = 0.000, C_p = 0.678) and Mail-order store (χ^2 = 0.037, C_p = 0.546);
- Return of old products is depended on General retail (χ^2 = 0.000, C_p = 0.614), Specialized retail (χ^2 = 0.025, C_p = 0.552), Market (χ^2 = 0.004, C_p = 0.581) and Mail-order store (χ^2 = 0.000, C_p = 0.624);

- Return of old products is depended on General retail (χ^2 = 0.000, C_p = 0.614), Specialized retail (χ^2 = 0.025, C_p = 0.552), Market (χ^2 = 0.004, C_p = 0.581) and Mail-order store (χ^2 = 0.000, C_p = 0.624);
- Reverse taking of packaging is depended on General retail (χ^2 = 0.007, C_p = 0.575), and Specialized retail (χ^2 = 0.000, C_p = 0.627).

All relations between purchase places and reverse logistics activities are designed in Figure 1. In these connection are showed dependency value and in brackets contingency coefficient. Contingency coefficient describes the intensity between individual variables.

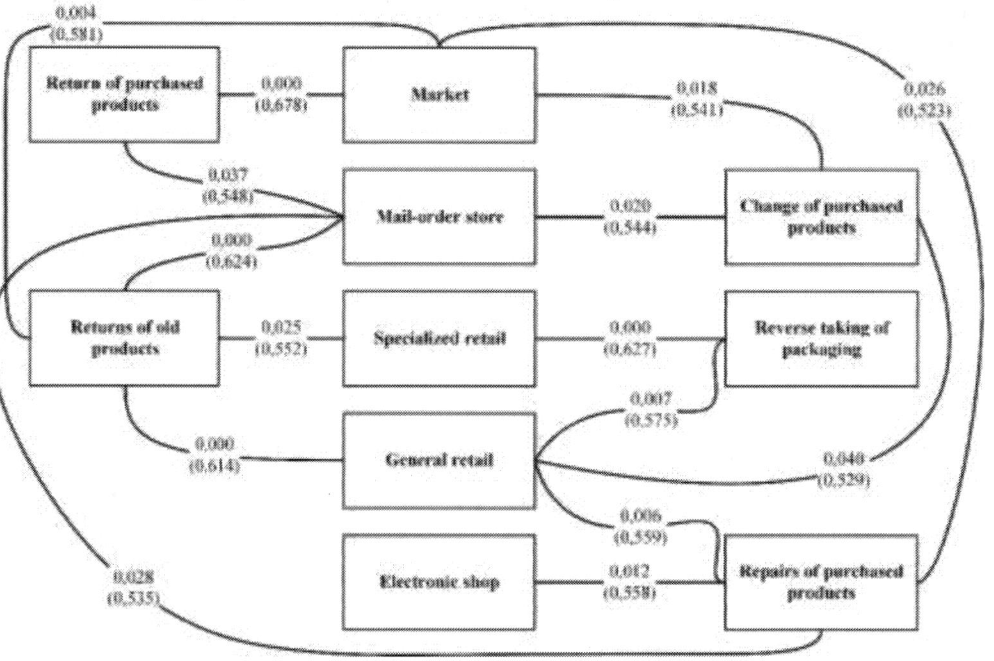

Fig. 1. Dependencies between reverse logistics and sell place

Source: Author.

5 Conclusion

Nowadays, reverse logistics has become one of the crucial aspect of business environment around the world. Several authors consider, that environmental issues in both of business and consumption life are closely connected within responsibility by individual chains in supply chain management (e.g. [38], [39]).

Zaman and Shamsuddin [38] discuss, that activities in supply chain processes are closely related to environmental quality and induce unhealthy environment (for business and living). As unhealthy they consider profit oriented approach in comparison with social oriented approach as healthy and sustainable way for the all countries.

Business activities and reverse logistics are closely interlinked areas that need to be addressed by businesses (manufacturers, retailers, customers) as well as end-users. The topicality of the whole topic reflects all requirements for the efficiency of enterprises in the use of production resources. The efficient use of production resources will minimize the generation of waste from the used production material and at the same time from the used packaging. The issue of used waste is a very serious problem, which results in a significant disturbance of the environment in the form of soil, water and air pollution (in case of landfilling) or in the case of combustion is the release of hazardous fumes [40]. Linking these business areas and reverse logistics activities can achieve the desired efficiencies at all levels (both customer and retailer).

References

1. T. Ambler, Market metrics: What should we tell the shareholders? *The Journal of Risk Finance*, **10**(1), 47-50, (2002)
2. V. Guide, L. N. Van Wassenhove, The reverse supply chain. *Harvard Business Review*, **80**(2), 25-26, (2002)
3. M. A. Lapré, A. S. Mukherjee, L. N. Van Wassenhove, Behind the learning curve: linking learning activities to waste reduction. *Management Science*, **46**(5), 597-611, (2000)
4. J. Hornungová, M. Klímková, The Relationship of Environmental and Economic Performance. *Scientia & Societas*, **9**(4), 207-217, (2013)
5. M. Pagell, Z. Wu, Building a more complete theory of sustainable supply chain management using case studies of 10 exemplars. *Journal of Supply Chain Management*, **45**(2), 37-56, (2009)
6. R. Dahlstrom, *Green marketing management*. Mason: South-Western Cengage Learning, (2011)
7. P. Pernica, *Logistika: Supply Chain Management*. Praha: Radix, (2005)
8. K. Govindan, H. Soleiman, D. Kannan, Reverse Logistics and Closed-Loop Supply Chain: A Comprehensive Review to Explore the Future. *European Journal of Operational Research*, **240**(3), 603-626, (2015)
9. G. Ferrer, C. D. Whybark, From Garbage to Goods: Successful Remanufacturing Systems and Skills. *Business Horizons*, **43**(6), 55-64, (2000)
10. R. Giuntini, K. Gaudette, Remanufacturing: The Next Great Opportunity for Boosting US Productivity. *Business Horizons*, **46**(6), 41-48, (2003)
11. S. E. Genchev, Reverse Logistics Program Design: A Company Study. *Business Horizons*, **52**(3), 139-148, (2009)
12. P. J. Daugherty, R. G. Richey, S. E. Genchev, H. Chen, Reverse logistics: Superior Performance Through Focused Commitments to Information Technology. *Transportation Research Pert E*, **41**, 77-92, (2005)
13. N. H. N. Abdullah, S. Yaakub, Reverse Logistics: Pressure for Adoption and the Impact on Firm's Performance. *International Journal of Business and Society*, 15(1), 151-170, (2014)
14. Y. Nikolaidis, *Quality management in Reverse Logistics: A Broad Look on Quality Issues and Their Interaction with Closed-Loop Supply Chain*. London: Springer, (2013)
15. G. Tomek, V. Vávrová, *Marketing: Od myšlenky k realizaci*. 3rd ed. Praha: Professional Publishing, (2011)
16. A. A. Hervani, M. M. Helms, J. Sarkis, Performance Measurement for Green Supply Chain Management. *Benchmarking: An International Journal*, **12**(4), 330-353, (2005)
17. L. C. M. Holanda, A. C. Francisco, Consumer's Perception of the Products Made with Recycled Components: Case Study from the Course of Administration of the Faculty of Campina Grande – FAC. *Espacios*, **34**(10), 1-9, (2013)

18. B. Lebreton, A. Tuma, A Quantitative Approach to Assessing the Profitability of Car and Truck Tire Remanufacturing. *International Journal of Production Economics*, **104**(2), 639-652, (2006)
19. Ch.-F. Hsueh, An Inventory Control Model with Consideration of Remanufacturing and Product Life Cycle. *International Journal of Production Economics*, **133**(2), 645-652, (2011)
20. S. S. Gan, I. N. Pujawan, Suparno, B. Widodo, Pricing Decision Model for New and Remanufactured Short-life Cycle Products with Time-dependent Demand. *Operation Research Perspectives*, **2**, 1-12, (2015)
21. C. K. M. Lee, J. S. L. Lam, Managing Reverse Logistics to Enhance Sustainability of Industrial Marketing. *Industrial Marketing Management*, **41**(4), 589-598, (2012)
22. A. El Korchi, D. Miller, Designing a Sustainable Reverse Logistics Channel: The 18 Generic Structures Framework. *Journal of Cleaner Production*, **19**(6-7), 588-597, (2011)
23. S. Wadhwa, J. Madaan, Intelligent agent technology for reverse logistics. *Information Control Problems in Manufacturing*, **12**(1), 215-220, (2006)
24. S. Wadhwa, J. Madaan, *Role of Quality Management Self-Assessment Model to Promote Reverse Logistics Operations*. Bangkok: International Conventions on Quality Control Circles, (2004)
25. A. Mihi Ramirez, L. Girdauskiene, Creation of knowledge and reverse logistics: Empirical analysis from perspective of the reverse based view theory. *Engineering Economics*, **24**(5), 478-487, (2013)
26. Y.-K. Lee, C.-T. Chang, P.-C. Chen, What sells better in green communications: Fear of Hope. It Depends on Whether the Issue Is Global or Local. *Journal of Advertising Research*, **57**(4), 379-396, (2017)
27. B. Schmitt. *Experiential Marketing: How to Get Customers to Sense, Feel, Think, Act, Relate to Your Company and Brands*. New York: The Free Press, (1999)
28. K. Alkilani, K. C. Ling, A. A. Abzakh. The Impact of Experiential Marketing and Customer Satisfaction on Customer Commitment in the World of Social Networks. *Asian Social Science*, **9**(1), 262-270, (2013)
29. S. Smilansky. *Experiential marketing: A practical guide to interactive brand experiences*. London: Kogan Page, (2009)
30. M. S. Lee, H. D. Hsiao, M. F. Yang, The study of the relationships among experiential marketing, service quality, customer satisfaction and customer loyalty. *The International Journal of Organizational Innovation*, **3**(2), 353-379, (2011)
31. S. W. O'leary-Kelly, B. E. Flores, The integration of manufacturing and marketing/ sales decisions: Impact on organizational performance. *Journal of Operational Management*, **20**, 221-240, (2002)
32. D. Pollard, I. Šimberová, Marketing in high-technology firms. *7th International scientific conference business and management 2012*, 465-471, (2012)
33. A. Toppinen, R. Toivonen, A. Valkeapää, A.-K. Rämö, Consumer Perceptions of Environmental and Social Sustainability of Wood Products in the Finnish Market. *Scandinavian Journal of Forest Research*, **28**(8), 775-783, (2013)
34. E. Ambrozová, V. Pokorný, Ke změnám v myšlení lidí v lidských systémech z hlediska management. *Scientia & Societas*, **8**(3), 139-152, (2012)

35. S. Rahmani, M. G. Lavasani, Gender Differences in Five Factor Model of Personality and Sensation Seeking. *Procedia - Social and Behavioral Sciences*, **46**, 2906-2911, (2012)
36. P. J. Costa, A. Terracciano, R. R. McCrae, Gender Differences Personality Traits Across Cultures: Robust and Surprising Findings. *Journal of Personality and Social Psychology*, **81**(2), 322-331, (2001)
37. W.-J. He, W.-C. Wong, Gender Differences in Creative Thinking Revisited: Findings from Analysis of Variability. *Personality and Individual Differences*, **51**(7), 807-811, (2011)
38. K. Zaman, S, Shamsuddin, Green logistics and national scale economic indicators: Evidence from a panel of selected European countries. *Journal of Cleaner Production*, **143**, 51-63, (2017)
39. C. Altuntas, O, Tunas, Greening Logistics Centers: The Evolution of Industrial Buying Criteria Towards Green. *The Asian Journal of Shipping and Logistics*, **29**(1), 59-80, (2013)
40. Eurostat. *Statistika odpadů*. [online] Available at: http://ec.europa.eu/eurostat/statistics-explained/index.php/Waste_statistics/cs (2017)

Energy Planning and Energy Efficiency in Smart City Areas

Yury R. Nurulin[1], *Inga V.* Skvortsova[1], *Olga A.* Kalchenko[1,*]

[1]Peter the Great Saint-Petersburg Polytechnic University, 29 Polytechnicheskaya, Saint-Petersburg, 195251, Russia

Abstract. The main value added of the approach, which is considered in this paper, is the joint development of an innovative concept for energy improvement city's areas, as well as methods and tools for its implementation. A new coordinated approach to energy planning and implementation at the district level within the framework of the concept of smart cities contributes to the efforts of consumers to improve energy efficiency. The research focuses on energy efficiency for existing built-up urban structures. They represent a large part of the built environment of European cities and face significant, often urgent energy challenges.
Key words: smart city, energy planning, energy efficiency, strategy

1 Introduction

For the last decade there has been a significant increase of interest in research and innovation the field of integration of information and communication technologies (ICT) for the management of urban property. Reflecting this interest is the Smart City concept that are designed to improve the level of comfort and quality of life in cities, reduce costs and resource consumption through the extensive use of ICT. This concept is a logical development of the sustainable development goals (SDGs), outlined in The United Nations report "Our Common Future" [1].

At the same time models and methods of energy planning applied by local and regional government bodies do not provide a comprehensive approach that could reduce the fragmentation of particular sectors of the city's energy economy and maximize the energy saving potential in urban areas. The study has been done in frame of CBC ENPI project EFEM shows that efforts to develop strategies that integrate the behavior of all key participants in the energy supply and consumption system are not sufficient [2]. This is due to the lack of cooperation between local and regional government authorities, energy companies and end-users of energy resources. In addition, there are no tools for understanding the motives for behavior and analyzing barriers to the adoption of effective management decisions by end users that can provide direct information exchange of consumers with energy suppliers and actively participate in strategic energy planning [3].

* Corresponding author: o.kaltchenko@mail.ru

2 Literature review

City areas are becoming "smarter," as governments, businesses, and communities increasingly rely on technology to overcome the challenges from rapid urbanization. Energy efficiency and the use of renewable energy are intimately linked with questions of energy security and energy supply and thus to sustainable development.

Sussman argues that local governments can have a positive impact on global warming by utilizing measures to foster energy efficiency and renewable energy use in government operations and by the general population [4].

Sustainable development cannot and must not be enacted top-down. A healthy balance between "bottom-up" and "top-down" is required to reach. "Think globally, act locally" may be a hackneyed phrase but it still resonates as a truth. Municipalities can have a substantial impact just by changing their own operations.

The transition towards renewable and sustainable energy is being accompanied by a transformation of communities and areas. Van Der Schoor and Scholtens investigated into the local community level towards realizing these ambitions from a social perspective [5].

Nilsson and Mårtensson, Palm studied energy systems in Swedish municipalities. Their analysis examined how municipalities promote oil reduction, efficient energy use, and the use of renewable energy [6, 7]. Brandoni and Polonara showed on the Italian example, that the site-specific nature of renewable sources, the need to involve citizens in the energy-planning process and the perspective of local governments, who are the first governmental stakeholders involved in the implementation of energy-saving initiatives, had emphasised the strategic role played by municipalities in the energy-planning process [8].

3 The problem definition

The above challenges are reflected in the Strategy of economic and social development of St. Petersburg until 2030 that is aimed to provide stable improvement of life quality of citizens and to increase global competitiveness of St. Petersburg based on strategic priorities of Russian Federation, sustainable economic growth and use of results of innovation and technologic industries [9]. Achievement of the general objective is possible on the basis of active managerial influence on the leading sectors of economic complex and social sphere. This implies the need for a coordinated and organized work of local and regional authorities, public property owners, municipalities, as well as appropriate participation of business, science institutions, public organizations and citizens of St. Petersburg. The system of state and municipal management at the regional level should be aimed at creating incentives for the implementation of strategic planning and budgeting tools, focused on the result – improving the level and quality of life. Improving urban development, providing sustainable economic growth, ensuring effective governance and civil society development, human capital development – are strategic directions of St. Petersburg.

Any strategy is effective only in conjunction with the methods and plans for its implementation. In this regard the research is needed to improve the energy efficiency of cities by developing, piloting and disseminating innovative approaches and tools for integrated planning and energy use at the urban level. In these studies, the focus of research should be shifted from an analysis of individual technical or organizational decisions in relation to individual buildings to form a holistic system perspective on energy efficiency and energy consumption. New approaches and tools should be developed for the interaction of local and regional public authorities with end-users of energy resources in the planning and implementation of energy efficiency decisions, which will allow using the synergistic effect to reduce the fragmentation of the energy consumption system at the level of urban areas, synergies at the district level and reducing fragmentation of sectors [10].

4 Materials and Methods

The system approach to the energy planning and energy efficiency in smart city areas opens energy planning to new actors to develop and implement more effective energy efficiency solutions, exploit synergies on district level and reduce sectoral fragmentation [11, 12].

Main outputs of the project aimed on realisation of above strategy will be as follows:
- the concept of integrated strategic approach to cooperative energy efficiency planning on district level;
- the guidance on cooperative energy planning on district level, that describes barriers and factors of success;
- the process model for the cooperation of local and regional authorities, energy utilities and public property owners in energy efficiency implementation on district level;
- new ICT-based tools for involving citizens and property users in energy planning.

5 Proposed solution for the problem defined

The main areas for research and innovation for realization of the strategy for energy planning and energy efficiency in smart city areas are the following:
- shift the emphasis in the mechanisms of solving urban problems to self-organization, communication, strengthening of self-government, search for innovative solutions, achieving public consensus;
- create a "digital city" with maximum data openness, access to information, formation of intelligent self-management systems, reduce the costs of access to information [13, 14].

The logical scheme of the project which is aimed on realization of the strategy for energy planning and energy efficiency in smart city areas is presented at the Table 1.

Table 1. Logical scheme of the project

Tasks (T), subtasks (ST)	The goal and outcomes	The method
T1. Social engagement and cohesion ST1.1. Identification of the key stakeholders ST1.2. Introduction and presentation of the key goals to decision makers with the support of the key supporters and local communities ST1.3. Preparation of the key challenges and the possible strategies of communication to the local communities ST1.4. Promotion of the digitalization approach to district energy supply system among key stakeholders	Reinforcement of local leaders. To recognize local/regional leaders, potential key local/regional supporters which will facilitate the communication within the community. Recognition of the potential for the project idea. An agenda of communication with local communities, a local working group based on the members of community and partners. Municipalities and other stakeholders empowered to become key actors in the promotion of digitalization of district	Meetings, local workshops, study trips, transnational meetings, panels and workshops between partners.

	energy system.	
T2. Selection of pilot districts for energy improvement ST2.1. Development of criteria and selection of objects for piloting ST2.2. Identification and analysis of barriers and factors for success in energy planning ST2.3. Involvement of public property owners and energy utilities	Energy improvement districts to be defined. Concepts and approaches for pilot district including energy utilities and public property owners.	Workshops, study visits, questioning and stakeholder analysis.
T3. Digitalization of district energy system for increasing energy efficiency ST3.1. Preparation of access points for digitalization of district heating ST3.2. Data collection through pilot testing under specific local conditions ST3.3. Analysis of collected data and allocation of district energy resources based on real demand ST4. District level cooperation model for local and regional authorities ST5. ICT tools development	To establish accompanying research for testing pilots in digitalization for increasing energy efficiency - Scientific research on legal framework for digital solutions and on implementation of digital solutions for energy resources use including interfaces to renewable energies. Data base with at least 3 years data from at least 5 access points ready for analysis. New ICT-based tools for involving citizens and property users in energy planning. The set of models (district energy smart greed, big data in district heating systems, motivation and behavior of consumers and stakeholders in district energy system).	Technical monitoring, best practice analysis, Big Data analysis, imitation modeling.

Source: Authors.

6 Discussion

Cities play a key role in reaching energy efficiency targets, but still face major challenges to achieve sufficient energy efficiency. Almost 50% of the entire energy is used for heating or cooling, and buildings account for around 40% of energy consumption [12]. The existing building stock and infrastructures in cities have a great impact on final energy consumption [15].

Approximately 75% of the Europe population live in cities and urban areas and they concentrate the largest share of current energy consumption. To face this situation, we suggest to pass from a supply-oriented market to one determined by demand, transnational awareness and social impacts. In order to do so, energy production, storage and consumption need to be customized (vertical principle) and harmonized (horizontal principle); in this sense, digitalization represents an excellent opportunity for data analysis, efficient energy allocation and reduction of CO2 emissions.

Individual behavior of citizens as energy consumers also plays a decisive role in improving the energy efficiency of cities. Modern approaches focus mainly on finding technical solutions that can't use the energy potential arising from behavioral changes. At the same time, according to date of the European Environment Agency, up to 20% of energy can be achieved through measures aimed at changing individual behavior. Therefore, the behavioral perspective should be an integral part of the approaches to improving energy efficiency. They must take into account the capabilities and needs of end users and involve energy users in the processes of energy planning, decision making and implementation.

7 Conclusion

The above measures for energy planning and energy efficiency in smart city areas focuses on the energy consumption of district energy systems including all Quadruple Helix stakeholders. With sensors installed in the regional grid and the connected buildings, collected digital data are used for control of energy supply system, and with predefined algorithms optimizes the whole energy system based on real demand. By using the virtual thermal inertia of the buildings, heat and energy can be saved without any change in the indoor temperature.

References

1. *Assembly UN General "The Future we want (Resolution adopted by the General Assembly on 27 July 2012)"* 53 [online]. Available at: http://www.un-documents.net/our-common-future.pdf (The United Nations, 2012)
2. *Project 2011-091-SE693 Efficient Energy Management – EFEM* [online]. Available at: http://acea.spb.ru/efem/www/ (2014)
3. *Boverket's building regulations – mandatory provisions and general recommendations, BBR. BFS 2011: 6, with amendments up to BFS 2016:* 6 [online]. Available at: https://www.boverket.se/globalassets/publikationer/dokument/2016/boverkets-building-regulations--mandatory-provisions-and-general-recommendations-bbr-23.pdf (2016)
4. E. Sussman, Reshaping municipal and county laws to foster green building, energy efficiency, and renewable energy. *NYU Environmental Law Journal*, **16**, 1, (2008)
5. T. Van Der Schoor, B. Scholtens, Power to the people: local community initiatives and the transition to sustainable energy. *Renewable and Sustainable Energy Reviews*, **43**, 666-675, (2015)
6. J. S. Nilsson, A. Mårtensson, Municipal energy-planning and development of local energy-systems. *Applied Energy*, **76**(1-3), 179-187, (2003)
7. J. Palm, Development of sustainable energy systems in Swedish municipalities: A matter of path dependency and power relations. *Local Environment*, **11**(4), 445-457, (2006)
8. C. Brandoni, F. Polonara, The role of municipal energy planning in the regional energy-planning process. *Energy*, **48**(1), 323-338, (2012)
9. *Strategy of economic and social development of St. Petersburg until 2030* [online]. Available at:

http://cedipt.spb.ru/media/acts/2015/12/09/Strategy_of_economic_and_social_development_of_St._Petersburg_until_2030.pdf (2018)

10. European Union (EU). *Energy Efficiency* [online]. Available at: https://ec.europa.eu/energy/en/topics/energy-efficiency (2017)
11. D. Popescu, S. Bienert, C. Schützenhofer, R. Boazu, Impact of energy efficiency measures on the economic value of buildings. *Applied Energy*, **89**, 454-463, (2012)
12. G. Escriva-Escriva, Basic actions to improve energy efficiency in commercial buildings in operation. *Energy and Buildings*, **43**, 3106-3111, (2011)
13. V. V. Glukhov, I. V. Ilin, A. B. Anisiforov, Problems of data protection in industrial corporations enterprise architecture. *Proceedings of the 8th International Conference on Security of Information and Networks*, 34-37, (2015)
14. V. V. Glukhov, I. V. Ilin, O. Y. Iliashenko, Improving the efficiency of architectural solutions based on cloud services integration. *Lecture Notes in Computer Science*, **9870**, 512-524, (2016)
15. J. Parasonis, A. Keizikas, Increasing energy efficiency of the translucent enclosure walls of a building. *Procedia Engineering*, **57**, 869-875, (2013)

The formation and development of social responsibility of business: Ukrainian experience in a context of decentralization

Andrew Pochtovyuk[1,*], *Victoria* Semenikhina[2], *Oksana* Onyshchenko[3] and *Bogdan* Ruban[4]

[1]PhD of Economics, Professor, Management Department, 39600 Kremenchuk Mykhailo Ostrogradsky national University, Kremenchuk, Pershotravneva str. 20, Ukraine
[2]senior lecturer of the Management Department, 39600 Kremenchuk Mykhailo Ostrogradsky national University, Kremenchuk, Pershotravneva str. 20, Ukraine
[3]PhD (Econ.), Associate Professor of the Department of Accounting and Finance, 39600 Kremenchuk Mykhailo Ostrogradsky national University, Kremenchuk, Pershotravneva str. 20, Ukraine
[4]Ph.D. student of the Management Department, 39600 Kremenchuk Mykhailo Ostrogradsky national University, Kremenchuk, Pershotravneva str. 20, Ukraine

Abstract. The current trends of using the practice of social responsibility (SR) and sustainable business in the world and in Ukraine in particular have been considered by the authors. The researches of business social responsibility policy in the field of sustainable development considered as a single international standard have been made. As well as the prospects of social responsibility in business development in the context of power decentralization in Ukraine have been considered.
Key words: corporate social responsibility, global compact and sustainable development, decentralization, stakeholders

1 Introduction

In conditions of the world economy development the numbers of companies are introducing a vector of socially responsible business into the corporate strategy. That development vector is one of the fundamental elements in the process of strengthening the company's competitive positions in the global market.

In recent years the approach has been applied in the Ukrainian companies' activities, whose management is aimed toward the introduction of socially responsible components in various areas of activity, in order to strengthen their competitive advantages, both in domestic and international markets. The works devoted to social responsibility development of business are written by G. Bowen [1], K. Davis [2], A. Carroll [3], P. Drucker [4] and Ukrainian scientist: Kolot [5], Krepak [6] and other.

* Corresponding author: andrey.pochtovyuk@gmail.com

2 Data and Methods

Dialectical approach was used for studying the definition of "social responsibility", which gives the opportunity to consider this definition as a dynamic category of society development. The historical approach was used in the analysis of the corporate social responsibility development. The system approach was used to assess the possible results in the formation of a transparent and democratic system of social responsibility with the definition of clear rules and business conditions within the Ukrainian realities and adaptation to the European experience.

There also have been used the methods of scientific generalization – aimed to search the unresolved issues in the theory of social responsibility; abstraction – aimed to establish the causes of business structures social functions failure; analysis and synthesis-to form the foundations of social responsibility of all economy subjects. The research has been conducted on the basis of data systematization on the global and Ukrainian practices of corporate social responsibility (CSR) and sustainable business development. The object of the study was CSR and sustainable business development in the world market and the Ukrainian market in particular. The subject of the study was the approaches evolution to the implementation of CSR and sustainable development in various areas of the world's companies in the uncertain environment.

3 Results and Discussion

The modern development of the national economy is connected closely with dilemma solving: improving the economic efficiency of business and evening out social issues. It costs to mark, that world economic relations represent a combination of different forms of business interaction that occurs between countries of the world, and primarily include trade and economic cooperation, the flow of investment flows, scientific and technical ties, the exchange of technologies and intellectual property [7].

The World Business Council for Sustainable Development (WBCSD) defines social responsibility as "the business obligation to contribute to sustainable economic development, working with employees, their families, the local community and society at large to improve quality of life".

The basic principles of social responsibility are directly related to the need to ensure sustainable development by the balance of business, state and society interests. Today, the basic principles of using social responsibility for business are, first of all, a positive reputation, own beliefs, expanding markets, access to financial resources and capitalization increasing. There are different approaches to the definition of "social responsibility" in the Table 1.

Table 1. Modern approaches to the definition of "social responsibility"

The concept of "social responsibility"	Source/organization
Organization responsibility for the impact of its decisions on society and the environment taking into account the transparent and ethical attitude that: (a) promote sustainable development, including the health and well-being of society; b) takes into account the expectations of stakeholders and complies with the laws; C) complies with international standards of conduct and is integrated with the organization activities	International standard of social responsibility ISO 26000

Promoting responsible business practices that benefit business and society, helps to achieve economic, social and environmentally sustainable development by maximizing positive influence of business on society while minimizing its negative impact	International Business Leaders Forum – IBLF
The concept which allows to integrate social and environmental aspects into the daily activities of commercial enterprises, as well as taking them into account in the process of their interaction with stakeholders and groups of influence	European Commission Green Paper
Companies ' long-term commitment to ethical policies and economic development, and improving the quality of workers' life and their families, communities and society at large	The World Business Council for Sustainable Development (WBCSD)
Achieving commercial success by means based on ethical standards and respect for people, society, environment	United States corporations "Business for Social Responsibility"

Source: According to the data [1, 2, 3, 5].

Well-known definition of social responsibility of business was proposed by Peter Drucker [4] – "to make your own interest what is essentially a public good". Thus, business really becomes responsible only when it equates its interests and goals with the society's interests and goals There are four main types of social responsibility of business: basic responsibility, economic (job creation and tax payments), legal (business compliance with local and international laws), ethical (compliance by employees and companies with universal norms of behavior in interaction with each other, with competitors, with the state and society. The highest type of social responsibility is a voluntary one, which involves investing in projects that are of particular society benefit but have no direct economic benefits (environmental protection, investments in health and literacy, infrastructure, poverty alleviation) [6].

In the fundamental article "The Case for and against Business Assumption of Social Responsibilities" K. Davis [2] summed up the main arguments for and against corporate social responsibility existing in the scientific literature at that time. Studying these aspects the author contrasted the concept of "social responsibility" and "economic efficiency", but this contrast was not absolute. The work of K. Davis [2] prompted further research in this area. Thus, in subsequent periods appears the model of A. Carroll [3], which includes four types of social responsibilities – economic, legal, ethical and so-called discretionary. It has been later transformed into a pyramid-shaped methodological model that applies management terminology and defines that "a socially responsible company should strive to make a profit, obey the law, be ethical, and be a good corporate citizen" [3].

In countries with developed economies, state programs which provide socially responsible companies with significant preferences in the field of taxation, public procurement and lending are successfully carried out. The strategy of promoting the development of social responsibility of business in Ukraine should be aimed at the implementation of the principles and tasks the state policy in the field of social responsibility of business and determine the activities of the Executive authorities and local authorities related to the creation of the necessary conditions for the development of social responsibility of business in Ukraine [8].

Social responsibility of Ukrainian business is rapidly growing despite the lack of legal framework, lack of awareness, lagging behind global trends and the lack of state support. It can be stated with confidence that socially responsible business will become the main component of the development strategy of Ukrainian companies the nearest future only with the state support. These goals may be achieved by using the mechanisms that stimulate social responsibility of business and include the development of a regulations set aimed at increasing the social responsibility of business structures; attracting substantial business potential contributing to stable employment, poverty eradication, rational use of energy and environmental protection, strengthening national competitiveness, effective management and fight against corruption; provision of the state budget in the coming years with special sources of funding for programs aimed at the development of social responsibility of business in Ukraine.

The corporate social responsibility – the enterprises' responsibility for their impact on society is not regulated by law in the world. Companies voluntarily improve the quality of employees and society life; reduce the burden on the environment. Nowadays the CSR in the EU is moving to a new development level. It will no longer be just a business desire to take into account the interests of all parties of the process of its activities so as to be responsible for potential risks to employees and the environment, to take measures aimed at improving the environment and protecting human rights and freedoms. Since 2017 CSR has received the legal status of the business obligation to the society. The French Parliament passed the corporate social responsibility business act [9].

The law provides that multinational firms that are carrying out all or part of their activities in French territory shall establish mechanisms to prevent human rights violations and environmental damage throughout their supply chain. This law applies to large French companies. In particular, companies covered by the law include companies operating wholly or partly in France and counting (within their own company and their direct and indirect subsidiaries) at least 5,000 employees when their head office is in France or 10,000 employees when their head office is abroad. For example, one hundred leading French enterprises are obliged to identify potential risks to employees and the environment and to take measures aimed at preserving the environment and protecting human rights and freedoms. Corporations are required to establish annual special plans, where will be prescribed responsibility. In accordance with these documents they also should monitor the actions of the parent and subsidiary companies, as well as related financial relations of contractors and suppliers. According to the new law, a company will be fined up to €10 million if it has not developed a plan. If the organization does not have a plan but it causes damage to the environment or violates the rights of employees, the amount of the fine will increase to €30 million. Lucia Ortiz – coordinator of the economic justice program in the international office of the organization "Friends of the Earth" believes that "members of the European Union should follow the example of France and begin to regulate the sphere of business within their country and in neighboring states" [10].

Many countries of the European Union are planning to follow this example. The European Commission is already making steps in this direction. On 16 June 2011, the United Nations Human Rights Council approved the guidelines for states and enterprises to "Protect, respect and remedy" (Protect, Respect and Remedy), which set standards for business control, as a prerequisite for this process development. In 2014, at the 26th session in Geneva, the UN Council opened an intergovernmental working group on Ecuador and South Africa initiative. It is developing an international legal instrument that will oblige transnational corporations and other business enterprises to respect human rights, in view of the increasing violations in the business sector [10]. Let's study the modern directions of corporate social responsibility development in Table 2.

Table 2. The main directions of CSR development and their content

The direction of CSR	Content
Business ethics	Rules and values that companies and their employees are guided. Corporate ethics codes and business conduct codes implemented in the leading companies of the world regarding sexual harassment or informal relationships, as well as conflict of interests, security issues, environment, career advancement, etc.
Corporate leadership	Profit as a means of production of goods and services to society in the format of sustainable development on the principle of "three-way outcome": economy (production), environment and social development. Quality, safe and affordable goods and services produced under conditions of maximum care for society and the environment are the main responsibility of business.
Promotion of social and environmental responsibility through suppliers and contractors companies	Requirements to their suppliers and contractors are not limited to the quality of the products supplied, safety or working conditions of employees, excluding any discrimination. They include additional requirements for human rights, environmental production, measures to reduce the negative impact on climate change, disclosure of offshore beneficiaries, etc.
Non-financial reporting	More and more countries are introducing mandatory non-financial reporting, mainly for large public companies. Thus, the European Union in 2014 introduced mandatory reporting for companies that employ more than 500 people. The most recognized today is the international standard of the Global reporting initiative for sustainable development GRI G4.
Environmental efficiency	Acceptable for business is the production with the so-called decoupling, when with the growth of production the enterprise impact on the environment is reduced. More and more companies-leaders of CSR are moving from the policy of eco-efficiency to the policy of eco-innovation, designed to reduce the volume of consumed resources, primarily energy, to make revolutionary changes in the technological process.
Communication with stakeholders	CSR leaders use social media and crowdsourcing channels to discuss CSR and sustainable development activities with stakeholders. Public dialogues and hearings using social networks in the development of non-financial reports have been firmly established as platforms for constructive exchange of views.

Source: According to the data [5, 6].

It should also be noted the impact of CSR on the formation and implementation of new large-scale projects and programs in the field of sustainable development, alternative energy, social development of regions, etc. Thus, in France at the world forum Lille for a responsible economy, an official presentation of the Nord-Pas-de-Calais region was held, which showed that in France, following Germany, Italy, the United States and the Netherlands, the concept of mobilizing and combining the efforts of regional authorities and socially responsible business with the involvement of large-scale resources of the public financial sector is already being implemented. Within a year, 150 projects in the field of alternative energy, infrastructure, industry and agriculture were launched in this Northern region of France. There are five directions:1. transition to alternative energy; 2. transformation of building structures into micro-power plants for collecting alternative energy at the local level; 3. using of hydrogen and other storage technologies in each building and in the entire infrastructure for intermediate energy production; 4. using Internet technology to transform energy flow into energy Internet (when buildings produce small amounts of alternative energy locally and can sell their surplus green energy through a common stream and share in this way with their neighbors) and 5. transfer of the vehicle fleet to electric traction and cars that can buy and sell green energy through high-tech interactive energy flow. The synergetic effect of simultaneous implementation of all components will lead to a fundamental economic, social and environmental renewal of the region, and by 2050 – to a reduction in energy consumption by 60% and the creation of 150 thousand new jobs. The projected level of future annual costs for the implementation of the plan is 6 billion euros at the initial volume of the commitments of the authorities, business and the banking community in 1 billion euros for 2014.

In 2000 the United Nations launched an initiative of promoting social responsibility of business and reporting on the implementation of this policy called the Global compact. As of February 2018 more than 13 thousand organizations from 188 countries have joined the Global Treaty. The movement for social responsibility appeared long before the Treaty signing, in the late 1940s. During this time the strategies of social responsibility have become a must-have for large corporations. In the US and EU annual investment in CSR reaches $2 trillion.

In Ukraine, the Soviet times society got used to the fact that all problems are solved by the state. At the same time, social investments in infrastructure, such as energy audit and heating systems reconstruction are almost imperceptible by people and do not affect the reputation of the investor company.

In the modern world of capitalism, there are still disputes about need of CSR. Accordingly business does not find it necessary to spend money on social problems solving. And if it does it perceives them as they already are the CSR actions. In most cases, CSR acts as a component of public relations work. It does not focus on the systemic sustainable development of regions, but on charitable or even one-time actions. The defenders of CSR ideas claim that there is economic justification for the social responsibility: companies receive numerous benefits from the fact that they do the work for the long run. But critics say that CSR takes business away from its fundamental economic role and becomes only an embellishment of reality. The only thing that the opposing parties agree with is that CSR is a voluntary, non-mandatory business contribution to the society development in the social, economic and environmental spheres and is directly related to the main activities of the company. For example, an automobile enterprise that pollutes air should be involved in environmental protection projects, and the pharmaceutical industry should pay attention to educational programs for the prevention of dangerous diseases.

So there is a logical question: why should a large business, which has always been distinguished by healthy pragmatism, bear social responsibility to the society in which it works?

One answer is the claim that CSR affects customer loyalty and brand value. For example, a recent report by Trendwatching, a consumer trend researcher, shows that 53% of consumers refuse to buy goods and services from companies that have a negative impact on the environment or society and 73% believe that business should not only earn money but also do more for society.

Socially responsible business is an activity with such main participants:

- responsible manufacturer which produces high-quality, safe products for consumers and does not overstate the price for it, provides truthful information to consumers, operates in accordance with the rules of environmental law, worries about the environment, introduces the latest technology to reduce the negative impact, the elimination of hazardous waste and preventing their emissions;

- responsible employer acts in accordance with the rules of labor law, takes care of working conditions and social welfare of its employees;
- responsible participant of social relations participates in the maintenance of society welfare , is manifested in charity for orphans, disabled and other socially vulnerable groups, supports spirituality, education, science, culture, promotion of the region;
- responsible participant of economic and political relations with the state, who pays taxes and social insurance contributions in good faith, operates in accordance with the current legislation, avoids corruption, effectively conducts business without benefits and subsidies from the state, ensures transparency of corporate finance, supports the rule of law, independent justice and political competition;
- responsible business partner, which forms its relations with partners on the principles of compliance with contracts, agreements and professional standards, financial responsibility.

Modern business is interested in the formation of a transparent and democratic system of social responsibility with the definition of clear rules and conditions of business within the national economy and the gradual adaptation to the European experience. Awareness of social responsibility role contributes to the information and communication ability of the business- environment in achieving tactical and strategic goals of commercial activity.

The reform of the territorial power organization and local self-government on the basis of decentralization has become one of the key directions of systemic social transformations in Ukraine, since it is one of the basic conditions for independent and effective activities of local authorities: the decentralization of decision-making processes increases the opportunities for the participation of local authorities in the development of the territory under its control; contributes to the effective provision of public services through careful coordination of government expenditures with local needs and priorities. Many companies invest in infrastructure development, medical and educational spheres, which are facilitated by the implementation of reforms in Ukraine, decentralization and increasing the role of local communities in solving existing socio-economic and environmental problems within certain territories.

In Ukraine, the process of decentralization began in 2014 with the adoption of the "Concept of reform of local self-government and territorial organization of power" (01.04.2014), Laws "On cooperation of territorial communities" (17.06.2014), "On voluntary association of territorial communities" (05.02.2015) and changes to the Budget and Tax codes on financial decentralization. This process has allowed forming a significant effective and capable institution of local self-government at the basic level – united territorial communities, according to the provisions of the European Charter of local self - government.

In 2014 at the time of the concept approval of local self-government reformation and territorial power organization in Ukraine [11] the existing problems of local governance have accumulated enough. Among them are:
- excessive fragmentation of territorial communities and, as a consequence, the inability to perform all the powers of local governments. In Ukraine, until 2014, about 12 thousand territorial communities have been formed, in more than 6 thousand communities the number of residents was less than 3 thousand people, among them in 4809 communities - less than 1 thousand people, and in 1129 communities - less than 500 people;
- most of the local budgets were subsidized. 5419 budgets of local governments were subsidized; it is over 70%, 483 territorial communities were contained for 90 % at the expense of the state budget;
- the development of local self-government was actually carried out at the level of territorial communities of cities of regional importance. Since 1991, the rural population has decreased by 2.5 million and the number of rural settlements has decreased by 348. However, the number of village councils increased by 1067 units [11];

- the absence in most territorial communities of a favorable living environment necessary for the comprehensive development of the person, his self-realization, protection of his rights. In most of them, the executive bodies of the relevant village councils have not been established; there are no institutions and municipal enterprises. The local self-government bodies of such communities are practically unable to exercise the powers granted to them by law;
- deterioration of infrastructure and networks;
- low professional level of local government officials due to the low competitiveness of local governments in the labor market, which led to low efficiency of management decisions.

Decentralization aims to overcome existing problems by enabling each local community to choose its own, optimal management of local economy and financial resources, the taxes and fees that remain in the local budgets. The goal of the reform in Ukraine is to decentralize so that each Ukrainian community has enough financial resources and opportunities to put all the responsibility for what is happening at the local level to the local authorities. The experience of four years of decentralization reform in Ukraine shows its success.

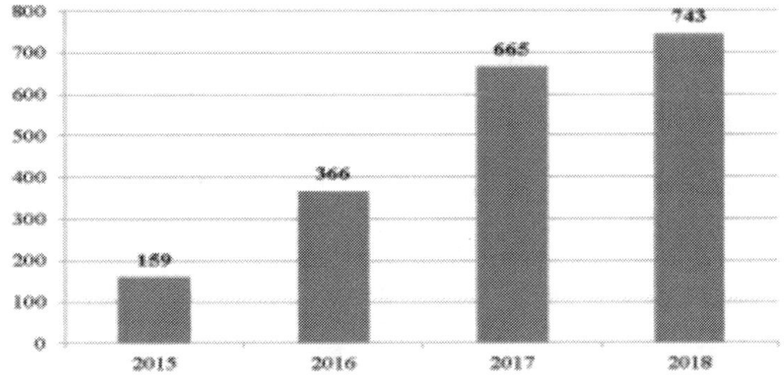

Fig. 1. Number of united territorial communities in Ukraine (2014-2018)

Source: According to the data [12].

According to the data [12] from 2014 to 2018, 3,443 territorial communities were united into 743 united territorial communities, they are home to 6.4 million people (18.2 % of the total population of Ukraine) (Fig. 1).

Own revenues of local budgets has increased from 68.6 billion UAH in 2014 to 192.7 billion UAH in 2017 [12]. However, the main purpose of the changes that are now taking place in most Ukrainian communities is to create conditions for effective development. This also will provide high-quality and affordable services, comfortable conditions for people to live in each community – regardless of whether it is a big city or a small village. Communities are not only united to gain more power and resources, they are consolidating their efforts to become stronger, to have resources and to be able to independently solve most of the issues that concern their residents.

But, if the company wants to invest in the development of infrastructure, medical or educational sphere, to finance the project of a charity fund or public organization, then, according to Ukrainian law, only part of this amount can be attributed to its expenses in taxation. At the moment, there are no direct incentives for the development of CSR in the state. The decentralization reform changed the tax distribution formula, which affected the balance of interests between business and government.

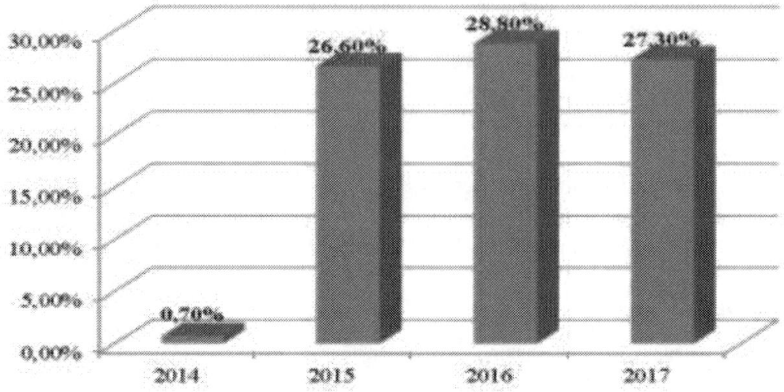

Fig. 2. Part of local taxes and fees in revenues of local budgets (2014-2018)

Source: According to the data [12].

According to the data [12] part of local taxes and fees in revenues of local budgets in 2016 has grown to 28,8% (Fig. 2). Communities are interested in attracting business to their territory, as a significant part of taxes remains in local budgets. At the same time, the communities should also increase their level of responsibility to the business: the investor should not doubt that his funds can be used for other purposes. The state could provide enterprises with a choice – to pay taxes to the state or to finance projects for the development of society directly. The main benefit in the implementation of CSR should not be in tax benefits, but in the stable and sustainable development of the community and business.

Among the main reasons for the failure of economic activities of their social functions can be identified:

- on the part of business: the systematic violation of the legislation, it finds its manifestation in the evasion of taxes and other payments of social purpose; the lack of actions for the enforcement of minimum wages established by the legislation of the living wage for able-bodied individuals; the lowering of wages to employees.

- on the part of households: population passivity in their own income increasing;

-on the part of regional authorities and local self-government: failure to fulfill their powers in terms of creating new jobs and lack of interest in the growth of population income at the expense of their own work due to the threat of transfers reducing to local budgets from the state;

- on the part of the state: the shift of the organizational functions of the state in the direction of distribution functions, the failure of the government approved at the legislative level of state social standards and the slow development of the institutional framework for their implementation.

The introduction of such measurement tools in the evaluation of economy subjects social responsibility as: indicators of manifestation of the socially responsible aspects of individual; indicators of social responsibility manifestation of territorial and local authorities; indicators of social responsibility manifestation will identify the reality disparity to composed stereotypes.

The formation of the social responsibility foundations should not be based on fragmentary but on the integrative social responsibility of all economic entities (state, enterprises, territorial communities). Social responsibility must be well consolidated and make effective function of the activity of all national economy entities. Implementation of social responsibility measures has certain advantages for each of the participants in Table 3.

Table 3. Benefits of implementing social responsibility measures

Subjects of economy	Advantages of SR introduction
For business structures	labor productivity and competitiveness of products increasing; ensuring stable operations in the short and long term; creating a favorable investment climate; improving the company's image; providing advertising, including in the media
For the personnel	Stability of wages; professional development; creation of attractive workers places; provision of appropriate working and rest conditions; social insurance; health promotion; material assistance and support in difficult situations; introduction of basic ethical principles in enterprises
For local communities	charity and sponsorship events; support for education, science, culture and national traditions; the environment care; care for socially vulnerable segments of the population; preservation of social stability; providing quality products and services; increasing the role in solving problems of society and so on.

Source: Authors.

The organization of social responsibility of business practice in regards to the community will ensure the rapid and effective development of the administrative unit through the belonging of employees-members of the community to the enterprise and accounting by economic entities and investors of the interests of public groups and promote the development of the territory in which the enterprise is located and operates. This approach will contribute to the implementation of the main areas of social responsibility of business aimed at sustainable society development and the formation of an effective mechanism of territorial communities financial support through decentralization.

4 Conclusion

The CSR allows companies to lobby their interests effectively and transparently by increasing the population trust and support. For example, western monopolistic electric companies invest in economic development of territories and non-profitable "green" energy sources. And when there is a need to increase tariffs for the population, receive the necessary support of society and politicians. Studies have shown that companies that pay attention to CSR, attract more motivated staff, participation in the development of the territories where the company operates, and the solution of its problems helps to retain specialists.

There are two main goals pursued Ukrainian companies in the field of CSR. This is reducing and mitigating risks and creating a positive image for customers, society and the state. It should be noted that there are practically no structures able to implement CSR practice in Ukraine. These are either international companies that are represented in Ukraine, or the largest domestic business. As for small and medium-sized businesses, it pays little attention to CSR issues.

A slight interest in sustainable development can be explained by the fact that in Ukrainian realities CSR does not play a significant role in the reputation of the company as it does abroad. In addition, the effect of conducting CSR activities in Ukraine for the company is still expressed in non-financial indicators and does not lead to direct additional income. Companies that manage a wider range of risks (and this is one of the objectives of CSR), look more stable for investors, with a better forecast, and therefore are more investment-attractive.

The work in the direction of CSR and in the quotes of many foreign companies shares is taken into account and that stimulates them to develop this direction. There is no such practice in Ukraine, but experts hope that everything will change over time. The country's economy, albeit with a delay, but follows the global trends, which means that changes are inevitable.

References

1. H. Bowen, *Social Responsibilities of the Businessman.* New York: Harper and Row, (1953)
2. K. Davis, The case for and against business assumption of social responsibilities. *The Academy of Management Journal*, **16**(2), 312-322, (1973)
3. A. Carroll, The pyramid of corporate social responsibility: toward the moral management of organizational stakeholders. *Business Horizons*, **34**(4, 39-48, (1991)
4. P. Drucker, *Management, Tasks, Responsibilities, Practices.* New York: Truman Talley Books/E.P. Dutton, (1973)
5. A. Kolot, Corporate social responsibiliтy: the evolution and development of the theoretical views. *Economic theory*, (4), 5-26, (2013)
6. D. Krepak, *Work on society: what is the corporate social responsibility of business* [online], Available at: http://forbes.net.ua/ua/opinions/1413277-pracyuvati-nasuspilstvo-shcho-take-korporativna-socialna-vidpovidalnist-biznesu (2016)
7. O. Kratt, K. Pryakhina, M. Bilyk, Ukrainian-Chinese collaboration: Prospects of development. *SHS Web of Conferences – Innovative Economic Symposium (IES2017)*, (2017)
8. *Strategy for promote the development of social responsibility of business in Ukraine until 2020.* [online], Available at: http://www.svb.ua/sites/default/files/201309_strategiya_spriyannya_rozvitku_svb_v_ukrayini.pdf (2018)
9. *French National Assembly Adopts Corporate Duty of Vigilance Law* [online], Available at: http://www.csrandthelaw.com/2017/02/28/french-national-assembly-adopts-corporate-duty-of-vigilance-law (2018)
10. *UN treaty is about binding rules not corporate social responsibility* [online], Available at: https://www.foei.org/features/un-treaty-binding-rules-not-corporate-social-responsibility (2018)
11. *Concepts of reforming local self-government and territorial organization of power in Ukraine* [online], Available at: http://zakon1.rada.gov.ua/laws/show/333-2014-p (2014)
12. *Decentralization of power* [online], Available at: http://decentralization.gov.ua/uploads/library/file/274/Monitoryng_11.06.2018.pdf (2018)

Determinants of real decision making on productive components of consumption under the current economic conditions of the Slovak Republic

Marianna Psárska[1,*]

[1]Institute of Technology and Business in České Budějovice, School of Expertness and Valuation, Okružní 517/10, 37001 České Budějovice, Czech Republic

Abstract. The aim of the contribution is to determine the main determinants of the formation of their productive components in consumption under the real conditions of the selected country, using the decision-making mechanism of the individual and of the households. The benefit of the article is in particular applied bonding and verification of the relation between the current theories and the practice that is realized through a questionnaire survey. By recognizing these determinants the partial decision-making issues will be clarified and it will open the space for the eventual desired behavior of household behavior in specific Slovak Republic conditions. The conclusions of the article point out that household decision-making is influenced by several factors in parallel, but psychological factors play a very important role. It is clear from the conclusions that Slovak households, by their decision, confirm theories of behavioral economists even when deciding on productive components in consumption.
Key words: productive components of consumption, rational decision-making, irrational decision-making, quality of life, determinants of consumption

1 Introduction

If we want to monitor the selected components or certain categories of the consumer basket of households, it is necessary for us to first understand their decision making mechanism from a theoretical and practical point of view. This is the key to understanding the shaping of consumption that is inevitably built on the individual he is looking for in the consumer sense. We understand households as recipients and consumers of pensions, but also as creators of savings. In this category, basic descriptive characteristics such as who is at the forefront of household preference, the number of dependent children where the household lives, its size, structure, age, and so on are important. This explains in part the motives of individual population groups, their basic tendencies in decision making and the tendencies of the future development of the studied components of consumption.

* Corresponding author: psarska.m@gmail.com

The choice of the category of household and individual is linked to the methodological individualism of the neoclassical school. It is an approach that explains all economic categories and laws to the rational behavior of an isolated individual, "Economic human" (homo oeconomicus) [1].

In principle, at present, households use products, but individuals usually buy them. There are different models of household decision-making (family), but they are dealt with in detail in marketing. We are going back to the decision-making mechanism and the powers that make the most of it. The decision process can be explained by a general logical description of the consumer's decision-making situation. The basic decision model is actually a system model, and it consists in practice of ten basic elements [2].

1.1. Rational decision-making

One of the axes of economic theory is the claim that a person decides rationally. Rationality is understood as a reason-based action. However, setting criteria for rational decision-making is rather complex because individuals perceive rationality differently. The basic concept of rationality is the "homo oeconomicus" paradigm, which has remained the focus of economists' attention to rationality, even though it has been the subject of constant criticism [3].

The pillars of rational decision-making consist in the fact that:

- a person acts in such a way to achieve a better situation than the one he is in;
- the paths that lead people to their goal are important. And here a person chooses the most effective ways to reach their goal with minimal distress and maximum benefit;
- the basic principle of optimal behavior is that a person continues to operate while their costs are below his marginal yield (or borderline utility);
- a person can only explain the behavior of people in a large group;
- In a large group there are interchange relationships based on people, that is, the market and the core coordinator is the price;
- a person makes mistakes but learns and does not repeat them;
- A rational person does not get all the information, he does not want to be fully informed. Rational behavior is characterized by rational ignorance, which is, in fact, an optimal level of awareness when, the borderline yield from information equals the borderline cost of information;
- the principle of optimization is based on the claim that a person uses their limited resources in an optimal way when deciding. Microeconomic analysis is based on the ability of a person to choose or prioritize one solution from a set of permissible solutions at the limit of their consumer options [4];
- The "homo oeconomicus" paradigm is based on the fact that the purpose and goal of human behavior is to best satisfy its individual needs while ignoring the consumption of other people. It is therefore based on egoistic preferences and the ability to always find the best optimal solution to maximize their own usefulness. This subtle rationality is also referred to as complete or equivalent or unlimited [5];

Pillars of rational decision-making also include two basic standards that serve to analyze rational human behavior. Formal theory of rational behavior consists of the rationality standard of self-interest and rationality, of the current target [6].

1.2 Alternative approaches to rationality

Alternative approaches to rationality can be considered a modern development of the neo-classical homo oeconomicus paradigm. They point to the weaknesses of this paradigm, the current economic and non-economic issues related to consumer decision-making.

Alternative approaches ponder rationality as limited (imperfect). The limited rationality is connected and explained through the principle of satisfaction, cognitive imperfection and the concept of altruism.

The principle of satisfaction lies in the fact that one does not seek an optimal variant indefinitely, but the evaluation process ends when he finds a satisfactory solution. He does not look for an optimal solution, but an acceptable one. Another one is the principle of imperfection that describes the consequences of human imperfections such as hesitancy, evasiveness, and so on. The concept of altruism captures the situation when one also includes the benefits for other people in their decision-making motives [7].

A limited rationality is manifested by the fact that a person is not always able to apply the expenditure and income calculations immediately, and decides to be uncertain. his decision may be influenced by the cost of obtaining information, or it is realized under the influence of emotions. Psychology, which is influenced by behavioral economics, also has a major impact. It realizes an optimally satisfactory solution to its situation.

Behavioral Economics† combines psychology with economics to provide answers to many challenging questions about economic decision-making in the current times. It deals with the circumstances of individuals' failures and the causes of crises. Behavioral economics justifies the emergence of crises by the fact that all behave equally irrationally. As a result, there are bubbles in the markets. In a summarized form, research into behavioral economists has shown that:

- emotions greatly affect a person's behavior, and a person is acting differently as an individual and differently in a crowd;
- a person prefers choosing an alternative that brings change of wealth, but when choosing from alternatives, the most important role is not played by static wealth, but its relative expression and relative change over a reference point, plays the most important role.
- Human behavior is motivated by two major systems. One is based on looking for rewards and the other on resistance to loss;
- intuitive judgment belongs to the same place as deliberate thought activities;

people misrepresent themselves as they act reasonably and deliberately, but they are actually influenced and vulnerable and repeatedly making misconceptions in bad shopping and investment decisions. These decisions are therefore not only irrational but even predictably irrational;

- we avoid comparing things that need to be very thought out;
- the so-called effect of herd behavior that occurs when we form an opinion based on the opinion of other people;
- the endogenous preferences for which habits, customs, reference group pressure, parental influence on children's preferences, advertising, love and affection have an effect on the maximalization of utility. Thus, the model of dependence of current and future preferences on past personal experiences or social influences on preference making and many others is working.

In our article, given the complexity of the issue, we will only focus on a selected category of productive components of individual consumption, specifically households.

† Behavioral Economics applies the results of research on human cognitive and emotional tendencies in explaining economic decisions and explaining how these tendencies affect market prices, revenue and resource allocation. Behavioral models integrate the knowledge of psychology into neoclassical economics.

1.3 Productive components in consumption and their key determinants

If we want to examine the selected product aspects of household consumption theoretically, we must use the alternative approach to rationality. Economics of Productive Consumption according to R. Valencik [8] considers the household behavior and the growth of the present value of the expected future income from the creation and operation of assets forming their "family portfolio". Under the productive components of consumption we mean those items of the consumer basket of households, which by their nature are investments in human and social capital. These are therefore components of consumption that are not necessary for survival, and their importance in the consumer basket is increasing only after reaching a certain standard of living for households. The fact is that in developed countries the importance of productive components of household consumption is growing especially when economies are at the peak of the economic cycle when these spending groups should reach the highest values. The question, however, is why it is not, for example, in Slovakia. In the context of decision-making about their consumption, it is necessary to know the process of implementing these decisions [2].

In general, we distinguish the four main content groups of determinants affecting consumption and consumer behavior. They are economic, social, demographic and other determinants. We consider economic factors as the primary premise for their application as the most important and as the determinants of the productive components.

But let us go back to the process of making decisions about these components of consumption. From the point of view of behavioral economy preferences [9-13], one prefers to choose an alternative that brings a change of wealth. Productive components of consumption definitely bring this possibility of change of wealth. The question, however, is whether the real households are really aware of this fact. Or is it just applicable to a tiny percentage of households?

Because the behavioral approach is based on the assumption that human behavior is motivated by two main systems, namely seeking rewards or avoiding losses. Is it therefore necessary to consider what rewards do productive components of consumption bring and whether there is any kind of loss? And also the question of how can the productive components of household consumption be protected from risks? Questions of this type lead us to the motives that lead people to spend their money on any kind of expenditure.

According to D. Kahneman [14], the current view of consumer decision-making is focused on three main themes, namely heuristics of judgment, risky choices and framing. In all three areas thoughts, intuitions and preferences were studied, which a person thinks about quickly and without long pondering. D. Kahneman and A. Tversky argue that intuitive judgment belongs to the same place as deliberate thinking, but focuses on the existing discrepancies between intuition and knowledge especially in statistics. Are the productive components of household consumption relevant to intuitive or deliberate thought processes, or does the human nature play a major role in these issues? D. Kahnemann also deals with the formalization of intuition and the issue of change of wealth or function of values.

Another author of behavioral economics is D. Ariely [15], who has attributed an important role in the decision making process to the so-called accidental dependence and anchors from the past. Are productive components of consumption also decided in this way? The analysis of D. Ariely, which focuses on the magic word "FREE", is a very exciting one, as we often make a decision which we did not want in the first place and which is not the best from the point of view of rational decision-making. Whatever is free, it seems to us a lot more valuable than what it really is. And why? Because human beings have instinctive fear of loss [3]. The success of this word is based on this fact and is often used, for example, in pharmacies for the sale of supplementary medications, and thus falls under household spending on healthcare and productive components of consumption.

Another point of view on this issue is the G. S. Becker theory, which, for example, considers habit (rationality) to be rational if the increase in past consumption increases current consumption [16]. For reflection on the rationality of custom behavior, the key point is the knowledge that the extent of addictive activities is relatively high in humans, so we can not exclude them from economic analyzes. Is it possible that just the habit of free public education within our economic system and family is just the necessary minimum spending that hampers the possible change from the point of view of their development and the growth of well-being in the long run? Or is there another cause?

If we knew the answers to these questions, it would be possible not only to better understand people, but also help them in the sense of better choice of purchased goods and services supporting productive components of household consumption, which in the long run would bring about the growth of their wealth and their role in the company has the ability to create much more added value.

We will end this part of the article with R. Thaler's work, analyzing how people can be encouraged to make better decisions. It outlines how governments can use behavioral models to increase the efficiency and quality of a wide range of public services. By which it points out some sort of possible positive interaction or correct motivation to change behavior even in our conditions. R. Thaler's research can also be applied to everyday human economic decisions as well as to the behavior of the whole market. In his work, he incorporated psychologically realistic thinking into economic decision-making. His research into the impacts of limited human rationality, social priorities and lack of self-control shows how these human characteristics systematically influence economic decision-making of individuals [17, 23, 24, 25].

2 Data and methods

The examined data for the purpose of this contribution were surveyed through an anonymous questionnaire survey on the productive componens of consumption and the implementation of the decision-making mechanism of specific individuals.

Overall, the questionnaire survey consists of two parts and has 25 questions. The first part is focused on the descriptive characteristics (see Table 1) in terms of determining the status of household preference, the number of dependent children in households, the level of net income, or the basic characteristics of the person surveyed. The second part deals with the issue of consumption, where the questionnaire survey contained categories of spent expenditure on healthcare, education, recreation and culture, and also included open questions, which gave us the opportunity to highlight the motives of decision-making and gained many insights into the selection of other perspective components of the expenditures of productive character.

Table 1. Descriptive characteristics of the target sample (Survey questions 1-3)

Question (1.)	**Position in household**	Question (2.)	**No. of children in the household**	Question (3.)	**Net income**
Retiree	2.53%	Without children	50.63%	below 500 Eur	50.63%
Maternal leave	2.53%	With 1 child	22.78%	from 500 to 1000 Eur	3.80%
Self-employed person	11.39%	With 2 children	22.78%	from 1000 to 1500 Eur	22.78%
Employee	83.54%	With 3 and more children	3.80%	above 1500 Eur	22.78%
Total sum	100.00%	**Total sum**	100.00%	**Total sum**	100.00%

Source: Author.

In the conclusions drawn from the collected data, we are of course aware of the fact that it is a smaller sample and unevenly distributed in proportion to the number and structure of the population, but to verify certain predicted tendencies, determinants or directions in the development of consumption of these components, our approach for this purpose is sufficient .

3 Results

Productive components of consumption are typical that they are not indispensable for survival, but their importance in the consumer basket is increasing only after reaching a certain standard of living for households. In the long run, their importance should increase significantly in order to improve the quality of life and overall development [18].

3.1 Key real determinants forming the productive components of consumption in SR

The first and partly conditional determinant is income, which has been increasing in Slovakia over the last period, yearly (in line with the rise in the minimum wage), which should create very favorable conditions in relation to the categories of expenditure pursued. Here, in the first step, we will think about the causes of income growth. According to the OECD list, in 2015, we were among the most working-class countries in terms of hours worked, and the average Slovak employed about one hour more in the day and excelled in work in difficult conditions, at night and over weekends where there are currently approved overtime raises. At present, the situation has changed slightly and overtime hours have fallen. Our survey (Table 2) has shown that overtime hours are least worked by people with the lowest income, the higher the income, the greater the willingness to work more. We did not assess attitude towards work. From the point of view of the willingness to undertake the risk, the ratio of the answers is balanced and in the context of the income there is no dependence, so we consider this aspect considerably individual. However, very interesting were the answers to the question about the decision making and the attitude towards it in relation to the monitored groups of expenditures. Up to 72% consider themselves to be people who think and plan things thoroughly rather than to spend on the expenses of that nature. The remaining 27.85% of respondents are impulsive. In relation to their income, low-income categories do not allow themselves to behave impulsively, on the contrary, with the increase in the income, the amount of people with impulsive attitude rises. In the highest income category, the ratio is almost 1: 1.

Table 2. Descriptive characteristics of the target sample (Survey questions 4-5)

Question (4.)	**Average overtime**	Question (5.)	**Do you like to undertake risk?**
Work nonstop	2.53%	Yes	39.24%
On average up to three hours	7.59%	I cannot judge	24.05%
On average up to two hours	6.33%	No	36.71%
On average an hour	18.99%	**Total sum**	100.00%
On average a houlf hour	36.71%	Otázka č. (6.)	**Opinion on own decision-making on spending?**
No overtime	26.58%	Acts impulsively more often	27.85%
Currently unemployed	1.27%	Thinks and plans more often	72.15%
Total sum	100.00%	**Total sum**	100.00%

Source: Author.

In the second step relating to income, we might notice that if households work on average more, they have less time to develop productive components of consumption and therefore have less time for education, culture, relaxation. During this period of time, if they would like to change their spending on productive aspects of consumption, they would probably have to spend less time at work, which would be a loss of income in the first place (provided they have paid or otherwise rewarded overtime). This view is supported in part by Poštová Banka's analysis of 2014, when it examined the proportion of employees working at night, evening, Saturday and Sunday. At the top of the list were Slovaks, 16% of them working at night, 25.2% in the evening, 24.4% on Saturday and 19.2% on Sunday [19]. This fact affects not only the level of income through mandatory surcharges but also the level of quality of life in households in which the provider is missing in the dedicated common time, which in turn limits the application of the aforementioned expenses. Our survey showed that up to a third of the respondents spend 1-3 hours of overtime a day, a third spends on average a half an hour, and about a third of the respondents does not work any overtime.

The **determinant of wealth** is largely influenced by looking at debt issues, where appropriate, the nature or type and nature of goods or services purchased by the individual or household debt. The **determinant price level** and **interest rate** are also closely linked to this issue. For both determinants, the change in the structure of consumption in terms of long-term sustainability and continuous targeted improvement of the income situation is the consequence or key to addressing. The survey revealed, based on the last open question, that most of the people surveyed do not think about the productive components of consumption at all. Of the total number of respondents, 72% said so. From this, up to 50% of the respondents given this question (not asked) simplified their answer, and repeated only the offered examples to this question. Only 18% of respondents given this question seriously considered and captured the basic idea of the survey and attempted to elaborate on it.

Allocation and redistribution of pensions, and **tax policy** are a very important determinant of the formation of consumption and, in particular, of productive components or the creation of savings. The main idea and reason for mentioning this determinant is the fact that the state affects the income of the tax payers, changes the relative prices of goods and services (excise duty) and also consumer choices, choosing employment, health care, choosing a place of residence (county town, village, etc.) and waste management etc. At present, for example, a reduced VAT rate from 20% to 10% is applied for medicines, medical devices and books, which directly falls under the monitored issue.

Another determinant is the **behavior of households**, which is largely influenced by their optimistic or pessimistic expectations. Whether optimistic or **pessimistic**, they are always associated with a certain amount of uncertainty or mistrust, which is also reflected in consumer decision-making in households. Household decision-making aims to keep the household at at least a constant level of its assets, and spend only on the incomes of individual assets. Savings are mainly taken as a policy to mitigate variations in retirement. Nowadays, according to our survey, the expectations of people that are too optimistic, which distort their decisions in the direction of less caution when deciding on productive components of consumption, are at the top of our survey. In addition to expectations, behavior and decision-making influence other factors, depending on the focus of specific types of expenditure.

Other influential determinants include **demographic and structural characteristics** relating to individuals or households, **psychological determinants**, and last but not least **cultural influences and social determinants** [3].

3.1.1 Main determinants forming expenditure on education

The analysis of the educational structure of the population in relation to the income level revealed that the amount of income is determined by the degree of education achieved only with a few exceptions [21]. Education is therefore something essential for development and success in the knowledge society and every economy. It represents a kind of investment for every individual, thanks to which they can gradually increase not only income, but also the knowledge to make it more efficient and to work more efficiently.

He level of spending on education for individual types of households in Slovakia is, on average, around 0.3-0.6% of the total net cash expenditure of households, which is very low. The expenditure on education was higher in households with a self-employed provider. The reason for the low spending on education can be busyness from the point of view of work responsibilities that go beyond normal working hours, it may also be a lack of motivation for self-education, other spending priorities (eg spending on other members of the family, children) or still a misconception about the fact that one is able to complete one school and no further training is needed. All of these factors were confirmed in the survey and the variability of the responses was very large even though 97% of the respondents considered spending on education to be beneficial and useful. Which is heavily opposed to the answers they developed in the questionnaire survey subsequently. In the following Table 3, the answers to the question of the benefits of education are what their point of view was. From the answers, it is obvious that the respondents noticed the reaction of the environment and felt the positive and the negative response that came out of it. This table also shows the benefits of spending on education based on higher income, non-financial benefits, knowledge and contacts, and the good feeling they had with themselves. All of these variables are of great importance in the long run and also have an impact on the level of their monetary and mental wealth.

Table 3. If you saw any benefits in education, what were they? (Survey question 9)

Respondents' answers, question 9	**In %**
I refuse to answer, it's private	2.53%
I got the job of my dreams, and that pleases me	3.80%
I have gained valuable business contacts for people with whom I will continue to share my work experiences	15.19%
I have received other non-financial benefits through education	3.80%
I have benefited from other non-financial benefits through education, I have improved my knowledge / abilities through education which has been positively evaluated by my environment	7.59%
I have gained other non-financial benefits, accelerated, streamlined and improved my work without any feedback from my environment	1.27%
I improved my knowledge / abilities through education which was positively evaluated by my environment and I had a good feeling about it	35.44%
I accelerated, streamlined and improved my work without any feedback from the environment	12.66%
I have gotten a raise in my salary through education	7.59%
Several of the benefits mentioned together (increased salary, contacts, non-financial benefits)	10.13%
Total sum:	**100.00%**

Source: Author.

3.1.2 Main determinants forming healthcare expenditure

For this group of expenditure, it is necessary to say that the health of the population is not improving at all and the number of reported occupational diseases is increasing. In fact, mediated factors are declining in terms of disease prevention, for example the consumption of vegetables, partly fruits and the production of own food [22]. So, what form of prevention is being implemented and therefore what influences real savings in healthcare spending? We found the answer to this question in part in the questionnaire survey, but the question of the objectivity of the answers is questionable as the questionnaire results partly contradict some statistical data. However, it can be said that, from the point of view of the amount of prevention, the respondents seek it from 34% 1-3 times a year, 55.7% realize it on a monthly basis and 8.8% seek it on weekly or daily basis, indicating the lack of intensity of prevention, although in terms of the content and attitude of the respondents, the structure and the satisfaction (86% consider the prevention of their health problems as sufficient) are almost ideal in the following table.

An interesting finding is the fact that Slovakia has a leading position within the OECD countries when considering the consumption and spending on medicine. Why is it? According to the survey, the respondent considered this form a suitable type of prevention and protection of health as a reason for their purchase. This belief is probably due to marketing and advertising, which is richly represented in several forms.

Table 4. How do you prevent disease and improve your health? (Survey question 11)

Answers to question 11 (method of disease prevention)	**In %**
I'm looking for alternative forms of prevention and treatment from experts	1.27%
I buy enough fresh fruit, vegetables, legumes and everything that is necessary for a varied diet	31.65%
I buy herbal and natural remedies in specialized stores	7.59%
I use multiple options of prevention at once	22.78%
I buy minerals and vitamins in pharmacies, and herbal and natural remedies in specialized stores	17.72%
I irregularly go to the sauna, swim, douse in cold water, etc.	2.53%
I do not have any health prevention expenses, it is not necessary	5.06%
I regularly go to baths, regularly go to the sauna, swim, douse in cold water, etc.,	6.33%
I grow my own fruits, vegetables + raise own poultry	2.53%
No response	2.53%
Total sum:	**100.00%**

Source: Author.

3.1.3 Main determinants of recreation and culture expenditure

Expenditure on recreation and culture is closely related to the time spent on individual spheres of quality of life. They cross through each other. If a person spends more time in the workplace, and in the sphere of one-off or out-of-work expenses, the sphere of free time, in which these expenditures in particular belong, shortens. This trend was confirmed by our survey. With this expenditure, the notion of motivation comes first, as the attitude of an individual towards spheres depends directly on it. As we know, motivation expresses the process, while the motif expresses a hypothetical disposition to this process (for example, the need to relax, to relax, to enjoy a cultural experience or other similar primary or secondary motives), so we should focus on the motives and the process separately and answer the question as to why is there a difference between Slovak households in this spending group and what forces the households to reduce or cut this share of expenditure.

In the first step, we determined whether or not the households spent any expenditure on recreation at all. Of the respondents, 88% answered positively, the remaining responded negatively, and most of the reasons stated a lack of time. What is also confirmed by the next question is that we have focused on the number of holidays, where up to 81.6% of respondents said that they go on holiday once or twice maximum. In terms of expenditure, we consider this value to be highly distorted and very variable, so I do not mention it. On the question of satisfaction, 83% responded that they were satisfied with the holiday. Reasons for dissatisfaction with the others were related to insufficient holidays, dissatisfaction with the location or a busy household due to young children. It is very interesting to note that if we asked whether they considered after their holiday that they have to change something the next year, 62.8% of the respondents answered yes. Those who answered yes, we asked whether they really did. Here 22.8% of respondents answered that they did not. And as a reason they have mentioned, for example, insufficient preparation or lessons from mistakes, little time or small children, which make it impossible to plan and organize things. For questions about this type of spending, it also happened that 79% of the respondents did not read the question at the end and their response was not related at all to what we asked about. This fact also speaks about how people perceive and focus very little. Often they think quickly, intuitively, and do not know what the person is asking specifically, or they make questions up. This also influences decision-making. If they are absent mentally, it also affects their decision-making, behavior, and action.

The second group of expenditure is spending on culture. 93.7% of the respondents from our sample spend expenses on it. The rest of them said they did not do it, and the reason is again the lack of time, money or inadequate and uninteresting possible choice of cultural events. Very interesting was the answer that it is difficult to change the habit, you just get used to some sort of thing and even incorporate some cultural action into their schedule is extremely difficult, even if they wanted to incorporate it into the schedule. Which is confirmed by the theory of behavioral economists and the unomittable role of habit.

4 Conclusion

The aim of the article is, using the decision-making mechanism of the individual and related to that, the household, to determine the main determinants of the formation of their productive components of consumption in real conditions. By recognizing these determinants, the partial decision-making issues were clarified and opened space for possible options to shape household behavior in specific SR conditions. Besides the basic determinants of consumption such as income, wealth, interest rate, price level, tax conditions, demographic, societal and cultural determinants, our survey confirmed the severity of the so-called psychological factors in terms of behavioral economists' conclusions, but focusing on specific conditions of Slovakia. Even with productive components of consumption and spending on them, people behave emotionally, intuitively, habitually and comfortably. People avoid unnecessary thinking, deceiving themselves and choosing in the most simple and short ways. Which is exactly what is a big risk with these components of consumption. As a result, productive components have a positive effect on the lives of individuals from an economic and mental perspective in the long run, but in the short term, the spending of resources now represents a reduction in the available resources in the household budget, which is partly done with impunity, and these expenses are for a number of reasons uninteresting.

It is generally known that the current knowledge economy is based on four basic pillars: education, corresponding to the information structure, clear rules of the game and the existence of an innovation system [3]. The justification of the consumed productive components of consumption coincides with several pillars of the knowledge economy. The core is not only to economically evaluate expenditure groups but also to know their themes, expectations and personal profile of people living in that country and only then is it possible to reliably predict the development of consumption or its selected components.

From our survey, many respondents have surprisingly focused their view on ecology, which is also productive in nature, but it is the so-called "environmentally conscious consumption", which is the expression of qualitative individualisation, which we can regard as voluntary or intentional modesty in the behavior of some households. However, it is debatable whether this environmentally-friendly lifestyle is really reaching out to potential consumers or it is just a modern trend to talk about.

5 Recommendations and Suggestions

The conclusions of this article point out that household decision-making mediated by individuals is influenced by several factors in parallel, and each of them represents a sort of channel through which action can be taken. Of particular importance in the decision-making mechanism are the psychological factors, which we have proven in our article.

From the perspective of economic policy recommendations, emphasis should be placed on taking these facts into account when generating government economic policies. Very interesting and now often used is the term "libertarian paternalism" which can be used not only in the sense of better information but also in the sense of targeted measures by governments in order to guide people's decision in a desirable direction without being limited [23-25].

In addition to these positives, it is criticized in a negative sense that it directs the decision-making and behavior of an individual, regardless of the description of what complex human spirit it should rely on and relies heavily on economic and psychological factors alone. At the same time, the multiplication of researches and comparisons or the new "paternalism" is beneficial and, compared to the old one, makes real progress [26]. Finally, there is another worrying tendency that is associated with the growth of some experiments in connection with "paternalism," and it is questionable to what extent they are known and their consequences are addressed. Also, with regard to the academic sphere, they are increasingly preferring and appreciating academic knowledge of psychological, neurological and behavioral sciences, with the science of social studies potentially deteriorating and being marginalized [27]. This discussion obviously brings many interesting points of view, but also opportunities to develop a moral, ethical and perspective correct form of "paternalism".However, in addition to this discussion, we are going back to the first possible form of action, in raising awareness as a possible policy measure, also here are limitations, for example in the limited attention of people who can not use all available information efficiently in the event of some kind of attention already exhausted, for example by fatigue, or by the overload of information, then decision making becomes intuitive and shallow. It therefore stops taking into account important aspects of the situation, and such decision-making can lead to suboptimal results by overestimating the importance of the most pronounced stimuli and ignoring the less visible ones. This could be avoided by some kind of help in terms of targeted databases or institutions through which people could learn the most accurate information. In general, this is precisely the aforementioned aid by the so-called "nudging", which aims to complement decision-making processes with elements that help people make the same intuitive decisions as they would if their cognitive resources were not restricted [28]. At the same time, we are at odds with the possible negative "paternalism" in keeping with ethical and moral standards or some sort of targeted manipulation.

From the point of view of a specific type of measure aimed at a particular individual, it can be, for example, a reform of education that would include content with due importance, depth and thorough preparation of a subject focusing on increased awareness of health and healthy lifestyles, financial literacy or targeted support of critical and logical thinking and decision-making psychology, which would improve decision-making over the longer term (health, education, consumption, marketing, etc.). In so-called financial literacy under the current condition, certain savings schemes that would allow to systematically generate savings or solve the financial situation of individuals and households in the long run in due time, would be appropriate. For example, it is possible to create savings for education and to motivate people to think ahead, and such products by banks and the state could also be favored and supported. There are, of course, many choices.

The already established and partly used tools include tax policy and the use of lower tax rates, for example, VAT (Value Added Tax) on selected types of supported household expenditure. In this measure, it would be appropriate to focus on the productive ingredients in consumption and to favor those with the appropriate disclosure of reasons. This type of measure can be used across all of the monitored spending groups and is just a selection of several options that could help to improve the benefits of such measures, with the fact that they would be thoroughly evaluated and published for possible corrections in cooperation with the academic sphere that deals with the issue in detail.

References

1. J. Lisý a kol., *Dejiny ekonomických teórií.* Wolters Kluwer, (2003)
2. L. Madarász, J. Vaščák, R. Andoga. T. Karoľ, *Rozhodovanie, zložitosť a neurčitosť, teória a prax.* Košice: Elfa, (2010)
3. M. Psárska, *Mikroekonomické aspekty spotreby domácnotí SR v nových ekonomických podmienkach.* Dissertation thesis, (2013)
4. J. Hlaváček, M. Hlaváček, *Zobecnená mikroekonomie* (2010)
5. P. Sirůček a kol., *Hospodářské dějiny a ekonomické teorie.* Prague: Karolinum, (2007)
6. R. H. Frank, *Mikroekonomie a chování.* Prague: Svoboda, (1995)
7. V. Holková., Ekonomická vzácnosť a racionálne rozhodovanie z pohľadu mikroekonomickej teórie. *Proceedings of the EAPG workshop 2011: Economic analysis and Policy Group,* 1-8, (2011)
8. R. Valenčík, Ekonomie - Jak ji chápat a k cemu je? (Doplnující text ke studiu mikroekonomie na magisterském stupni). *Marathon*, (53), (2004)
9. D. Ariely, *Jak drahá je intuice.* Práh, (2011)
10. E. Andrade, D. Ariely, The Enduring Impact of Transient Emotions on Decision Making. *Organizational Behavior and Human Decision Processes*, **109**(1), 1-8, (2009)
11. P. Singer, Famina, Affluence, and Morality. *Philosophy and Public Affairs,* **1**(1), 229-243, (1972)
12. D. Small, G. Loewenstein, The Devil You Know: The Effects of Identifiability on Punishment. *Journal of Behavioral Decision Making* **18**(5), 311-318, (2005)
13. D. Small, G. Loewenstein, Helping a Victim or Helping the Victim: Altruism and Identifiability. *Journal of Risk and Uncertainty,* **26**(1), 5-13, (2003)
14. D. Kahneman, *V jeho prednáške z udeľovania Nobelovej ceny za ekonómiu, z 8.decembra 2002, ktorej názov je „Mapy ohraničenej racionality: Pohľad na intuitívny úsudok a voľbu",* (2002)
15. D. Ariely, *Jak drahé je zdarma.* Práh, (2009)
16. G. S. Becker, K. M. Murphy, The Theory of Rational Addiction. *Journal of Political Economy,* **96**(4), 675-700, (1986)
17. R. H. Thaler, C. R. Sunstein, *Nudge (šťouch): Jak postrčiť lidi k lepšímu rozhodovácní o zdraví, majetku a štěstí.* Kniha Zlín, (2010)
18. M. Psárska, Produktívne zložky spotreby domácností SR a ich aktuálny vývoj v kontexte blahobytu / Marianna Psárska. - VEGA 1/0570/11. *QUAERE 2012 Reviewed Proceedings of the Interdisciplinary International Scientific Conference of PhD Students,* Hradec Králové, Czech Republic, (2012)
19. *The Post Bank* [online], Available at: https://openiazoch.zoznam.sk/cl/160648/Slovaci-opat-na-cele-rebricka-Stvrtina-pravidelne-pracuje-vecer-a-kazdy-siesty-chodi-do-prace-v-noci (2018)
20. *Zdroj údajov; ŠÚ SR, ČŠÚ* (2018)
21. E. Čitáryová, T. Chrenko, Analysis of the educational structure of Slovak population in terms of income levels. *Medzinárodný seminár: Kvantitatívne metódy v ekonómii,* (2011)

22. *Štatistická ročenka Slovenskej republiky 2016*. Veda, (2017)
23. R. H. Thaler, *Neočekávané chování, Příběh behaviorální ekonomie*. Argo, Dokořán, (2015)
24. A. Belayanin, Richard Thaler and behavioral economics: From the lab experiments to the practice of nudging. *Nobel Memorial Prize in Economic Sciences 2017,* **1**, 5-25, (2018)
25. R. H. Thaler, Nudge, not sludge. *Science*, **361**(6401), 431-431, (2018)
26. B. Pickett, The New Paternalists. *POLITY*, **50**(2), 300-329, (2018)
27. R. Jones, M. Whitehead, Politics done like science': Critical perspectives on psychological governance and the experimental state. *Environment and planning d-society & space*, **36**(2), 313-330, (2018)
28. P. Houdek, P. Koblovský, D. Šťastný, M. Vranka, Consumer Decision Making in the Information Age. *Society*, **55**(5), 422-429, (2018).

Theoretical and practical view of development and differences in the application of productive consumption components in Czech and Slovak households under current economic conditions

Marianna Psárska[1,*]

[1]Institute of Technology and Business in České Budějovice, School of Expertness and Valuation, Okružní 517/10, 37001 České Budějovice, Czech Republic

Abstract. The aim of the contribution is to point out the possible causes of changes and differences in the behavior of current households in the context of new consumption theories and current conditions by means of selected categories of net cash expenditure of households[2] embodying the productive aspect of consumption. The monitored data is based primarily on Family Account Statistics for the period of 2010-2017. The benefit of this contribution is in particular the application and clarification of the links between the current theories and practice which helps to understand the economic behavior of households. The conclusions of the article point out that household decision-making is largely influenced by the economic situation of households, but also by other factors which we also deal with in the contribution. It is clear from the conclusions that Czech and Slovak households apply a different approach to expenditures on productive components of consumption, which in the long run will deepen the differences between countries. In the context of the evolving direction of the Behavioral Economy, it would be helpful to include the monitoring of psychological factors in Family Account Statistics.
Key words: household consumption, productive components of consumption, net cash revenue, net cash expenditure, household savings, quality of life

1 Introduction

If we want to examine the productive aspects of household consumption from a theoretical point of view, we must abstain from some constraints imposed by traditional microeconomic theories, such as the principle of unlimited rationality, or the optimization of utility understood as a subjective phenomenon. If we assume this optimization assumption, it means that it is possible to earmark an absolute final consumption, which

* Corresponding author: psarska.m@gmail.com

[2]A privately-owned household consisting of persons sharing a permanent dwelling in one home and joint spending on important household expenses. A household member is also a person who is not related to other members of the household, but lives with others in one home and participates in the payment of expenses. Definition according to ŠU SR, family accounts statistics.

ends definitively with the prospect of utility and therefore does not have a productive character. In the real economic system, however, this does not work and there are demonstrably many examples of a functioning productive aspect. Economics of productive consumption by R. Valencik [1] considers the household behavior and the growth of the present value of the expected future income from the creation and operation of the assets forming their "family portfolio" as a target orientation. Another prerequisite from which we abstain is the so-called exogenous character of utility in relation to the economic system, which means that the relationships and mechanisms covered by the classical microeconomic theories in which human subjects interact with each other have no role in creating their subjectively perceived benefit.

Traditional microeconomic theories follow the so-called modern theories of consumption, which emerged in response to the theory of consumption expenses and savings by J. M. Keynes [2]. In contrast to Keynes' theory of purchasing consumer goods, modern theories focus on explaining the use of consumer goods services, which is a step closer to the issue we are dealing with. They are therefore trying to capture the real flows of consumption of services provided by consumer goods purchased by households. This approach is called actual consumption. For example, the work of J. S. Duesenberry, I. Fisher, F. Modigliani, R. E. Brumberg, A. Anda, M. Friedman, T. Sargent, and many others [2] is a reference to modern consumer theories.

Especially interesting is the theory of consumption based on the hypothesis of a permanent pension developed by Milton Friedman. It was based on criticism of the simple consumer function of J.M. Keynes, and follows the work of Irwing Fisher, who in his consumption function considers the impact of wealth and interest rate on consumption. M. Friedman's specific contribution is that he transferred the Fisher idea into a consumption-permanent relationship and indicated how the entity anticipates future pensions that are the determinants of its wealth and permanent retirement. Consumption in this theory is understood as a function of permanent retirement. M. Friedman considers wealth and retirement as identical concepts. According to M. Friedman's more detailed specification of nominal wealth, its five components can be distinguished, namely money, bonds, shares, physical goods, and human capital. Here he points out and proves a very interesting idea, that whether or not the households save or consume, they create their own asset portfolio from the mentioned components of nominal wealth and act in such a way as to maximize the present value of the future income or, of the return on the portfolio [3].

Also G. Becker introduced a microeconomic analysis of consumption to examine human behavior, interpersonal relationships, and off-market relationships. He dealt with the link between quality of life and the economy. His work had a great theoretical benefit because he expanded the perception of consumption and contributed to the formation of contemporary theories. According to G. Becker: "The choice of consumption and other activities of an average person depends on his or her childhood, life experiences, social interactions and cultural influences" [4]. Which, in connection with the productive aspects of consumption, talks about possible motives for their full application in the lives of individuals or households from the view of the environment, where a certain amount of room for corrections is also created by the applied economic policy.

Classical, or traditional and new theories of consumption are the theoretical background of past consumption trends. Each of these consumption theories emerged in a particular environment and period.

This development has not occurred and is still continuing mainly thanks to Behavioral Economists. New views on consumption and the direction of its development are conditioned by the decision-making mechanism, which is the key and basis of understanding the formation of current consumption. This is ultimately visible also in the form of realized expenditures, whether individuals or the whole household. In addition to this basic mechanism, when looking at households and their behavior, it is necessary to perceive the situation in the income level, ie the existing stratification of society and the differences in the economic situation of individual types of households, which tells us much about their decision-making approach.

Behavioral Economics combines psychology with economics to provide answers to many challenging questions about economic decision-making in the current times. It deals with the circumstances of individuals' failure and the causes of crises. Behavioral economics justifies the existence of some group irrationality. As a result, for example, the emergence of bubbles in markets or crises. The founders of this new direction were D. Kahneman, R. Thaler and A. Tversky[†].

R. Thaler also analyzes how people can be encouraged to make better decisions. It outlines how governments can use behavioral models to increase the efficiency and quality of a wide range of public services. By pointing out some sort of possible positive interaction or correct motivation to change behavior even in our conditions. R. Thaler's research can also be applied to everyday human economic decisions as well as to the behavior of the whole market. In doing so, he incorporated psychologically realistic thinking into economic decision-making. His research into the effects of limited human rationality, social priorities and lack of self-control shows how these human characteristics systematically influence the economic decision-making of individuals.

Modern research [5-9] has also highlighted the importance of emotions that greatly influence a human's behavior. From the perspective of behavioral economy preferences, one prefers to choose an alternative that brings about a change of wealth. Productive components of consumption definitely bring this possibility of change of wealth, but it must be borne in mind that choosing the alternatives does not play the most important role of static wealth but its relative expression and relative change towards a reference point. The question, however, is whether actual households are really aware of this fact. Or is it only applicable to a tiny percentage of households?

The behavioral approach is based on the assumption that human behavior motivates two major systems. One is based on looking for a reward and the other on resistance to loss. The reward system creates values, priorities, and motivates individual behavior. Avoidance of risk is, on the contrary, a protective function aimed at the survival of an individual. So, here is the need to reflect on what rewards the productive components of consumption bring, is there a threat of any kind of loss? Or the question of how to protect the productive components of household consumption from risks? Questions of this type lead us to the motives that lead people to spend their money on any kind of expenditure.

According to D. Kahneman [10], the current view of consumer decision-making is focused on three main themes, namely heuristics of judgment, risky choices and framing. In all three areas thoughts, intuitions and preferences were studied, which one thinks quickly and without long thinking. There is a distinction between intuitive and deliberate thought processes. Together, D. Kahneman and A. Tversky argue that intuitive judgment belongs to the same place as deliberate thinking, but focuses on the existing discrepancies between intuition and knowledge especially in statistics. Are the productive components of household consumption an intuitive or deliberate thinking process? D. Kahnemann also deals with the formalization of intuition and the issue of change of wealth or function of values.

Another author of behavioral economics is D. Ariely. Through many experiments, he has come to the conclusion that "people deceive themselves by their own imagination as they

[†] At the time of its inception, this theory was still called "prospect theory". The breakthrough work is the well-known Prospect theory: An analysis of Decission under Risk. So it's a new theory. However, its principles have been proven and defended in the past.

act reasonably and deliberately, but are in fact impressionable and vulnerable. They are prone to make misconceptions about bad shopping and investment decisions. These decisions are not only irrational but even foreseeably irrational" [11]. According to him, in practical life, we compare things that lie side by side, things that are easy to compare. On the contrary, we avoid comparing things that need to be very thought out. In human decision-making, accidental addiction and the anchor of the past also play an important role. So, if we decide on the old anchors and memories, it can not be said that the trade will give us all the greatest personal happiness and benefit.

The analysis of D. Ariely, which focuses on the magic word "FREE", is a very exciting one, and we often make our decision and we are making a decision we did not want and which is not the best from the point of view of rational decision-making. Whatever is free, it seems to us a lot more valuable than what it really is. And why? Because human beings have instinctive fear of loss [12]. The success of this word is based on this fact and is often used, for example, in pharmacies for the sale of supplementary medications, and thus falls under household spending on health and productive components of consumption.

Another point of view on this issue is the theory of G. S. Becker. It uses the economic outlook on rationality in both non-economic and non-market areas. He applied the economic concept of rationality to all forms of human decision making. G. S. Becker, for example, considers habit to be rational if the increase in past consumption increases current consumption [13]. For reflection on the rationality of habitual behavior, the key point is the knowledge that the extent of addictive activities is relatively high in humans, so we can not exclude them from economic analyzes. Is it possible that just the habit of free public education within our economic system and family is just the necessary minimum spending that hampers the possible change from the point of view of their development and the growth of well-being in the long run? Or is there another cause?

2 Data and methods

The data base for this contribution consists mainly of outputs from the family accounts statistics where data for the Slovak Republic for 2011, 2012 and 2015 are obtained through a network of randomly selected households. The figures for 2013, 2014, 2016 and 2017 were simulated. Methodology in the Czech Republic is more unified by 2016 and the research is carried out on a set of 3000 households selected by deliberate quota selection, which limits the possibility of generalizing data to the whole population, but allows for a deeper analysis of the differences between household types, which I consider very positive for the purposes of this article. According to the data of the Czech Statistical Office, the methodology will change in 2017 (data for 2017 are not available). According to the Family Accounts Stats, the partial sourcing from the EU-SILC sample was selected by selecting from existing household samples and changing from quota to random selection.

Consumption is generally understood as the value of consumer goods and services purchased by households. The Family Account Statistics is covered through the amount and structure of net cash expenditures, depending on the status of household preference and the number of dependent children. The reason for choosing these household groups is to ensure that net cash expenditure is comparable to net cash income and at least partially capture the differences between households in both countries. Thanks to this link, we can see the economic situation of individual households and how they react to changing conditions.

The method of a scientific paper is the form or procedure by which the intended result is reached from a certain initial state. Therefore, the methods used to produce this contribution are chosen to lead to the achievement of the stated primary goal. The theoretical and practical definition and subsequent possible comparison of productive consumption components in Czech and Slovak households, as well as the availability of data and the

relatedness of the methodology in both countries, are a key issue for us in connection with data collection. To do this, it is necessary to choose the thematically appropriate groups of net cash expenditures, whose choice was in terms of content suitability, value in the consumer household basket and our discretion in relation to the importance of status within the category of productive components of expenditures. In order to best assess not only the economic situation of households in both countries but also their overall quality of life, we have used reporting according to the spheres of quality of life. This reporting is based on the fact that household consumption is one of the components of quality of life and is literally its material component. The following spheres of quality of life are known: [14]:
- The out-of-work sphere consists of expenditure aimed at satisfying basic material needs, therefore on food and non-alcoholic beverages, alcoholic beverages and tobacco, clothing and footwear, housing, water, gas and electricity, furniture and furnishings and health.
- The sphere of leisure time is understood as the part of the day when one acts up to his own discretion, choices and possibilities of activities that contribute to his relaxation and self-realization. It arises after deduction of work and out-of-work time, that means that the spheres of the way of life are both conditioned and limited by each other. The structure of expenditure consists of expenditure on recreation and culture, education, hotels, cafes and restaurants.
- The cross-sectional sphere of expenditures takes up activities that blend in with other spheres of life. We include travel, communications, and various services to modernize one's way of life. But we also include some of the complementary equipment and commodities of households that have the character of luxurious estates, and thus denote the modernization of the way of life, of the reached level of living standards.
- The fourth sphere of quality of life is the sphere of work, which is not embodied in expenditure.

From the development of the share of spheres of quality of life in households, the so-called consumption models are derived. According to I. Laluhu, there are three models of consumer behavior of our households, namely the economical, luxury and conservative models of consumption. The starting point of this breakdown is the amount of spending going towards the three spheres of quality of life [15]. Consumption and quality of life are also linked to prosperity [16]. According to S. Mc. Calla it is the starting point of linking the notion of quality of life with the terms "satisfaction of needs" and "happiness" [17].

In fulfilling the aim of this contribution, we use the method of economic analysis where we analyze the individual consumption theories and the productive components of consumption and analyze in detail how we examine these parts and select the key facts. At the same time, we use a comparison method and the critical analysis of related literature, which is chosen because of the necessity of comparison, such as a comparison of consumption theories, their essence and benefits. From this section, a number of questions emerged on the basis of which we have chosen to select the data base and to process it appropriately using mathematical methods and several other commonly used methods, such as the method of observation, comparison, synthesis, induction, deduction, or the overall historically-logical process of examining the monitored variables with a consideration to the possibilities of the future.

3 Results

Under the productive components of consumption, we mean those items of the consumer basket of households, which by their nature are investments in human and social capital and ultimately have a retroactive effect on the income level of household members. These are therefore components of consumption that are not necessary for survival, but their importance in the consumer basket is increasing only after reaching a certain standard of living for households. In the long run, their importance should increase in the interest of improving the quality of life and overall development. Typical examples are, for example, educating members of the household, maintaining good health and overall physical and psychological well-being, including spending on health, creating and maintaining social contacts, creating the conditions for the most effective use of leisure time, cultivating the image of the household, recreation [18].

3.1 The current economic situation and the quality of life of households in the Czech Republic and Slovakia

As the monitored countries were once a part of one state, they have many similar "character" features. But nowadays, there are many differences between them that we will focus on in terms of economic situation and quality of life. To begin with, it is essential to say that the monitored countries do not have the same currency. For this reason, for the sake of comparability, the following tables are given in percentages, allowing easier comparison between the two countries [19]. From Table 1, we see that the level of net monetary incomes, expenditures and savings in terms of the difference between earnings and expenditure (in % of income) is increasing in both countries, creating favorable conditions for increasing the share of productive components in household consumption. It is also clear from the table that households are not only earning income but also spending. The positive thing is that it grows more slowly. In view of the absolute height of the monitored quantities, the best position of the households of SZČO and the households of employees. In the worst economic situation are the households of others, and especially in Slovakia (eg. unemployed) [20].

Table 1. Overview of Percentage Changes in Monthly Net Cash Income, Expenses, and Savings in Czech and Slovak Households from the view of Household Preference

		2010		2012		2015		2016	
		SR	**ČR**	**SR**	**ČR**	**SR**	**ČR**	**SR**	**ČR**
Households of employees	Change in income			2.0%	2.5%	19.6%	5.0%	4.3%	5.5%
	Change in expenses			3.2%	1.3%	11.5%	1.9%	0.8%	0.4%
	Income savings	12.0%	14.2%	11.0%	15.2%	17.0%	17.8%	19.8%	21.8%
Households of self-employed persons	Change in income			0.8%	7.7%	16.3%	-0.7%	4.0%	-0.3%
	Change in expenses			1.6%	7.5%	18.4%	-6.7%	0.8%	0.3%
	Income savings	14.4%	3.9%	13.7%	4.1%	12.1%	9.9%	14.8%	9.4%
Households of retirees	Change in income			8.9%	7.9%	-3.8%	3.7%	2.8%	0.2%
	Change in expenses			5.5%	4.9%	-7.2%	2.2%	0.9%	0.3%
	Income savings	12.0%	3.5%	14.7%	6.2%	17.7%	7.4%	19.3%	7.3%
Households of others	Change in income			3.4%	5.6%	-21.7%	-1.1%	5.0%	0.0%
	Change in expenses			2.8%	3.4%	-15.4%	-0.5%	2.4%	0.0%
	Income savings	6.2%	4.1%	6.7%	6.1%	-0.7%	5.5%	1.8%	5.5%

Source: Author.

In Table 2, we can see that in Slovakia, households with two or more children have a smaller and even negative income-saving ratio, which speaks about their worse economic situation than in the Czech Republic with similarly structured households, although in 2017 there are noticeable tendencies for improvement. We also see clearly that incomes in the Czech Republic are growing more slowly than in Slovakia. [20] However, here we are going to think about the causes of income growth. According to the OECD, in 2015 we were among the most working-class countries in terms of hours worked. Slovakia was in eighth place with the number of hours worked with 1906 per year, and the Czech Republic was in 11th place with the number 1867. When we count this data into the working time fund of that year, we deduct the average number of holiday days we find that the average Slovak worked approximately daily an hour more and an average Czech for a quarter hour more. Currently, the situation is different, both countries have gone down in the rankings and tend to work less. From the point of view of psychological factors, this is warranted, as there is a high demand for workers on the market, employment is on the increase, and it would grow if the labor force were not available on the market, and there was no need for extreme effort or performance. At present, Slovakia moved to the 20th place with 1740 hours. The Czech Republic is higher, on the 17th of July with the number of 1770 hours. We have paused for one important reason and it is that if households work on average more, have less time to develop productive components of consumption, and therefore have less time for education, culture, relaxation. During this period of time, if they would like to change their spending on productive aspects of consumption, they would probably have to spend less time at work, which would be a loss of income for the first time (provided they are paid or otherwise rewarded for overtime) . This view is supported in part by Poštová Banka's analysis of 2014, when it examined the proportion of employees working at night, evening, Saturday and Sunday. At the top of the list were Slovaks, 16% of them working at night, 25.2% in the evening, 24.4% on Saturday and 19.2% on Sunday. In the Czech Republic, 4.4% of the population work in the night, 9.1% in the evening, 18.2% on Saturdays, and 12.2% on Sunday [21]. This fact affects not only the level of income through mandatory surcharges but also the level of quality of life in households which, in the case of time-consuming for the common time, does not, in effect, detract from the limitation of the application of the aforementioned costs.

Table 2. Overview of Percentage Changes in Monthly Net Cash Income, Expenses, and Savings in Czech and Slovak Households in terms of Number of Dependent Children in the Household

		2010		2012		2015		2016	
		SR	CR	SR	CR	SR	CR	SR	CR
Households without children	Change in income			7.4%	5.2%	14.0%	2.9%	3.8%	1.5%
	Change in expenses			6.0%	2.6%	6.9%	1.3%	0.9%	1.0%
	Income savings	14.5%	11.3%	15.6%	13.5%	20.9%	14.8%	23.1%	15.2%
Households with 1 child	Change in income			2.8%	0.4%	13.6%	6.3%	4.6%	3.8%
	Change in expenses			5.6%	-0.7%	2.9%	2.3%	0.8%	1.0%
	Income savings	11.0%	12.2%	8.6%	13.2%	17.2%	16.4%	20.2%	18.7%
Households with 2 children	Change in income			1.0%	6.4%	14.4%	1.9%	4.0%	3.9%
	Change in expenses			0.9%	3.3%	12.7%	-2.2%	0.7%	1.0%
	Income savings	9.0%	6.5%	9.1%	9.1%	10.4%	12.7%	13.3%	15.2%
Households with 3 + Children	Change in income			11.0%	3.2%	2.1%	-5.8%	4.4%	0.5%
	Change in expenses			10.0%	29.1%	18.8%	-23.1%	2.4%	1.0%
	Income savings	6.9%	16.2%	7.8%	-4.8%	-7.3%	14.3%	-5.3%	13.9%

Source: Author.

A very interesting and complementary look is provided by the spheres of quality of life, which are made up of individual categories of monitored net cash expenditure per household per person per year. The out-of-work sphere is defined in terms of content as a sphere designed to meet basic material needs, whose share in the growth of quality of life should logically decrease. It also includes health expenditure, but they are not normally monitored in the context of prevention. The sphere of leisure time is understood as the cut of the day when one devotes to himself, according to his own discretion, to the choice and possibilities of activities that contribute to his relaxation. This means that the individual spheres, both in the context of the sphere of work and the cross-sectional expenditure, are both conditional and limited. The expenditure structure is a leisure time sphere consisting of spending on recreation, culture, education, hotels, cafes and restaurants. It is largely made up of productive aspects of household consumption and is therefore key to us. The last sphere is the sphere of one-off expenditure, which includes activities and activities that go hand in hand with other spheres of life, such as transport, communications, various modernization services [12].

In Table 3 we see the development of spheres of quality of life. It is obvious that in Slovakia, by 2016, the quality of life has increased compared to 2010, for example by the fact that the non-work sphere spending has decreased by 5%, however, the leisure expenses also by 0.7%, indicating that the households besides improvement of their situation from the point of view of investing in their education, relaxation in the form of recreation or expansion of horizons thanks to culture, have lost their time and means in the direction of cross-section areas, which grew by 5.7%. A different situation is occuring in the Czech Republic, where households spent 2% more on the non-work sphere, but the leisure expenses could increase by 0.6% and thus have a 3% lead in 2016. This expenditure in the long run creates favorable conditions for the growth not only of their quality of life but also of prosperity [20].

Table 3. Overview of spheres of quality of life and comparison of achieved levels in both countries

Year / Country	**2010**		**2012**		**2015**		**2016**		**Change**	
Spheres of Quality of Life:	**SR**	**CR**	**SR**	**CR**	**SR**	**CR**	**SR**	**CR**	**SR**	**CR**
on out-of-work time	57.8%	51.6%	57.4%	51.5%	53.0%	52.6%	52.8%	53.7%	-5.0%	2.0%
on cross-section areas	29.6%	34.2%	29.8%	34.9%	35.5%	33.3%	35.3%	34.2%	5.7%	0.0%
on free time	12.5%	14.2%	12.7%	13.6%	11.6%	14.2%	11.8%	14.7%	-0.7%	0.6%

Source: Author.

3.2 Education expenses

The analysis of the educational structure of the population in relation to the income level revealed that the amount of income is determined by the degree of education, up to a few exceptions [22]. Education is therefore something essential for development and success in the knowledge society and every economy. It represents a kind of investment for every individual that can gradually increase not only income, but also the knowledge to make use of at home, in leisure time, whether to solve problems and to better decide or repeat mistakes. For example, by recognizing the role of habit, emotion, or risk attitude in our article, knowing that, if one wants to achieve a goal, he must create positive repetitive habits in terms of the continuous fulfillment of the chosen goal or by monitoring economic reports and the economic and political situation to minimize the risks associated with bad decisions, and also to not make decisions in an emotional distress or in anger. All this is related to education and making decisions in the direction of specific individuals.

Table 4 below shows the level of spending on education for individual types of Slovak and Czech households monitored according to the preference of the household. On average, they are around 0.3-0.6% of the total net cash expenditure of households, which is very low, although compared to 2010 in Slovakia, with the exception of the households of pensioners where their share has not changed. Regarding the expenditures on education, the households of self-employed persons stand out the most. The cause of ow spending on education may be busyness from the point of view of work responsibilities that go beyond normal working hours, it may also be insufficient motivation for self-improvenet, other spending priorities (eg, significant increase in spending on long-term consumption goods, other services) or the still existing misconception that one is able to finish one school and no further training is needed.

Table 4. Net cash expenditure for education by household preference (% of total net cash expenditure)

Year/Country	**2010**		**2012**		**2015**		**2016**		**2017**	**Change**	
Type of monitored household	**SR**	**CR**	**SR**	**CR**	**SR**	**CR**	**SR**	**CR**	**SR**	**SR**	**CR**
Households of employees	0.4	0.7	0.5	0.7	0.8	0.6	0.8	0.7	0.8	0.4	0.0
Households of self-employed persons	0.6	0.8	0.6	0.7	1.1	0.7	1.1	0.7	1.1	0.5	-0.1
Households of retirees	0.1	0.0	0.0	0.2	0.1	0.1	0.1	0.0	0.1	0.0	0.0
Households of others	0.2	0.5	0.2	0.3	0.4	0.6	0.4	0.6	0.4	0.2	0.1
Average value	0.3	0.5	0.3	0.5	0.6	0.5	0.6	0.5	0.6	0.3	0.0

Source: Author.

Table 5 focuses on the share of spending on education and its change in terms of the number of children in households. It is obvious that the number of children is increasing education expenditure, which is related to the education of individual children in families, not with the continuous education of parents, which we consider equally necessary. In the Czech Republic, these expenses in families with two or more children are growing faster, probably also due to their overall better economic situation (see Table 1-2).

Table 5. Net cash expenditure for education by number of children (% of total)

Year/Country	**2010**		**2012**		**2015**		**2016**		**2017**	**Change**	
Type of monitored household	**SR**	**CR**	**SR**	**CR**	**SR**	**CR**	**SR**	**CR**	**SR**	**SR**	**CR**
Households without children	0.1	0.3	0.1	0.2	0.3	0.2	0.3	0.1	0.3	0.3	0.1
Households with 1 child	0.5	0.7	0.6	0.9	1.1	0.7	1.1	0.8	1.1	1.1	0.8
Households with 2 children	0.7	1.1	0.7	1.0	1.0	1.1	1.0	1.1	1.0	1.0	1.1
Households with 3 + Children	0.6	1.3	0.9	1.0	1.4	1.8	1.5	1.7	1.5	1.5	1.7
Average value	0.5	0.9	0.6	0.8	1.0	0.9	1.0	0.9	1.0	1.0	0.9

Source: Author.

3.3 Healthcare expenses

In the introduction to this group of expenditure, it is necessary to say that the number of visits to outpatient clinics is increasing, as well as the number of hospitalized patients in neurology and psychiatrists, and the use of spa treatments stagnates and the number of reported occupational diseases is increasing. Also, the consumption of vegetables, partly fruits and the production of our own food, which we have available, especially in the sense of Slovakia [19], in terms of disease prevention is decreasing, which in many cases it is similar to the situation in the Czech Rebulic as well [20]. So, what form of prevention is being implemented and therefore has an impact on real savings in healthcare spending? An interesting fact is that in terms of consumption and medication spending, Slovakia and

Czech Republic have a leading position within the OECD countries. Why is this? Why does the spending of households of retirees or other households grow steadily? Is it caused by psychological or real causes in terms of growing morbidity and health problems?

Table 6. Overview of net cash expenditure on health by household preference (% of total)

Year/Country	**2010**		**2012**		**2015**		**2016**		**2017**	**Change**	
Type of monitored household	**SR**	**CR**	**SR**	**CR**	**SR**	**CR**	**SR**	**CR**	**SR**	**SR**	**CR**
Households of employees	2.5	1.9	2.6	2.0	2.4	2.0	2.5	2.1	2.5	0.0	0.2
Households of self-employed persons	2.2	1.9	2.2	2.0	2.2	2.0	2.2	2.1	2.3	0.1	0.2
Households of retirees	5.1	4.1	5.3	4.3	5.0	3.7	5.1	3.4	5.3	0.2	-0.8
Households of others	3.2	3.1	2.9	2.0	3.7	1.8	3.8	1.8	4.0	0.8	-1.3
Average value	3.3	2.8	3.3	2.6	3.3	2.4	3.4	2.3	3.5	0.3	-0.4

Source: Author.

The following Table 7 again shows the clear difference between households in the Czech Republic and Slovakia. From the point of view of tendencies, it speaks about the decrease in expenditures in Slovakia and in the Czech Republic and about their stagnation. At the same time, we can see a roughly half per cent permanent difference between the monitored countries. What causes it? For net healthcare expenditure, for example, we found that in the Czech Republic, spending on healthcare and pharmaceutical products and therapeutic devices account to around 71% of total health expenditure. Slovakia is up to 81% on average over the same period. What is the cause of their purchases? Is it marketing or an attempt to treat oneself with mistrust in public healthcare?

Table 7. Overview of Net Cash Expenditure for Healthcare by Number of Children (% of total)

Year/Country	**2010**		**2012**		**2015**		**2016**		**2017**	**Change**	
Type of monitored household	**SR**	**CR**	**SR**	**CR**	**SR**	**CR**	**SR**	**CR**	**SR**	**SR**	**ČR**
Households without children	3.9	3.0	3.8	3.1	3.5	2.8	3.5	2.7	3.6	-0.3	-0.3
Households with 1 child	2.4	1.9	2.6	1.9	2.4	1.9	2.4	2.3	2.5	0.1	0.4
Households with 2 children	2.4	1.7	2.5	1.8	2.2	1.8	2.2	1.8	2.2	-0.2	0.1
Households with 3 + Children	2.2	2.1	2.4	1.3	2.2	2.0	2.3	1.8	2.4	0.2	-0.3
Average value	2.7	2.2	2.8	2.0	2.6	2.1	2.6	2.1	2.7	-0.1	0.0

Source: Author.

3.4 Recreation and culture expenditure

Expenditure on recreation and culture is closely related to the time spent on individual spheres of quality of life. They cross each other. If a person spends more time in the workplace, and in the sphere of one-off or out-of-work expenses, the sphere of leisure time, in which these expenses in particular fall, is reduced. Here, the notion of motivation becomes the foremost, as the individual's attitude towards spheres is directly dependent on it. As we know, motivation expresses the process, while the motif expresses a hypothetical disposition to this process (for example, the need to relax, to enjoy a cultural experience or other similar primary or secondary motives) should focus on the motives and the process separately and answer the question why is there a difference between Czech and Slovak households in this spending group and which forces households to reduce or lower their share of expenditures (see Table 8-9).

Table 8. Overview of net cash expenditure on recreation and culture by household preference (% of total expenditure)

Year/Country	2010		2012		2015		2016		2017	Change	
Type of monitored household	SR	CR	SR	CR	SR	CR	SR	CR	SR	SR	CR
Households of employees	8.1	9.1	7.0	8.5	6.4	8.7	6.4	9.2	6.3	-1.8	0.1
Households of self-employed persons	7.4	9.9	8.2	9.2	6.9	9.3	6.9	9.3	6.8	-0.6	-0.6
Households of retirees	5.6	8.6	6.2	7.5	5.0	7.7	5.1	7.8	5.0	-0.6	-0.8
Households of others	5.0	7.4	5.7	6.6	3.5	6.5	3.5	6.5	3.6	-1.4	-0.9
Average value	6.5	8.7	6.8	7.9	5.5	8.1	5.5	8.2	5.4	-1.1	-0.6

Source: Author.

Table 9. Overview of net cash expenditure on recreation and culture by number of children (%)

Year/Country	2010		2012		2015		2016		2017	Change	
Type of monitored household	SR	CR	SR	CR	SR	CR	SR	CR	SR	SR	CR
Households without children	6.4	8.6	6.4	8.2	6.2	8.1	6.2	8.0	6.2	-0.2	-0.6
Households with 1 child	7.6	9.2	6.9	8.5	6.1	8.7	6.2	9.3	6.1	-1.5	0.1
Households with 2 children	8.7	9.7	7.4	8.9	6.4	9.3	6.5	9.7	6.4	-2.3	0.0
Households with 3 + Children	6.1	12.0	8.2	8.1	5.2	9.4	5.3	9.7	5.3	-0.8	-2.2
Average value	7.2	9.8	7.2	8.4	6.0	8.9	6.1	9.2	6.0	-1.2	-0.7

Source: Author.

4 Conclusion

At the moment, the probability of another recession is often mentioned, and from the point of view of household psychology it may be a kind of self-fulfilling prophecy, the more people start to expect it, the sooner it comes, and at least even by starting home in such an atmosphere to save more economically, slows down and reduces spending on key productive components of household consumption. Nowadays, although the willingness to shop is growing, it is already a paradoxical situation. This is the case if interest rates are lower than consumer inflation, which means that banks actually pay mortgage borrowers real interest for taking their debt, which leads to misleading household decisions and excessive debt, which is considered to be very the current problem affecting consumer decision-making [23-26].

However, the assumption of rationality or irrationality of decision-making and awareness of these risks is not a guarantee of stability under the current conditions. Household income may not evolve as expected by households. Family Account Statistics also neglects the qualitative aspect and the problem of the shorter lifecycle of products, which would have a direct impact on the level of consumption and quality of life of households in both countries. From the point of view of consumption and its productive components, it is necessary to monitor them in the context not only of economic, but also of psychological and social factors influencing households, which are embodied in other indicators. Therefore, it is necessary to be aware of the links between them and thus not to exclude them, for example, from the total time fund that households and individuals have available to the individual real spheres of quality of life. At the same time, it is clear from the contribution that Czech and Slovak households apply a different approach to expenditures on productive components of consumption and thus they get a different trend of development and point to different household conditions. This is mainly the trend of household expenditure growth, which is generally faster in Slovakia than in the Czech Republic, which is also confirmed by their increasing indebtedness. Expenditure on productive components of consumption points to the existing differences between countries and their households and the different trends in their development as a result of several factors discussed in the article. These, in the long run, create qualitative differences.

With regard to the evolving Behavioral Economics and in connection with the issues examined, we conclude that it would be beneficial to include psychological factors and decision-making motives in Family Accounts Statistics, which would allow for a more thorough overall analysis, a better understanding of the behavior of households in current conditions and last but not least advocate a better prognosis of household consumption as one of the key factors of GDP.

5 Recommendations and Suggestions

The conclusions of the paper point out that household decision-making is largely influenced by the economic situation of households but also by other factors such as lack of time to implement them or the fact that at present the probability of another recession that would reduce spending on key components of household consumption and has slowed down their implementation in the coming years. Another very important fact is that Europe is at the start of the new industrial revolution. This is considered to be the fourth jump forward of that kind, which has therefore been given the label "Industry 4.0." The basis of this term is mainly the change in so-called socio-economic behavior of people and human society as such, which also directly affects households through different channels of influence as it is a matter of change. There are a number of areas ranging from industry, technical standardization, security, education, legal framework, science and research to the labor market and the social system. These channels in the context of households create not only an environment full of risks but also opportunities to handle these challenges. For positive incentives, the possibilities of development from the point of view of the individual and the household can be considered as productive components of consumption, which create space to adapt to incoming changes.

However, in terms of economic policy recommendations in both countries and individuals, greater emphasis should be put on better awareness of these changes, as well as the qualitative side of the manufactured and purchased consumer basket components, and taking into account the long-term horizon when deciding on upcoming changes and in doing so, to implement the applied economic policy. It is also necessary to focus on consumption research in terms of monitoring and explaining the implications of emerging changes, together with targeted and meaningful redirection of money to more perspective spheres enhancing the quality of life of households. Very interesting and often inflexible is the term "libertarian paternalism[‡]" which can be used not only in the sense of better information, but also in the sense of targeted measures by governments in order to guide people's decision in a desirable direction, without limiting them in any way [27-29]. From our point of view, it could be a reform of education that would include content with the appropriate importance and depth of subjects aimed at increased awareness of health and healthy lifestyles, financial literacy or targeted support of critical and logical thinking and

[‡] We perceive the concept as trying to help people achieve their own goals, in other words, it is a sort of "poke" towards better decisions with regard to the situation we are in.

psychology of decision-making in the long term to improve decision-making in a number of problematic areas (health, education, consumption, marketing, etc.).

References

1. R. Valenčík, Ekonomie - Jak ji chápat a k cemu je? (Doplnující text ke studiu mikroekonomie na magisterském stupni). *Marathon*, (53), (2004)
2. D. Rozborilová, *Teórie spotreby, úspor, investícií vládnych výdavkov 8.* Lura Edition, (2005)
3. M. Friedman, *Theory of the Consumption Function.* Princeton University Press, (1957)
4. G. S. Becker, *Teorie preferencií.* Liberální institut, (1997)
5. D. Ariely, *Jak drahá je intuice.* Práh, (2011)
6. E. Andrade, D. Ariely, The Enduring Impact of Transient Emotions on Decision Making. *Organizational Behavior and Human Decision Processes*, **109**(1), 1-8, (2009)
7. P. Singer, Famina, Affluence, and Morality. *Philosophy and Public Affairs*, **1**(1), 229-243, (1972)
8. D. Small, G. Loewenstein, The Devil You Know: The Effects of Identifiability on Punishment. *Journal of Behavioral Decision Making,* **18**(5), 311-318, (2005)
9. D. Small, G. Loewenstein, Helping a Victim or Helping the Victim: Altruism and Identifiability. *Journal of Risk and Uncertainty,* **26**(1), 5-13, (2003)
10. D. Kahneman, *V jeho prednáške z udeľovania Nobelovej ceny za ekonómiu, z 8.decembra 2002, ktorej názov je „Mapy ohraničenej racionality: Pohľad na intuitívny úsudok a voľbu",* (2002)
11. D. Ariely, *Jak drahé je zdarma.* Práh, (2009)
12. M. Psárska, *Mikroekonomické aspekty spotreby domácností SR v nových ekonomických podmienkach.* Dissertation thesis, (2013)
13. G. S. Becker, K. M. Murphy, The Theory of Rational Addiction. *Journal of Political Economy,* **96**(4), 675-700, (1986)
14. M. Psárska, Nerovnosti v kvalite života domácností Slovenskej republiky z hľadiska ich príjmov a jednotlivých sfér kvality života. *Sociálny rozvoj a kvalita života v kontexte makroekonomickej nerovnováhy*, Velký Meder, Slovakia, 571-587, (2012)
15. I. Laluha, Kvalita života a dynamika socialno-ekonomickeho pohybu. *Socialna politika,* (2008)
16. M. Bunge, What is Quality of Life indicator. *Social Indicators Research*, **1**(2), (1975)
17. S. McCall, *Quality of Life Social Indicator Research*, **2**(2), (1975)
18. M. Psárska, Productive components of household consumption and their current development in the context of well-being / VEGA 1/0570/11. *QUAERE 2012 Reviewed Proceedings of the Interdisciplinary International Scientific Conference of PhD Students,* Hradec Králové, Czech Republic, 359-369, (2012)
19. *Štatistická ročenka Slovenskej republiky 2016.* Veda, (2017)
20. *Zdroj údajov; ŠÚ SR, ČŠÚ,* (2018)
21. *The Post Bank* [online], Available at: https://openiazoch.zoznam.sk/cl/160648/Slovaci-opat-na-cele-rebricka-Stvrtina-pravidelne-pracuje-vecer-a-kazdy-siesty-chodi-do-prace-v-noci (2018)

22. E. Čitáryová, T. Chrenko, Analysis of the educational structure of Slovak population in terms of income levels. *Medzinárodný seminár: Kvantitatívne metódy v ekonómií*, Nitra, Slovakia, (2011)
23. M. Lehuta, Kedy a odkiaľ príde ďalšia kríza. *TREND, Týždenník o ekonomike a podnikaní*, 30-31 a 46-47, (2018)
24. A. Strachotová, Zadlženosť domácností na Slovensku. *BIATEC, Odborný bankový časopis*, **18**, (2010)
25. P. Staněk, *Globálna kríza- hrozba alebo výzva?* Sprint dva, (2010)
26. P. Staněk, *Cena prežitia* [online], Available at: http://www.cez-okno.net/clanok/hospodarstvo/peter-stanek-cena-prezitia
27. R. H. Thaler, *Neočekávané chování, Příběh behaviorální ekonomie*. Argo, Dokořán, (2015)
28. A. Belayanin, Richard Thaler and behavioral economics: From the lab experiments to the practice of nudging. *Nobel Memorial Prize in Economic Sciences 2017*, **1**, 5-25, (2018).
29. R. H. Thaler, Nudge, not sludge. *Science*, **361**(6401), 431-431, (2018).

Analysis of Consumption on the Goods and Services Market in the Cybernetic Model of Selected Country

Pavel Rousek[1,*]

[1]Institute of Technology and Business in České Budějovice, School of Expertness and Valuation, Okružní 517/10, 37001 České Budějovice, Czechia

Abstract. The paper analyses the consumption on the goods and services market in the cybernetic model of Czechia. The theoretical basis is given by the consumption function in the short-term, the macroeconomic multiplier of the two-sector economy and the cybernetic model with the goods and services market. All the above-mentioned theory is applied to the real conditions of the selected country, which is the Czech Republic.
Key words: cybernetic model, goods and services market, consumption, macroeconomic multiplier

1 Introduction

A real economy is a complex system that can be simplified into an economic model with inputs, outputs, and links inside. Exogenous inputs are most often considered as autonomous components of consumption, investment, government purchases, exports and imports. The key output is then the amount of product.

One of the most important input and output parts is consumption. There is the input autonomous consumption and the output total consumption as part of the gross domestic product. This article analyses autonomous and total consumption in a two-sector economy.

2 Data and Methods

2.1 Consumption in the Short-Term

The theory of consumption function is very well explored in the professional literature of the 20th and 21st centuries [1-3].

Consumption function in the short-term is based on Keynesian theory. It is a relation between consumption and macroeconomic product. Mathematically expressed, these two variables are dependent and independent variable. The equation of the function is as follows in the formula (1).

[*] Corresponding author: rousek@mail.vstecb.cz

$$C = C_0 + c \times Y \qquad (1)$$

Where:
C is the total consumption,
C_0 autonomous consumption,
$c \times Y$ induced consumption,
c marginal propensity to consume,
Y product.

2.2 Macroeconomic Multiplier of the Two-Sector Economy

The so-called Keynesian multiplication process uses the property of consumption function and especially of the post-induced consumption through the marginal propensity to consume. "The Keynesian multiplier process is the economist's paradigmatic positive feedback loop, in which an initial departure from full-employment equilibrium cumulates instead of being corrected" [4].

The principle of multiplication is infinite and weakening repetition of the original impulse, which is the change in autonomous expenditures. Mathematical evaluation of the strength of the effect is given by the formula (2).

$$\Delta Y = \Delta(C_0 + I_0 + G_0 + NX) \times \frac{1}{1-c} \qquad (2)$$

Where:
Y is the product (GDP),
C_0 autonomous consumption,
I_0 autonomous investments,
G_0 autonomous government purchases,
NX net export,
c marginal propensity to consume.

2.3 Cybernetic Model with the Goods and Services Market

"Cybernetics allows expression of the dependence of output change ΔY on exogenous input changes Δ(C0+I0+G0+NX)" [5]. Cybernetic model with the goods and services market has following parameters:

- Inputs
  - Autonomous consumption C_0
  - Other autonomous expenditures $I_0 + G_0 + NX$
- Internal feedback
  - Marginal propensity to consume c
- Output
  - Product Y

A constant price level is assumed, which can be achieved by calculations with input parameters at constant prices. The graphical interpretation of the above is shown in Fig. 1.

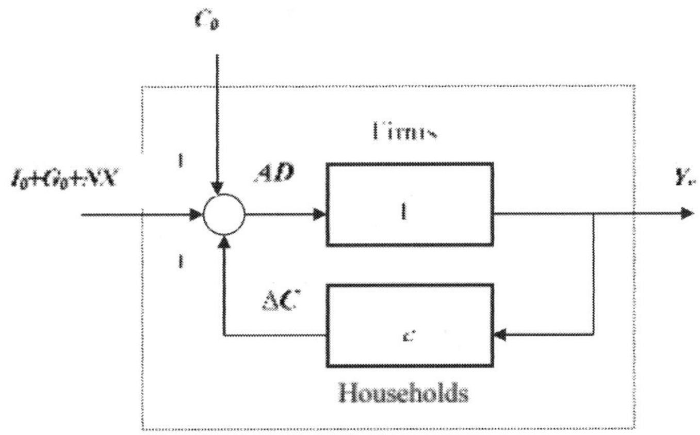

Fig. 1. Cybernetic Model of a Static Economy with the Goods and Services Market

Source: [6].

3 Results and Discussion

3.1 Consumption in Selected Country

Under the conditions of the Czech Republic, objective time series can begin in 1993 when Czechia and Slovakia became independent. In this case, the relevant data can be traced back to 1996 as shown in Table 1.

Table 1. Consumption and product data of the Czech Republic (in CZK million, constant prices) [7].

	Product	Consumption
2017	4612481	3052490
2016	4418198	2955606
2015	4306516	2865773
2014	4089400	2777475
2013	3981303	2734118
2012	4000653	2704547
2011	4032910	2743834
2010	3962464	2765219
2009	3874383	2741341
2008	4069840	2731172
2007	3963527	2667991
2006	3753246	2587425
2005	3512515	2516852
2004	3297100	2453097
2003	3142892	2408273
2002	3033592	2290145
2001	2984277	2193217
2000	2899925	2164111
1999	2781256	2113471
1998	2741968	2050586
1997	2751011	2074671
1996	2767468	2014077

Source: [7].

By fitting the regression line with these points, an accurate consumption function with $R^2 = 0.9781$ is obtained. The parameters of this consumption function are $c = 0.5137$ and $C_0 = 683545$ (Fig. 2). The complete consumption function is given in equation (3).

$$C = 683545 + 0.5137 \times Y \qquad (3)$$

Where:
C is the total consumption,
C_0 autonomous consumption,
$c \times Y$ induced consumption,
c marginal propensity to consume,
Y product.

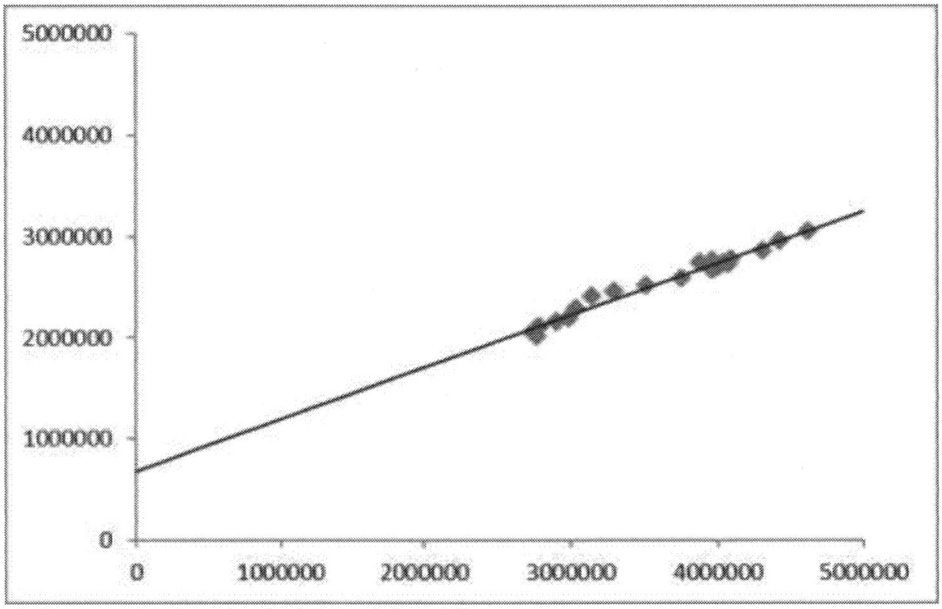

Fig. 2. Consumption Function in the Czech Republic.

Source: Author.

3.2 Macroeconomic Multipliers in Selected Country

The Keynesian multiplier is naturally based on the values of the consumption function, autonomous consumption and the marginal propensity to consume. For the Czech Republic, the calculation method (4) and the value of the multiplier (5) apply.

$$\Delta Y = \Delta(683545 + I_0 + G_0 + NX) \times \frac{1}{1-0.5137} \qquad (4)$$

$$\Delta Y = \Delta(683545 + I_0 + G_0 + NX) \times 2{,}0563 \qquad (5)$$

Where:
Y is the product (GDP),
C_0 autonomous consumption,
I_0 autonomous investments,
G_0 autonomous government purchases,

NX net export,
c marginal propensity to consume.

3.3 Cybernetic Model of Selected Country

The cybernetic model of the Czech Republic is supplemented by specific values of autonomous consumption $C_0 = 683545$, and marginal propensity to consume $c = 0.5137$. This means that the change in any autonomous expenditures (autonomous investments, autonomous government spending, net exports) will be 2.0563 times in the economy.

Fig. 3 shows an actual cybernetic model of the Czech Republic with the goods and services market.

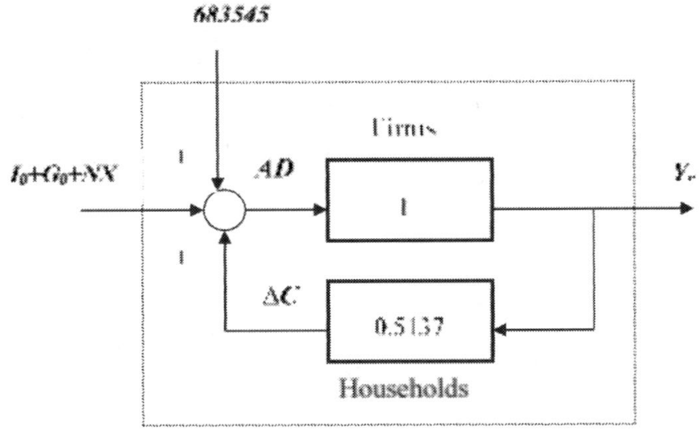

Fig. 3. Cybernetic Model of the Czech Republic with the Goods and Services Market

Source: Author.

3.4 Complex Dataset of Selected Country

Czechia can be described from the point of view of consumption, Keynesian multiplier and cyber model by the parameters in Table 2.

Individual quantities were determined as follows:

- Y – real data,
- C – real data,
- $C_0 + I_0 + G_0 + NX$ as the total input of cybernetic model,
- C_0 as one parameter of the consumption function,
- $I_0 + G_0 + NX$ as the difference between cybernetic model and consumption function data.

Table 2. Consumption and product data of the Czech Republic (in CZK million, constant prices).

	Product Y	Total Consumption C	Total Autonomous Expenditures $C_0+I_0+G_0+NX$	Autonomous Consumption C_0	Non-consumption Autonomous Expenditures I_0+G_0+NX
2017	4612481	3052490	2243050	683545	1559505
2016	4418198	2955606	2148570	683545	1465025
2015	4306516	2865773	2094259	683545	1410714
2014	4089400	2777475	1988675	683545	1305130
2013	3981303	2734118	1936108	683545	1252563
2012	4000653	2704547	1945518	683545	1261973
2011	4032910	2743834	1961204	683545	1277659
2010	3962464	2765219	1926946	683545	1243401
2009	3874383	2741341	1884112	683545	1200567
2008	4069840	2731172	1979163	683545	1295618
2007	3963527	2667991	1927463	683545	1243918
2006	3753246	2587425	1825204	683545	1141659
2005	3512515	2516852	1708136	683545	1024591
2004	3297100	2453097	1603380	683545	919835
2003	3142892	2408273	1528388	683545	844843
2002	3033592	2290145	1475236	683545	791691
2001	2984277	2193217	1451254	683545	767709
2000	2899925	2164111	1410234	683545	726689
1999	2781256	2113471	1352525	683545	668980
1998	2741968	2050586	1333419	683545	649874
1997	2751011	2074671	1337817	683545	654272
1996	2767468	2014077	1345820	683545	662275

Source: Author.

4 Conclusion

Consumption in the short-term, macroeconomic multiplier of the two-sector economy, and cybernetic model with the goods and services market have been used to define conclusions of this paper.

GDP increased by 67%, consumption increased by 52%, total autonomous expenditures increased by 67%, autonomous consumption did not increase or decrease, and non-consumption autonomous expenditures increased by 135%. This has happened over the past two decades.

References

1. J. M. Keynes, *The General Theory of Employment Interest and Money*. London: Palgrave Macmillan, (1936)
2. J. M. Keynes, The 'Ex ante' theory of the rate of interest. *Economic Journal*, **47**(188), 663-669, (1937)
3. M. Friedman, *A Theory of the Consumption Function*. Princeton: Princeton University Press, (1957)
4. P. Howitt, The microfoundations of the Kenyesian multiplier process. *Journal of Economic Interaction and Coordination*, **1**(1), 33-44, (2006)
5. P. Rousek, In printing, (2018)

6. M. Máče, P. Rousek. *Makroekonomie pro technické školy*. Prague: Grada, (2013)
7. *Czech National Bank*. Available at: https://www.cnb.cz/cnb/STAT.ARADY_PKG.PARAMETRY_SESTAVY?p_sestuid=28788&p_strid=ACCAAA&p_lang=CS (2018)

Environmental Footprint Evaluation of Eco-Friendly Products in the Global Economy

Pavel Rousek[1,*]

[1]Institute of Technology and Business in České Budějovice, School of Expertness and Valuation, Okružní 517/10, 370 01 České Budějovice, Czechia

Abstract. Ecological products, their terminology, and the process of certification are not unified within the global economy. This paper gives a comprehensive overview of organic products labelling. The novelty of the contribution is the effort to introduce the evaluation of the environmental footprint of eco-friendly products.
Key words: eco-friendly products, environment, evaluation, environmental footprint

1 Introduction

Environmentally friendly, environment-friendly, eco-friendly, nature-friendly, or green products have many different names and may be subject to many different certification processes to verify their environmental footprint.

Food, agricultural products, clothing, cars and other consumer products have a number of eco-labels or green stickers. Some supranational organizations are trying to unify it, but there is no global consensus.

Most of these labels use only a binomial scale (product meets or does not meet the conditions) or a discrete scale (products are grouped into several groups). A continuous scale (the product is rated by the value of an environmental footprint) is not widely accepted for the time being. The paper proposes just such a continuous scale, the units of which would be money.

2 Current State of Eco-Friendly Products Labelling

In the past, many non-governmental organizations (NGOs) have been involved in the labelling of organic products. The attempt is now for global harmonisation, unification, and standardization of these efforts. An example of such standardization is the International Organization for Standardization ISO and the European Union of the EU.

Standards for labelling procedures by the International Organization for Standardization ISO in the framework of ISO 14000 Environmental management [1-3]:

- Type I Environmental Labelling (ISO 14024:2018) [1];
- Type II Self-Declared Environmental Claims (ISO 14021:2016) [2];

* Corresponding author: rousek@mail.vstecb.cz

- Type III Environmental Declarations (ISO 14025:2006) [3].

The European Union is particularly involved in energy labels. The energy consumption labelling system is regulated by EU Council Directive 92/75/EEC [4], Directive 2010/30/EU of the European Parliament and of the Council [5], and Regulation (EU) 2017/1369 of the European Parliament and of the Council [6]. This applies to the following groups of appliances:

- Refrigerating appliances;
- Washing machines;
- Dishwashers;
- Ovens;
- Air conditioning;
- Bulbs;
- Television;
- Cars;
- Tires.

2.1 Organic Label

The organic label is assigned to food, clothing or wine that has been produced by organic farming methods and is thus certified according to organic farmer standards. [7] The organic label has a number of regional variants.

2.1.1 Eco Label and Bio Label

The sign of the inconsistency of labelling is the fact that in some countries "eco" or "bio" abbreviations have been used to designate organic food:

- Spanish "Eco";
- French "Bio";
- German "Bio" or "Öko".

2.2 Natural Label

The organic designation is natural. Here, in many countries, there is a lower level of regulation. Again, this is a marking that is or is not assigned. This marking does not bring any deeper information.

2.3 Energy Label

In terms of energy consumption labelling, the European Union has a very sophisticated system that puts appliances in categories from A to G. The most inefficient category G is not used in many products, yet the most efficient category A is divided into subgroups A, A^{+}, A^{++}, A^{+++}.

3 Environmental Footprint Evaluation

The idea of assessing the environmental impact is based on the fact that the impact of production, consumption and disposal of the product (or service) on the environment would be assessed. Unit outputs of this expertise would be the currency unit (Euro, US Dollar, etc.) Natural resources of the Earth are limited and production and consumption must take this into account [8].

3.1 Environmental Footprint of Production

The environmental footprint of production represents the positive or negative externalities of the production process. Positive externalities in production are external benefits; negative externalities in production are external costs. Both examples affect the offer of the product or service.

3.1.1 Positive Externalities in Production

When positive externalities are produced, private marginal cost (private supply) is higher than the social marginal cost (social supply). The consequence is that the actual market quantity is lower than the socially optimal quantity. Production reductions are accompanied by increased sales.

Benchmarking positive production externalities is part of a positive environmental footprint of eco-friendly products.

3.1.2 Negative Externalities in Production

This case is characterized by the fact that the private marginal cost (private supply) is lower than the social marginal cost (social supply). The consequence is that the market quantity is higher than the socially optimal quantity. The increase in production is accompanied by a sale at reduced prices.

For eco-friendly products, it is not possible to predict a greater expansion of negative externalities in production, but this case cannot be totally excluded.

3.2 Environmental Footprint of Consumption

In analogy with the above, the environmental footprint of consumption represents the positive or negative externalities of the consumption phase of product life. Positive externality in consumption is external benefits; negative externalities in production are external costs. Both examples affect the demand for a given product or service.

3.2.1 Positive Externalities in Consumption

If positive externalities are used in consumption, private marginal benefit (private demand) is lower than the social marginal benefit (social demand). The consequence is that the actual market quantity is lower than the socially optimal quantity. Reduction of production is accompanied by sales at reduced prices.

Positive consumption externalities are part of a positive environmental footprint of eco-friendly products.

3.2.2 Negative Externalities in Consumption

This case is characterized by the fact that private marginal benefit (private demand) is higher than the social marginal benefit (social demand). The consequence is that the market quantity is higher than the socially optimal quantity. Increases in production are accompanied by increased sales.

In the case of organic products, it is not possible to predict the widening of negative externalities in consumption, but this case cannot be entirely ruled out.

3.3 Environmental Footprint of Disposal

To the above, we also have to add the ecological impacts of the possible ecological disposal of products that are at the end of their life.

4 Proposed System of Eco-Friendly Products Labelling

The proposed system is based on the current Ecolabel and Energy Sharing Scheme, each product being characterized by the value of an environmental footprint expressed in money. The customer can compare the financial and environmental demands of the purchased products.

The new label might contain the following information:

- Product identification;
- Energy and environmental footprint of production;
- Energy and environmental footprint of consumption;
- Energy and environmental footprint of disposal;
- Total environmental footprint value.

5 Conclusion

Eco-friendly products are subject to many different certification processes with many different eco-labels or green stickers. There is no global unity. Attention is paid especially on agricultural products (such as food and clothing) and consumer products (such as cars).

There are different rating-approaches 1) Binomial Scale, 2) Discrete Scale, or 3) Continuous Scale. This paper suggests the use of a continuous scale with money units.

References

1. International Organization for Standardization. *ISO 14024:2018.*
2. International Organization for Standardization. *ISO 14021:2016.*
3. International Organization for Standardization. *ISO 14025:2006.*
4. *Council Directive 92/75/EEC.*
5. *Directive 2010/30/EU of the European Parliament and of the Council.*
6. *Regulation (EU) 2017/1369 of the European Parliament and of the Council.*
7. D. Conner, R. Christy, The organic label: How to reconcile its meaning with consumer preferences. *Journal of Food Distribution Research,* **35**(1), 40-43, (2004)
8. A. Y. Hoekstra, T O. Wiedmann, Humanity's unsustainable environmental footprint. *Science,* **344**(6188), 1114-1117, (2014)

Comparison of neural networks and regression time series in estimating the Czech Republic and China trade balance

Zuzana Rowland[1,*] *Petr* Šuleř[1] and *Marek* Vochozka[1]

[1]Institute of Technology and Business in České Budějovice, School of Expertness and Valuation, Okružní 517/10, 37001 České Budějovice, Czech Republic

Abstract. Foreign trade has been and is considered to be very important. Trade balance measurement provides one of the best analyzes of a country's external economic relations. It serves as a monetary expression of economic transactions between a certain country and its foreign partners over a certain period. The aim of this paper is to compare the accuracy of time series alignment by means of regression analysis and neural networks on the example of the trade balance of the Czech Republic and the People's Republic of China. Trade balance data between the Czech Republic and the People's Republic of China is used. This is a monthly balance starting in 2000 and ending in July 2018. First, a linear regression is made followed by regression using artificial neural networks. A comparison of both methods at expert level and experience of the evaluator, the economist, is performed. Optically, the LOWESS curve appears to be best out of the linear regression and the fifth preserved RBF 1-24-1 network seems the mot appropriate out of neural networks.
Key words: neural networks, regression analysis, comparison, trade balance, prediction

1 Introduction

Foreign trades have always been considered to be of the utmost importance and have always been given great attention. The balance of payments is recognized as an instrument that provides one of the best analyzes of a country's external economic relations and serves as a monetary expression of economic transactions between a country and its foreign partners over a certain period [1].

We can define the balance of payments as a systematic statistical record that captures all economic transactions between domestic and foreign currencies for a certain period of time. It includes transactions related to the movement of goods, services and earnings, operations with financial receivables and payables and operations referred to as transfers [2]. The balance of payments consists of two basic accounts, a current account and a capital account [3]. The current account of the balance of payments reflects the movement of real resources (eg goods, services, transfers, etc.), the capital account mainly depicts the international

* Corresponding author: rowland@mail.vstecb.cz

movement of capital - direct and portfolio investment, short-term capital, etc. [4]. The difference between a current account movement and a movement on a capital account is reflected in the change in foreign currency reserves. If the current account liability is not covered by the asset in the capital account, the central bank's foreign currency reserves will be reduced and vice versa [5].

Regarding the development of the balance of payments of the Czech Republic and China, the Czech Republic has several advantages from the commercial point of view of China. Thanks to its geographical position, the Czech Republic is very close to the Western markets, especially to neighboring Germany, the country is politically stable and is actively engaged in business transactions within the European Union, which gives the Chinese market additional opportunities [6]. The fragmentation of world production and the higher foreign value of Czech exports allow Chinese exports with added value to reach western markets through the Czech Republic [7].

China is the 4th largest trading partner for the Czech Republic and the second largest importer (since 2006). For the Czech Republic, China is the 18th largest export market. Czech exports to China have a growing trend over the long term. The largest increases were recorded in 2007 (by almost 60%) and in 2010 (53%). The increase was also high in 2011, when it was more than 30%. In 2012 and 2013 growth stood at 8-11%. Exports have tripled since 2007. In the past five years, they have grown by about 70% [8]. In 2017, exports from the Czech Republic to China reached 56 billion CZK, which represents a yearly increase of 20% and historically record exports from the Czech Republic to China. China is one of the largest but also the most demanding markets in the world [9]. Business is characterized by innumerable barriers, hard competition, and unusual environments for Europeans. Nevertheless, Czech exporters succeed in promoting themselves in the Chinese market [10]. In 2014, Czech exports to China amounted to 42.5 billion CZK. China imports to the Czech Republic in 2017 amounted to 473 billion CZK (y/y + 10%). The main position in Czech exports to China has long been occupied by machines and transport vehicles, which make up more than half of direct exports [9]. Evidence of a large increase in profits in the automotive industry is the latest investment of the Chinese automotive company Yanfeng in Planá nad Lužnicí, which has released 1.85 billion CZK for car production. The project is designed to create almost 600 jobs [11]. The software industry is also considered by the Chinese government to be necessary for economic progress and profit [12]. However, we must not forget the biggest problem of Czech-Chinese trade relations, which despite the economic progress remains at high deficit on the Czech side. According to the statistics mentioned above, the Czech Republic's deficit in 2017 amounted to International trade is an important factor in economic growth. In order for the Czech Republic to maintain its foreign relations with China, it is essential that it continue to invest in science and research. [13]. Sustainability of trade with China is mainly dependent on comparative advantages. The most cost-effective comparative advantage of China is the cheap cost factors in which we can include cheap labor in large quantities [14]. Thanks to the political transformation, the Czech Republic has incorporated the qualitative comparative advantage we can see in the industrial sectors [15]. The Czech Republic should invest in education, which will allow employees with knowledge of the Chinese market to be recruited. It should also look for opportunities in the development trends of China's economy and culture, and fight against the threatening factors of low market flexibility and high spending on research and development. The development of trade and economic cooperation between the Czech Republic and China depends, in particular, on the technical progress and innovation that the Czech Republic must continue to develop.

2 Data and methods

Data for analysis is available on the website of the World Bank etc. The data on the trade balance between the Czech Republic and the People's Republic of China will be used for the analysis. This will be the difference between total exports and imports between the two countries from the Czech point of view. The time period for which the data will be available is the monthly balance starting from January 2000 and ending in July 2018. This is about 223 input data. The unit is the Euro.

The descriptive characteristics of the data are given in Table 1.

Table 1. Characteristics of the data file

Samples	Month (Input variable)	Balance (Output (goal))
Minimum (Training)	36526.00	-1.050893E+09
Maximum (Training)	43252.00	-4.054250E+07
Average (Training)	39916.52	-3.661027E+08
Standard deviation (Training)	1949.82	2.289766E+08
Minimum (Testing)	36586.00	-9.687806E+08
Maximum (Testing)	43282.00	-5.088367E+07
Average (Testing)	39702.30	-3.679126E+08
Standard deviation (Testing)	2174.13	2.719910E+08
Minimum (Validation)	36951.00	-1.168102E+09
Maximum (Validation)	43040.00	-6.679944E+07
Average (Validation)	40047.88	-3.945945E+08
Standard deviation (Validation)	3096.16	3.098454E+08
Minimum (Overall)	36526.00	-1.168102E+09
Maximum (Overall)	43282.00	-4.054250E+07
Average (Overall)	39904.26	-3.705868E+08
Standard deviation (Overall)	1963.77	2.400326E+08

Source: Authors.

Interesting, of course, is price development over time. Figure 1 therefore shows the selected statistical characteristics in graphical form, including the histogram of the input data.

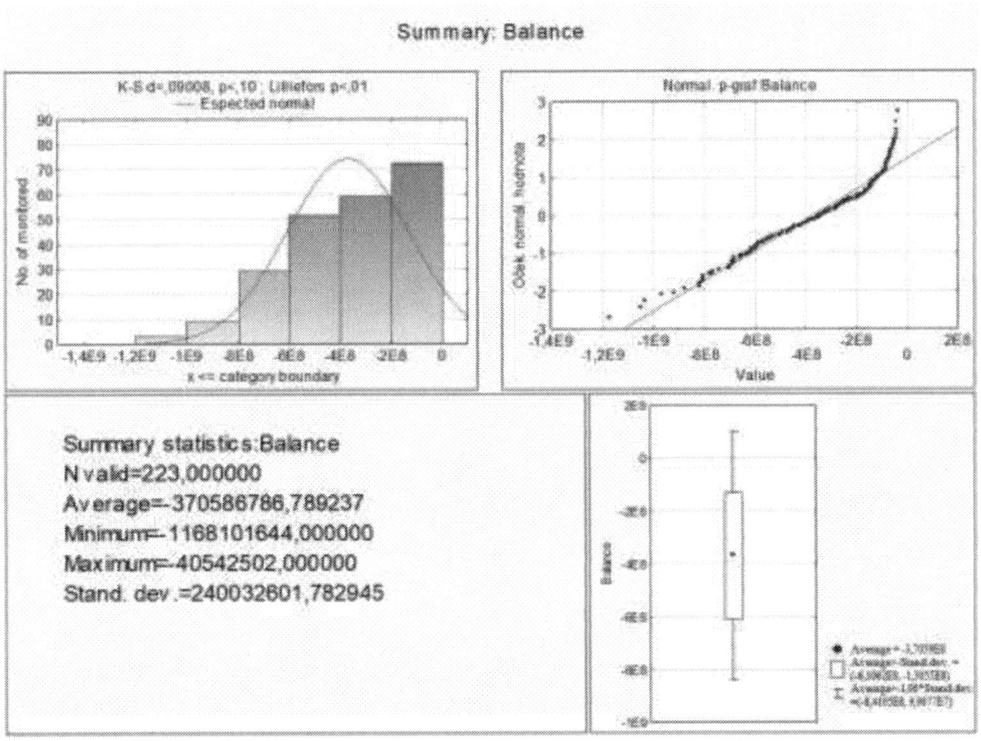

Fig. 1. Graph of basic statistical characteristics
Source: Authors.

For data processing, DELL's Statistica version 12 will be used. First, a linear regression will be performed. Afterwards, neural networks will be used for regression.

Linear regression will be performed on the monitored data sample for the following functions:

- Linear,
- Polynomial,
- Logarithmic,
- Exponential,
- Multiparameter of weighed distances,
- Multiparameter negative-exponential smoothing.

First, the correlation coefficient, ie the dependence of trade balance on time, will be calculated. Next, we will work with a level of significance of 0.95.

Then regression will be performed using neural structures. We will generate multilayer perceptron networks and neural networks of basic radial functions. Time will be the independent variable. We will determine the trade balance between the Czech Republic and the PRC as the dependent variable. We divide the time series into three sets – training, testing, and validation. The first group will have 70% of input data. Based on the training set of data, we generate neural structures. In the remaining two sets of data, we always leave 15% of the input information. Both groups will serve us to verify the reliability of the found neural structure, or the found model. The delay of the time series will be 1. We will generate 10,000 neural networks. We will preserve 5 of them with the best characteristics. We will be orientated using the smallest square method. We will terminate network generation if there is no improvement, ie to reduce the sum of squares. Thus, we will preserve those neural structures whose sum of squares of residue to the actual Czech

Republic and PRC trade balance will be as low as possible (ideally zero). In the hidden layer, we will have at least two neurons, at most 50. In the case of the radial basic function, there will be at least 21 neurons in the hidden layer, at most 30. For the multiple perceptron network we will consider these distribution functions in the hidden layer and in the output layer:

- Linear,
- Logistic,
- Atanh,
- Exponential,
- Sinus.

Other settings are left by default (using the ANS tool – automated neural network).

In conclusion, we compare the results of linear regression and regression using neural networks. Comparison will not take place in the form of residue analysis (minimum, maximum values, spread of residues, etc.), but at expert level and experience of the assessor, economist.

3 Results

Linear regression

The correlation coefficient is -0.8830, which represents a significant statistical indirect dependence of the trade balance on the development over time. The determination coeficient reaches 0.7797.

A point graph was constructed (see Figure 2), where the points were intersected by a regression curve, in this case linear. The line parameters are shown in the figure.

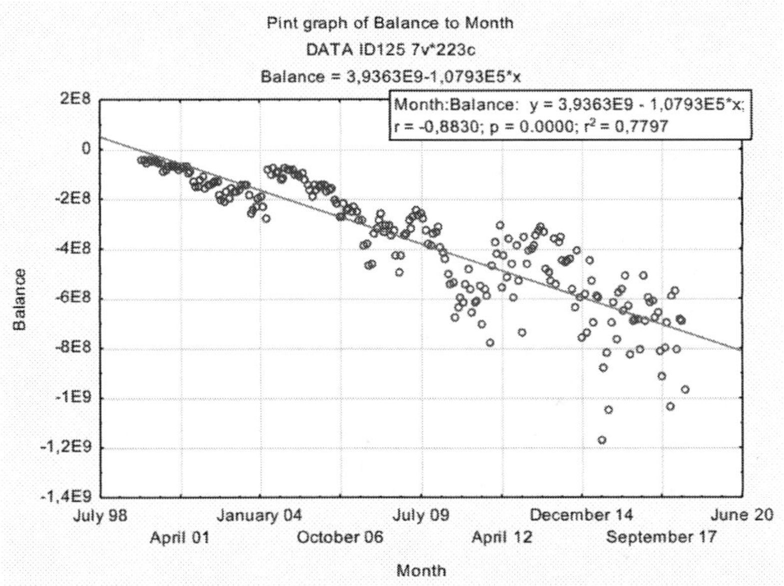

Fig. 2. The Czech Rep. and the PRC point graph intersected by the regression curve – linear curve Source: Authors.

The solid line represents the regression function. The linear functions, however, do not fully respect the trade balance of the Czech Republic and the PRC. Figure 3 represents the intersection of the point graph with a polynomial function.

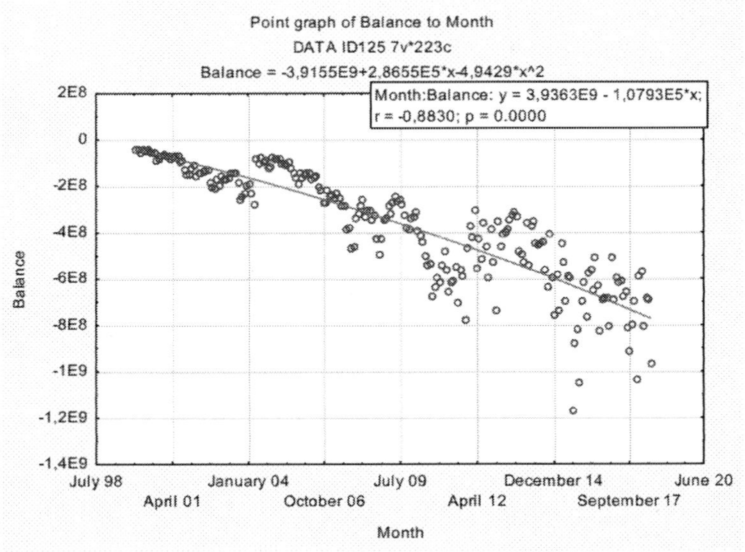

Fig. 3. The Czech Rep. and the PRC trade balance point graph intersected by a regression curve – polynomial function
Source: Authors.

Even in this case, the solid line represents a regression curve. Its shape resembles a straight line, a linear function. Even in this case, the trade balance of the partner countries is not fully represented.

Figure 4 shows a point graph intersected by a logarithmic function.

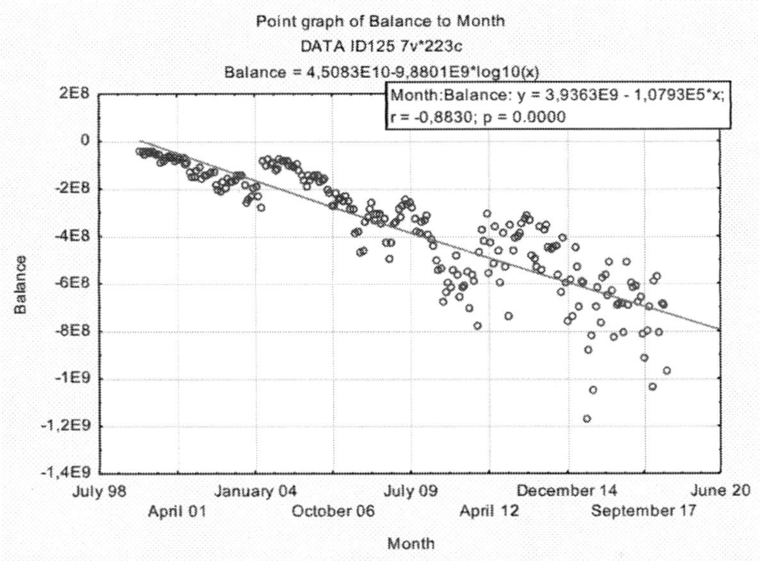

Fig. 4. The Czech Rep. and PRC trade balance point graph intersected by a regression curve – logarithmic function
Source: Authors.

The logarithmic function curve also closely resembles a straight line and therefore a linear function. From the location of individual points in the graph, it is obvious that the logarithmic function is not suitable for regression.

Figure 5 provides a graph of the trade balance of the Czech Republic and the PRC intersected by the LOWESS function.

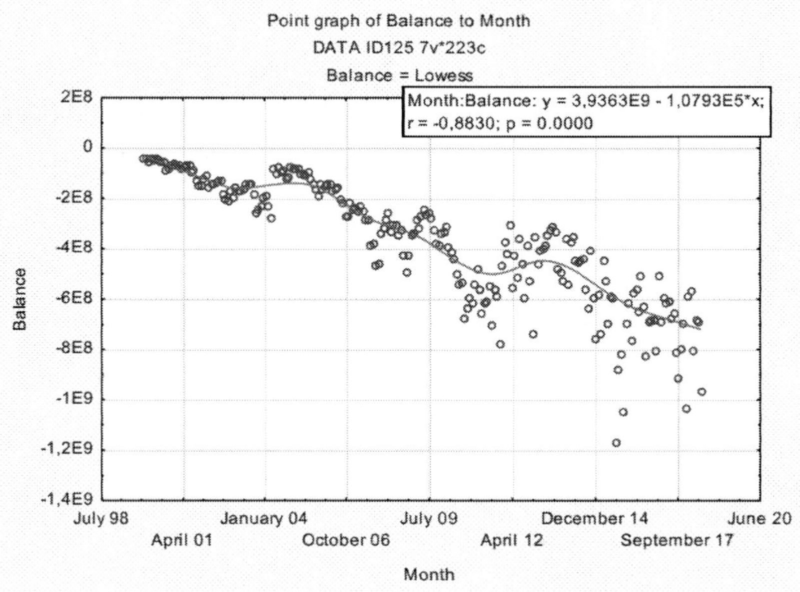

Fig. 5. The Czech Rep. and the PRC trade balance point graph intersected by a regression curve – LOWESS
Source: Authors.

The LOWESS (locally weighted scatterplot smoothing) function, or LOES (locally estimated scatterplot smoothing), is calculated as a regression function at partial intervals. The LOWESS curve aligns the time series and monitors its trend appropriately. The LOWESS curve intersection looks very interesting. It can fairly reveal the trend of the trade balance of the partner countries.

Figure 6 provides a point graph of the Czech Republic and PRC trade balance, intersected by smallest squares of weighed distances function.

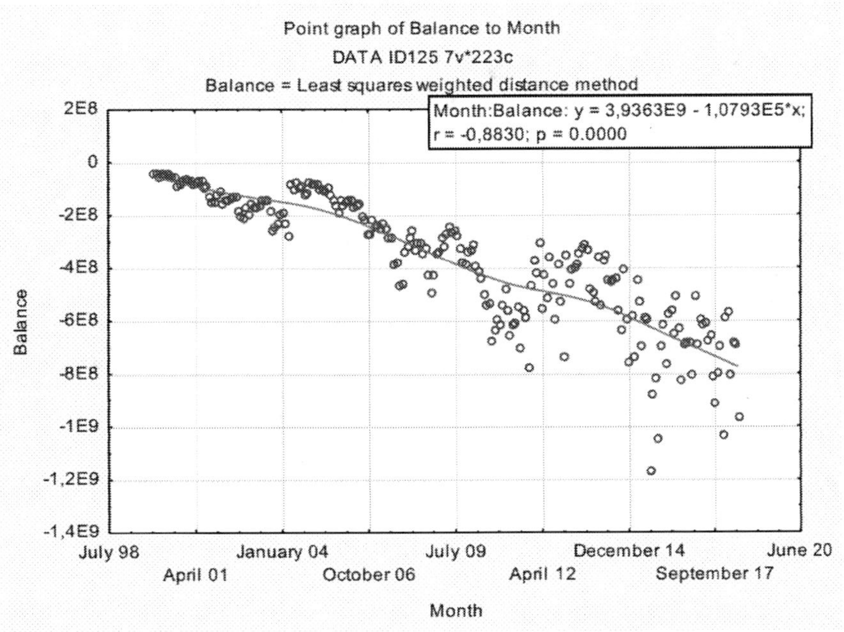

Fig. 6. Czech Rep and PRC trade balance point graph intersected by the regression curve – sm. sq. of weighed distances
Source: Authors.

The curve copies the development of trade balance over its entire interval better than linear functions, polynomial functions and logarithmic functions.

Figure 7 provides the intersection of the function obtained by smallest squares of negative - exponential smoothing.

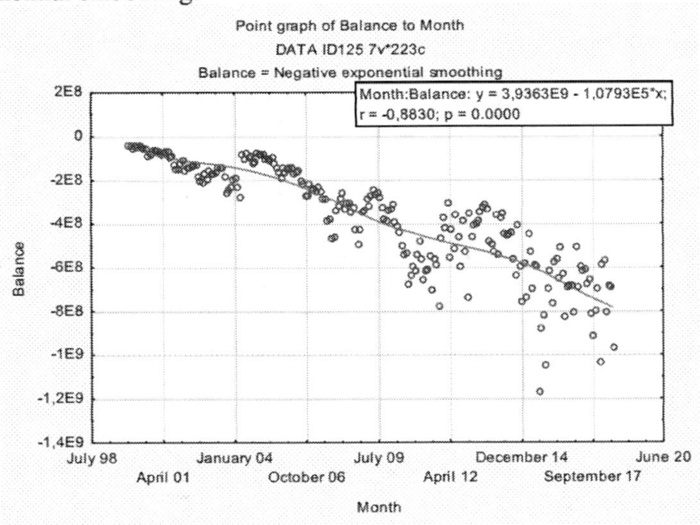

Fig. 7. CZ and PRC trade balance point graph intersected by a regression curve – sm. sq. negative exponential smoothing
Source: Authors.

Even this curve appears to be interesting and appropriate for a possible prediction. It also has better results than linear, polynomial, and logarithmic.

The correlation coefficient indicates significant statistical indirect dependence of the target variable on time development. Its value is -0.8830. The determination coefficient is 0.7797. If we only evaluated the results by comparing the trade balance between the Czech Republic and the PRC and the shape of the regression curve and taking into account a simple linear regression, we could certainly say that the LOWESS curve is closest to the development. Following it are curves obtained by the smallest squares method, namely negative-exponential smoothing and weighed distances. All three copy the basic trade balance between the Czech Republic and the PRC. At the same time, it should be recognized that the other functions (linear functions, polynomial functions, and logarithmic functions) are not suitable for time series alignment. It will simplify its course too much.

Neural structures

Based on the established procedure, 10,000 neural networks were generated. Five networks have been preserved, the ones showing the best parameters. Their overview is given in Table 2.

Table 2. Overview of preserved neural networks

Net. name	Train. perf.	Test. perf.	Valid. perf.	Train. error	Test. error	Valid. error	Train. algorit.	Error function	Activ. of hidd. lyr.	Output activ. funct.
RBF 1-22-1	0.9439	0.9426	0.9534	2.803986 E+15	4.4378 12E+15	4.5276 28E+15	RBFT	Sum of sq.	Gauss	Identity
RBF 1-24-1	0.8933	0.8560	0.9512	5.208266 E+15	1.0044 19E+16	3.6104 12E+15	RBFT	Sum of sq.	Gauss	Identity
RBF 1-30-1	0.9179	0.8705	0.9505	4.046748 E+15	8.9651 70E+15	5.6887 61E+15	RBFT	Sum of sq.	Gauss	Identity
RBF 1-27-1	0.9252	0.9042	0.9598	3.701526 E+15	6.9056 86E+15	4.2726 54E+15	RBFT	Sum of sq.	Gauss	Identity
RBF 1-25-1	0.9425	0.9289	0.9499	2.871419 E+15	5.7671 00E+15	4.7584 36E+15	RBFT	Sum of sq.	Gauss	Identity

Source: Authors.

These are only the neural networks of the basic radial function. The input layer has only one variable - time. In the hidden layer, neural networks contain from 22 to 30 neurons. In the output layer, we have a single neuron and the only output variable is the export balance of the Czech Republic and the PRC. For all networks, the RBFT training algorithm was applied. In addition, all neural structures used the same function to activate the hidden layer of neurons, namely the Gaussian curve. They also use the same function to activate the outer layer of neurons, and this function is identity (see Table 2).

Training, testing and validation performance is also interesting. In general, we are looking for a network that has the same performance in all sets of data (we remind that the data was randomly distributed). The error should be as small as possible. As far as the error is concerned, it should be recalled that we are working on the balance expressed in euros, ie with a great deal of detail, which will also correspond to the above-calculated error.

The performance of individual sets of data is expressed as a correlation coefficient. The values of the individual data sets according to specific neural networks are presented in Table 3.

Table 3. Correlation coefficients of individual data sets

	Balance (Training)	Balance (Testing)	Balance (Validation)
1.RBF 1-22-1	0.943854	0.942643	0.953409
2.RBF 1-24-1	0.893296	0.855999	0.951235
3.RBF 1-30-1	0.917891	0.870478	0.950481
4.RBF 1-27-1	0.925175	0.904176	0.959754
5.RBF 1-25-1	0.942461	0.928905	0.949845

Source: Authors.

The table shows that the performance of all stored neuronal structures is very high. Higher fluctuations in the performance of the individual datasets are shown in networks 2. RBF1-21-1 and 3. RBF 1-29-1. The correlation coefficient of all training data sets ranges from more than 0.89 to over 0.94. The value of the correlation coefficient of the testing data sets for all neural networks ranges from more than 0.85 to over 0.94. The correlation coefficient of the validation data set of all neural networks is above the 0.949 level. In order to select the most suitable neural structure, we need to analyze the results obtained. Table 4 provides the basic statistical characteristics of each set of data for all neural structures.

Table 4. Statistics of individual data sets according to preserved neural structures

Statistics	1.RBF	2.RBF	3.RBF	4.RBF	5.RBF
Minimum prediction (Training)	-898391801	-821663096	-857909633	-914757095	-905237832
Maximum prediction (Training)	-49086981	-40172975	-13537671	-4452348	-41323566
Minimum prediction (Testing)	-789576394	-804782911	-828736787	-706475950	-744790331
Maximum prediction (Testing)	-49635722	-66639725	-23270632	-73460448	-55363423
Minimum prediction (Validation)	-803294038	-842962423	-777155961	-816475698	-807210541
Maximum prediction (Validation)	-79169241	-72858834	-16165433	-68026809	-85082148
Minimum residues (Training)	-336438354	-427989158	-399848725	-377532214	-373817164
Maximum residues (Training)	258520249	277584103	188100987	225393996	210531734
Minimum residues (Testing)	-252876501	-563851527	-352153219	-283794827	-333348153
Maximum residues (Testing)	208414123	238241010	163190692	208652287	135282887
Minimum residues (Validation)	-364807606	-325139221	-390945683	-351625946	-360891103
Maximum residues (Validation)	146675635	131088493	100813927	75309342	141610295
Minimum standard residues (Training)	-6	-6	-6	-6	-7
Maximum standard residues (Training)	5	4	3	4	4
Minimum standard residues (Testing)	-4	-6	-4	-3	-4
Maximum standard residues (Testing)	3	2	2	3	2
Minimum standard residues (Validation)	-5	-5	-5	-5	-5
Maximum standard residues (Validation)	2	2	1	1	2

Source: Authors.

Ideally, the individual statistics of the neural network are cross-sectional in all sets (minimum, maximum, residue, etc.). However, in the case of the aligned time series, the differences are noticeable. They are apparent both between individual datasets of one neural network and between individual neural networks. Somewhat larger differences are reflected in the residue characteristics. However, we are not able to clearly identify which of the preserved neural networks has the most appropriate results.

Figure 8 is a graph showing the actual development of the trade balance of the Czech Republic and the PRC, as well as the development of predictions using individual generated and preserved networks.

The chart shows that all neural networks predict the development of the trade balance at different intervals differently. However, the similarity between predictions of individual networks, but similarity (or degree of consistency) with actual trade balance developments is not important.

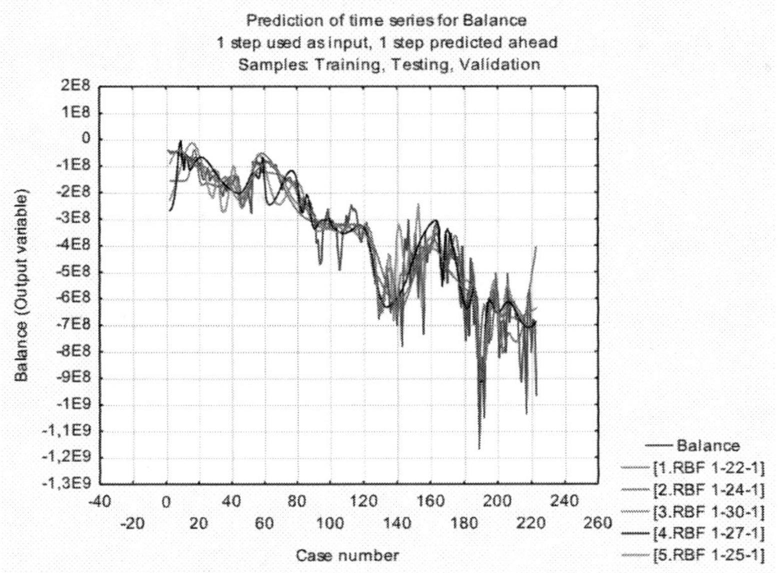

Fig. 8 Line graph – The Czech Republic and the PRC trade balance predicted by neural networks compared to the real trade balance in the monitored period
Source: Authors.

If we examine the course of the individual curves and their consistency with the real course of the trade balance between the Czech Republic and the PRC, we will come to the following conclusions:

- 1.RBF 1-22-1: aligns the time series fairly well. However, it can not register well local trade fluctuations.
- 2.RBF 1-21-1: offers a very similar pattern to network number 1. However, it exceeds the acceptable error value at the beginning and end of the period under review.
- 3.RBF 1-29-1: aligns time series in a rather interesting way. It tries to register local fluctuations. This, however, is predicted even in cases where there were actually none.
- 4.RBF 1-25-1: has a similar pattern to the curve constructed according to the first neural network. Only in the course of the trade balance at the beginning of the monitored period it shows a large fluctuation from the actual trade balance development.
- 5.RBF 1-24-1: shows the best results. Within a certain tolerance, the trade balance of the Czech Republic and the PRC is fairly well reflected.

4 Conclusion

The aim of the paper was to compare the accuracy of time series alignment using regression analysis and neural networks on the example of the trade balance between the Czech Republic and the People's Republic of China.

In general, each prediction is given by a certain degree of probability with which it is to be fulfilled. As we predict the future development of any variable, we try to estimate the future development of this variable on the basis of previous years' data. Although we can include most of the factors influencing the target variable in the model, we always simplify reality, and we always work with a certain degree of probability that some of the predicted scenarios will be fulfilled As with linear regression, as well as regression using neural networks, there is simplification - and quite substantial. We only work with two variables – input (time) and output (trade balance Czech Republic x PRC). So, we completely ignore other input variables, which often have a significant impact on the trade balance between the Czech Republic and the PRC (international political situation, taxation in both countries, production factor prices, state export support, natural resources and many others). The main determinant is then the purpose for which we perform the calculation. If we want to estimate the future development of individual countries' economies based on the trade balance between the Czech Republic and the PRC, it is likely that such simplification will be sufficient and the calculation will have adequate information value. If, for example, we are planning to plan for the shipping capacity of a particular company, it will not be enough to use such simplified time series. It is still true that we estimate aggregated variables better than partial ones.

At the same time, we can state that due to the great simplification of reality, it is not possible to predict the emergence of extraordinary situations and their impact on the Czech Republic and PRC trade balance (perhaps in the short term, yes, but certainly not long term). Ideal would be a prediction for the next few days, but for such a short prediction it is currently not possible to obtain data.

The trade balance of both countries can be determined on the basis of statistical methods, causal methods and intuitive methods. In this case, we have been comparing statistical methods. However, they only gave us a possible framework for the development of the monitored variable. It is important to work with information on the possible future development of the economic, political or legal environment. If we can predict their development, we can then project it into the monitored trade balance. At the same time, however, the personality of the evaluator comes into play – an economist who, on the basis of his knowledge and experience, corrects the price determined by the statistical methods and specified on the basis of causal links.

Optically, the LOWESS curve was the best out of linear regression, followed by the smallest squares curve by negative-exponential smoothing, and the curve obtained by the smallest square-weighed distance. From the neural networks, the network 5.RBF 1-24-1 proved to be the most usable in practice. If we look at performance from a correlation coefficient point of view, the neural network 5.RBF 1-24-1 remains usable. The aim of the paper was fulfilled.

References

1. J. Drábek, L. Lipková, M. Gress, Development of the balance of payments in the czech republic since 1993 with the emphasis on current account. *Economic Annals-XXI,* **7-8**(2), 4-7, (2015)
2. R. Martin, Central and eastern europe and the international economy: The limits to globalisation. *Europe - Asia Studies,* **50**(1), 7-26, (1998)
3. M. Engman, Expanding international supply chains: The Role of Emerging Economies in Providing it and Business Process Services. OECD Trade Policy Papers, (52), 66 p., (2007)

4. O. G. Golichenko, S. Samovoleva, The balance of externalities and internal effects in national innovation systems. *Proceeding of the 10th European Conference on Innovation and Entrepreneurship*, 223-230, (2015)
5. F. Novotný, Profitability Life Cycle of Foreign Direct Investment: Application to the Czech Republic. *Emerging Markets Finance and Trade*, **54**(7), 1623-1634, (2018)
6. *Czech Republic freight transport & shipping report - Q2 2017*, London: Business Monitor International, (2017)
7. T. De Castro, J. Vlčková, P. Hnát, Trade and investment relations between the czech republic and china: The czech republic as a gateway to the EU? *Society and Economy,* **39**(4), 481-499, (2017)
8. *Význam zahraničního obchodu s Čínou roste*, Czech Statistical Office [online]. Available at: https://www.czso.cz/csu/czso/vyznam_zahranicniho_obchodu_s_cinou_roste_20130627 (2013)
9. *Čína: Obchodní a ekonomická spolupráce s ČR,* Ministry of Foreign Affairs [online]. © Zastupitelský úřad ČR v Pekingu (Čína). Available at: http://www.businessinfo.cz/cs/clanky/cina-obchodni-a-ekonomicka-spoluprace-s-cr-19054.html (2018)
10. R. Gerard, The political economy of transition. *The Journal of Economic Perspectives,* **16**(1), 29-50, (2002)
11. I. T. Y. Chen, Balance of payments and power: assessing China's global and regional interdependence relationship. *International Relations of the Asia-Pacific*, **14**(2), 271-302, (2014)
12. Z. P. Hao, Regulation and organisation of accountants in China. *Accounting, Auditing & Accountability Journal,* **12**(3), 286-302, (1999)
13. A. Ortolani, Chinese begin paying by cellphone. *Wall Street Journal* [online], pp. 1. Available at: https://search.proquest.com/docview/399037632?accountid=133446 (2005)
14. S. Tseng, Trade flows between czech republic and east asia. *Review of Economic Perspectives,* **13**(3), 146-158, (2013)
15. *OECD science, technology and industry scoreboard 2007: Innovation and performance in the global economy* (2007). Paris: Organisation for Economic Cooperation and Development (OECD), 228 p.

The Significance of Internal Enterprise Resources from the Perspective of the Opinion Level of Top Managers of Czech Enterprises Operating in Service Sector

Jarmila Straková[1,*] and *Jan* Váchal[1]

[1]Institute of Technology and Business in České Budějovice, Department of Management, Okružní 517/10, 37001 České Budějovice, Czech Republic

Abstract. This contribution focuses on the issue of internal enterprise resources in the service sector. The level of opinion in top managers is being observed, respectively what significance is being attached by top managers of enterprises operating in service sector to internal enterprise resources from the perspective of their current and future situation. Statistically significant differences between the meaning of internal resources nowadays and in the future are being analysed and indicated via the pair Wilcoxon test on a testing set consisting of 255 enterprises within the service sector from the whole of the Czech Republic. Within the evaluation of the current situation top managers assign a high importance mainly to employee standards, financial resources, the name of the enterprise, management quality, service and marketing portfolio. Evaluating prediction of enterprise internal resources and their importance´s future development has pointed to a relatively high growth of significance in enterprise strategy, enterprise education, marketing, and information and communication systems. Statistically significant differences from Wilcoxon test have been proven in employee education, enterprise strategy, information and communication systems and marketing standards. Indicated internal resources will gain greater dominance in the future, and they may significantly influence the existence of enterprises operating in service sector.
Key words: top managers, service sector, internal enterprise resources, opinion level, evaluation

1 Literary Review

The economy of the Czech Republic has been characterized now for several years by permanent economic growth, shifting the Czech Republic among 5 EU countries with the best developing national economy. That is why it may be claimed that during the observed period (2016, 2017), during which a research survey of Czech enterprises focused on the area of strategic management and influence of enterprise environment was in process,

* Corresponding author: strakova@email.vstecb.cz

profitable businesses prevail not only within service sector, while loss enterprises occur there in much lower amounts. The current economic development makes enterprises feel the omnipresence of positive development, enterprises do not appreciate a range of resources and factors from enterprise surroundings, but also the influence of globalizing and internationalizing processes on the constantly growing competitiveness of enterprises and their prosperity. At the same moment it opens an assumption for business subject preparation for a period of time with an opposing economic cycle, related to mainly internal enterprise resources. A change of economic cycle may be expected, according to both local and foreign economists´ prognoses, within a time period of 3-5 years, however, the preparation for elimination of this change must be implemented in the current period of time, i.e. a time of economic growth.

The enterprise carries its activities out within certain environment which should be observed all the time. Enterprise surroundings may be defined as anything appearing beyond the fictional border of the enterprise as a technical and socio-economic system [13]. Dvořáček and Slunčík [2] as well as other authors classify enterprise environment into internal and external. According to Euremegbe [3], internal environment also includes situation factors within a specific enterprise. These factors are, largely, the result of a management process decision, mainly under management control. Pošvář and Erbes [11] consider harmonizing internal and external environment, with addition of management permanently analysing and observing its environment, determining important factors, possible future success, etc., to be the basic manager ability. The purpose of such analysis is to characterize the enterprise´s strengths and weaknesses and to determine the enterprise´s position on the market. Internal environment analyses contains the analysis of internal enterprise resources, on a general level we mean financial resources, human resources, material, and non-material resources (so-called resources of non-material character); at the same time, it also contains structure analysis in relation to environment change influence; system analysis, company culture analysis; analysis of its abilities, etc. Its result is determining the company´s strengths and weaknesses [1, 7, 15, 16]. Most local as well as foreign authors agree that optimal structure of internal resources is what becomes a base for creating competition advantage and reaching long-term prosperity [4, 6, 8, 16, 19].

As the environment becomes more differentiated, its dynamic changes create circumstances for enterprises with correctly formulated internal resources, and ability to survive and prosper, while other, less adaptable enterprises may stagnate, become obsolete or lose their sense of integrality and identity [10]. Some research surveys [14, 18] have proven that the enterprise´s ability to survive in competitive environment is affected straightaway by the value of its internal resources and ability to predict and use the opportunity, while others [5, 12] suggest that internal resources are even more important to keep competitiveness than external resources. Internal resources (such as human or technological capital) may, besides influencing competitiveness sustainability, may also become a source of knowledge for new product development [17]. Other research surveys warn about the fact that managers may be aware of new coming (or currently happening) changes in branches they are operating in, but they may not interpret these changes correctly. Often, they underestimate the meaning of given changes, and they may be waiting for too long as to react to them [9].

2 Material and Methods

Survey focused on the issue of strategical management, strategy, and influence of business environment on enterprises has been running since 2016. Data collection was done via a wide questionnaire survey. In 2017, the testing set consisted of 456 enterprises from the whole of the Czech Republic, enterprises were classified from the perspective of size categorization, sector differentiation, and spread of its influence. Nowadays, number of enterprises keeps growing, adding foreign companies from Russia and China, gaining a testing sample of Austrian and German enterprises is being planned.

For the purposes of this contribution, a testing set of 255 enterprises operating in service sector was chosen. A Wilcoxon´s pair test was used. It is the statistical method which is used to test whether the mean difference of the figures of paired data is statistically insignificant or this difference is statistically significant medium. The mean value estimate for this test is made by median. We test the hypothesis H_0: $x_{0,5}-y_{0,5}=0$ against the alternative H_1: : $x_{0,5}-y_{0,5}\neq 0$ at the significance level of 0.05 (i.e. reliability of 95%).

3 Results and Discussion

3.1 Significance of Internal Resources (current situation and prediction)

The lower-mentioned results determine the significance of internal resources, as from the perspective of their current situation, as of prediction, respectively what significance top managers of enterprises operating in service sector assign to enterprise resources nowadays and in the future.

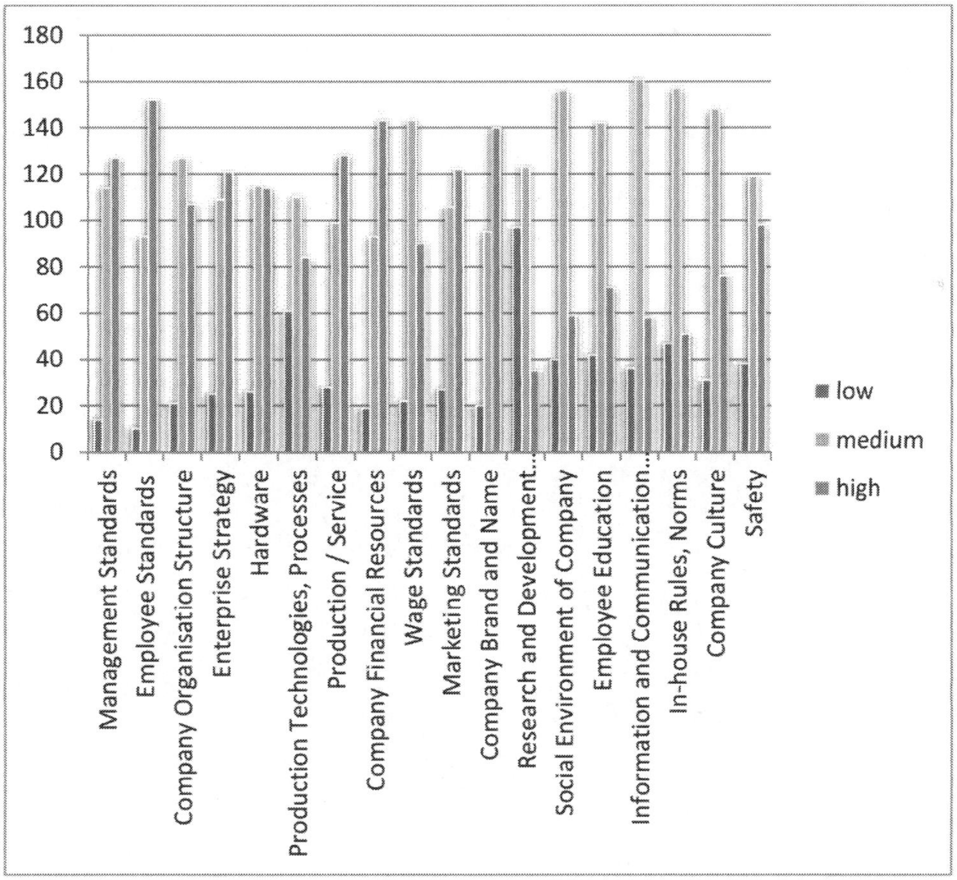

Fig. 1. Significance of Internal Resources from the Perspective of Level of Opinion of Managers of Enterprises Operating in Service Sector – the Current Situation.
Source: Authors.

Top managers of enterprises operating in service sector assign, within their internal environment, dominant significance to the following: employee standards, financial resources, brand, and company name. At a smaller rate (even so evaluated as highly significant) they further state service portfolio, management quality, marketing, and enterprise strategy. From the perspective of current situation, the following resources may be considered decisive as well as limiting: financial security level, employee, management quality, service portfolio, marketing, and an accurately set strategy. Good reputation, respectively the name of an enterprise may be understood as a significant competition tool that, specifically within service sector, reaches intensity, because enterprises here move within highly competitive environment.

On the contrary, a very low significance is assigned to research and development by top managers, this situation may be considered a very negative finding. Research and development is still not appreciated by a great percentage of Czech enterprise managers, not only in service sector, but also in production and industry sector, where this situation is even more alarming. Often, the Czech business sphere strongly criticizes support of science and research from governmental side, a low representation of research organizations including universities, is also being criticized. This critique, however, has two levels, one really existing, at which inclusion of scientifically-research basis into the service sector innovation process is objectively insufficient, the second level relates to the willingness and readiness of enterprises themselves, operating in service area, at which enterprises operating here do not appreciate their receiver and new-solution, process, and technology implementer roles.

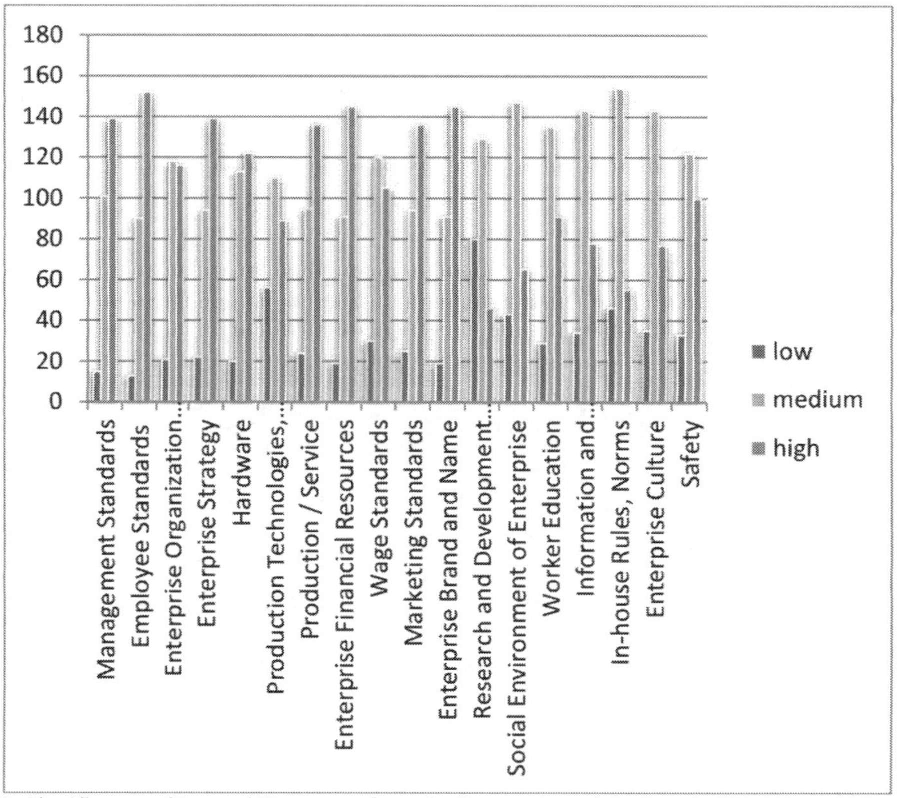

Fig. 2. Significance of Internal Resources from the Perspective of the Level of Opinion of Managers of Enterprises Operating in Service Sector – Prediction.
Source: Authors.

From the perspective of prediction in the development of significance of internal resources from the perspective of their inclusion into a value-creating enterprise process, significant differences were not proven, compared to the current situation. Results are almost identical, the influence of positive economic development may be expected, as well as an underestimation of the option of a change in economic cycle, thus an underestimation of inner resource significance. A growth in enterprise strategy, marketing, and management quality growth should be noticed, and considered a positive finding. A relatively high growth of significance in enterprise education, information and communication systems is happening. It is a necessity to highlight the fact that internal enterprise resource analysis was already happening within the mentioned period of economic growth, which probably significantly influenced top manager perception, including their lowered ability of predicting expected changes in the future development of enterprise environment, and their impact on the structure of internal enterprise resources.

3.2 Determination of Differences in the Significance of Internal Resources between the Current Situation and Prediction using Wilcoxon´s Pair Test Analysis

Table 1. Significance of Internal Resources – Current Situation (Service Sector).

Current Significance	Arithmetic mean	Median	Standard deviation	Modus
Management Standard	2.44	2	0.60	3
Employee Standard	2.55	3	0.59	3
Enterprise Organization Structure	2.32	2	0.67	2
Enterprise Strategy	2.36	2	0.71	3
Hardware	2.33	2	0.71	2
Technology	2.03	2	0.87	2
Service Portfolio	2.36	3	0.75	3
Financial Resources	2.47	3	0.68	3
Wage Standard	2.25	2	0.65	2
Marketing Standard	2.35	2	0.73	3
Brand and Name	2.45	3	0.70	3
Research and Development Standard	1.71	2	0.76	2
Social Environment of Enterprise	2.05	2	0.67	2
Employee Education	2.09	2	0.70	2
Information and Communication Systems	2.07	2	0.64	2
In-house Norms, Rules	2.00	2	0.66	2
Enterprise Culture	2.16	2	0.67	2
Safety	2.21	2	0.75	2

Source: Authors.

Table 2. Internal Resource Significance – Prediction (Service Sector)

Future Significance	Arithmetic mean	Median	Standard deviation	Modus
Management Standard	2.48	3	0.62	3
Employee Standard	2.54	3	0.61	3
Enterprise Organization Structure	2.35	2	0.69	2
Enterprise Strategy	2.44	3	0.72	3
Hardware	2.38	2	0.69	3
Technology	2.06	2	0.88	2
Service Portfolio	2.42	3	0.71	3
Financial Resources	2.48	3	0.67	3
Wage Standard	2.27	2	0.73	2
Marketing Standard	2.42	3	0.71	3
Brand and Name	2.47	3	0.69	3
Research and Development Standard	1.83	2	0.76	2
Social Environment of Enterprise	2.07	2	0.68	2
Employee Education	2.23	2	0.68	2
Information and Communication Systems	2.16	2	0.68	2
In-house Norms, Rules	2.02	2	0.68	2
Enterprise Culture	2.15	2	0.69	2
Safety	2.25	2	0.71	2

Source: Authors.

Table 3. Results of Wilcoxon´s pair test between the significance of internal resources nowadays and in the future

Future Meaning	V Value	p-Value
Management Standard	382.5	0.1334
Employee Standard	688.5	0.793
Enterprise Organization Structure	927.5	0.4128
Enterprise Strategy	662.5	0.01339
Hardware	961	0.09913
Technology	1757	0.4653
Service Portfolio	900	0.1449
Financial Resources	614	0.8053
Wage Standard	706.5	0.5566
Marketing Standard	690.5	0.06898
Brand and Name	718.5	0.4726
Research and Development Standard	-	-
Social Environment of Enterprise	1262	0.7519
Employee Education	558.5	0.0004054
Information and Communication Systems	880.5	0.02955
In-house Norms, Rules	-	-
Enterprise Culture	573.5	0.7005
Safety	554	0.3772

Note: Factors (research and development standard and in-house norms, rules) were not tested because the average value of the current level of these internal resources was lower than the value 2 (significant internal resource).
Source: Authors.

Table 3 contains the results of Wilcoxon´s pair test between the given significance of internal enterprise resources nowadays and in the future. Significant statistical differences have been indicated in employee education, enterprise strategy information and communication systems, less significant in marketing standards. All of the stated resources are significant for enterprises operating in service sector, and they significantly influence their current as well as future existence within more and more intense and sharpening competition environment. These internal resources are going to gain still greater significance, especially during the period of economic decrease in national economy performance. It may be claimed that the results obtained fully correspond to the service sector characteristics.

4 Conclusion

Research survey has brought and suggested a few valuable findings that may be taken into account within enterprise strategic management in business practice for service sector. The solution outcomes allow the following conclusions:

- Internal enterprise resources may be considered a starting base for the existence itself, development and competitiveness of enterprises within service sector, and that is why enterprise top managers should be paying corresponding, i.e. greater attention. A range of current managers is not, unfortunately, aware of this fact, the causes of this may be observed in the fact that, as it has already been suggested in Results and Discussion, Czech economy has found itself, now for a few years, in a period of economic growth, and internal resources, as well as mid-environment and macro-environment factors, remain usually unrecognized. A change in economic cycle is, however, inevitable, and managers should definitely not underestimate its preparation, they should take into account the fact that internal resources, their quality and optimal structure might mitigate and eliminate negative impacts related to this future change.
- Employee standard, financial resources, market brand and enterprise name may be considered the ′core′ internal resources in service sector in the current period, service portfolio, management quality, marketing, and enterprise strategy should be considered as additional, but still important. The absence of research and development factor is warning, and it requires a solution from the state′s side as well as from the side of enterprises themselves.
- Within significance testing in differences between the current and future situation among internal resources in service sector enterprises, statistically significant differences were proven in enterprise education, strategy, information and communication systems, and marketing. The named internal resources respond to the service characteristics.
- The service sector requires relatively ′little capital′, the more is its attention focused on human resources and their development. In the Czech Republic, unemployment has historically been the lowest. Enterprises get into trouble with lack of qualified available workforce, they start to look for reserves within internal resources and pay more attention to enterprise education.

References

1. C. Capon, *Understanding the business environment: inside and outside the organisation*. Prentice Hall/Financial Times, NY, (2009)
2. J. Dvořáček, P. Slunčík, *Podnik a jeho okolí: jak přežít v konkurenčním prostředí* C. H. Beck, Prague, (2012)
3. G.O. Euremegbe, Impact of business environment on organization performance in Nigeriaa study of union bank of Nigeria. *ESJ, Special edition*, 478-494, (2015)
4. R.T. Frambach, E.J. Nijssen, *Creating Customer Value Through Strategic Marketing Planning*. Kluwer Academic Publishers, London, (2001)
5. G. Hawawini, V. Subramanian, P. Verdin, Is Performance Driven By Industry- or Firm-Specific Factors? A New Look at the Evidence. *Strategic Management Journal*, **24**, 1, (2003)
6. Ch.W.L. Hill, G.R. Jones, *Essentials of strategic management*. South-Western/Cengage Learning, Mason, (2009)
7. T.R. Jain, M. Trehan, R. Trehan, *Business Environment*. VK Global Publication, New Delhi, (2014)
8. D. Jakubíková, *Strategický marketing*. Grada Publishing, Praha, (2008).
9. Z. Jovanovic, Management and changes in business environment. *Ekonomika, [Економика]*, **61**, 2, 143-151, (2015)
10. K. Motamedi. Resilience in dynamic complex environments, 131-141. *6th Business & Management Conference*, Geneva, (2017)

11. Z. Pošvář, J. Erbes, *Management I.* MZLU, Brno, (2008)
12. R. Rumelt, How Much Industry Matter? *Strategical Management Journal*, **12**, 167-185, (1991)
13. M. Synek, *Manažerská ekonomika.* Grada Publishing, Praha, (2003)
14. J. Wang, Y. Ying, Internal resource, external environment and competitive advantage of enterprises - Empirical study based on Hubei province manufacturer, *Proceedings of the International Conference on E-Business and E-Government, ICEE 2010,* 2930-2933, (2010)
15. I. Worthington, Ch. Britton, *The business environment.* Pearson Education Limited, UK, (2010)
16. M. Zamazalová, *Marketing*. C. H. Beck, Praha, (2010)
17. J. Zhang, W-P., Wu, Social capital and new product development outcomes: The mediating role of sensing capability in Chinese high-tech firms. *Journal of World Business*, **48**, 539-548, (2013)
18. J. Zhang, W-P., Wu, Leveraging internal resources and external business networks for new product success: A dynamic capabilities perspective. *Industrial Marketing Management*, **61**, 170-181, (2017)
19. R. Zuzák, *Strategické řízení podniku.* Grada Publishing, Praha, (2011)

Participation in the sharing economy in the Czech Republic

Jaroslav Stuchlý[1] and *Michal Šuta*[1,*]

[1]Institute of Technology and Business in České Budějovice, Faculty of Corporate Strategy, Okružní 517/10, 37001 České Budějovice, Czech Republic

Abstract. The sharing economy currently forms a really important aspect of an economy. The authors set out to determine the size of the population participating in the sharing economy through the use of social networks and other channels with the purpose of identifying and describing individual factors (variables) affecting this type of market segment. The objectives of this paper were to determine the age gap between male and female participants in the sharing economy and to identify specific factors that may affect their participation in it, in particular among respondents who actively use social networks. It was concluded that gender did affect participation in the sharing economy, whilst financial position and permanent address did not have any influence on this market segment. The authors also set out to determine the level of education of the population that is dominant on this market. In terms of methodology, the information presented in this paper is based on the results of an empirical questionnaire survey on participation in the sharing economy in the Czech Republic and the application of general statistical methods.
Key words: sharing economy, p2p lending, age identification

1 Introduction

The sharing economy is among the macroeconomic aspects of the state. It is an issue which is very important for the state, companies and individuals due to the high degree of globalization. Many works have been written on this topic in general; however, the narrowly specified topic of this contribution appears very scarcely in professional literature.

A new type of economic interaction appeared at start of the 2008-2009 financial crisis. Initially originating on the western coast of the United States, it gradually rolled out to many developed countries, spreading particularly fast in municipal areas. This new type of economic interaction is called many different names, including joint consumption, access economy, peer-to-peer economy, collaborative consumption or on-demand economy. However, it is best known as the sharing economy [5].

The sharing economy is an important element of shared consumption. It can be considered a hybrid economic system built on property lending, sharing or rental without the need to own the assets. This type of economy is based on state-of-the-art technologies which make it possible to reduce costs, in particular transaction costs and overheads.

* Corresponding author: suta@mail.vstecb.cz

As early as 1978, Felson and Spaeth [11] defined collaborative consumption as follows: "Events in which one or more persons consume economic goods or series in the progress of engaging in joint activities with one or more others". Russel Belk [3] disagreed with this definition by claiming it is very general and does not provide a satisfactory description of the acquisition and distribution of economic resources. Belk [4] therefore defines collaborative consumption as "people coordinating the acquisition and distribution of a resource for a fee or other compensation". Belk's definition therefore assumes that the sharing economy cannot exist without some form of compensation.

Teffer [18] states that the sharing economy is a remarkable trend that is disrupting traditional economic sectors, but that this phenomenon is very difficult to define and that empirical evidence of its impact is scarce. At present, two markets exist in the sharing economy, namely the peer-to-peer (P2P) and business-to-customer (B2C) markets [7]. The sharing economy provides temporary access to idle resources. The unused capacity of goods is financed by those who own them and consumers rent access to those goods without acquiring the assets themselves [13].

Botsman and Rogers [6] differentiate between three system subtypes of temporary access. The leading criterion is not the type of ownership, but in what form assets are shared. The first type of sharing is called "Product Service Systems". It is a model that allows companies to offer goods as services rather than sell goods as products. Goods being private property can be shared and lent peer-to-peer. These systems appeal to the growing number of people arriving at the opinion that they want to have an advantageous product but do not need to own it. The second type of system is called "Redistribution Markets". These markets redistribute used or previously owned goods from one place where they are not needed to another place where they are needed. In some markets, goods can be shared free of charge, in other markets in exchange or for a consideration. The third type of system is called "Collaborative Lifestyles". Under this system, people with the same needs or interests jointly share and exchange intangible assets such as time, skills, or location. For the most part, this exchange is done at the local or neighbourhood level, whereby people share working space, gardens or parking space. However, lifestyles are also shared globally through activities such as peer-to-peer lending and the rapidly growing peer-to-peer travel. Unlike the aforementioned authors, Lisa Gansky [12] only distinguishes between two types of sharing based on the type of ownership. The first is the so-called Full Mesh Model, whereby a company owns some assets and rents them out via an application. The customer orders/books the given asset(s) through an application, thereby getting access to the specific asset(s) without the need to visit a registered office. The second is the so-called Own-to-Mesh Model. This model is based on the principle that a company does not own assets and that sharing is mediated by a company.

A study published by PricewaterhouseCoopers [16] claims that global revenues of the sharing economy across five key sharing sectors (travel, carsharing, finance, staffing and music / video streaming) have the potential to exceed USD 335 billion by 2025. Currently, 19% of US adults have undertaken sharing economy transactions, whereas only 7% of US citizens are providers in the sharing economy.

Peer-to-peer market

The peer-to-peer market involves the sharing of both tangible and intangible assets and human skills, whereby the owner shares them with someone else either free of charge or for a consideration through an internet platform. There are many forms of sharing in the peer-to-peer market, including carsharing, peer-to-peer lending or couch surfing, as well as examples of the Own-to-Mesh model described by Gansky [12].

Carsharing is popular in the large cities of North America and Europe because people do not want to bear the additional costs of running their own car. Modern carsharing, which was pioneered in Switzerland, has spread around the globe ever since. Europe is one of the largest world regions to have adopted carsharing. Other world regions are showing considerable growth in this sector, and although this growth is much higher than in Europe and North America, the total numbers are still quite low. At present, the success of carsharing is mostly linked to densely populated areas [10]. Peer-to-peer carsharing is an innovative approach to carsharing and involves owners temporarily renting their cars to other people in their neighbourhood. Peer-to-peer carsharing is part of the larger sphere of the sharing economy, namely the economic model based on the concept of joint consumption as opposed to ownership [2]. Carsharing reduces the number of cars on the roads because the players in this market segment do not have to own their own car. Hampshire & Gaites [14] highlight the many studies that show that carsharing reduces transportation costs for a large segment of society. However, the authors note that the traditional carsharing model is difficult to measure geographically in areas with a lower population density because the operator must know the fixed costs for the car purchase or rental in advance. Shaheen & Cohen [17] analysed the carsharing market and established that in 2006, more than 346,000 people around the globe used 11,501 shared cars. Over the next eight years, the number of people making use of carsharing rose by more than 4.4 million and the number of shared cars increased by more than 90,000. This change is illustrated in Table 1.

Table 1. Carsharing analysis

Year	Users	Shared cars
2006	346,000	11,501
2014	+4.4 million	+ 90,000.

Source: Shaheen & Cohen [17]; authors.

Peer-to-peer lending (P2P) involves the provision of loans between individuals via online platforms, whereby financial institutions only act as intermediaries as required by law. The first commercial online platform for P2P lending (Zopa) was established in 2005 by groups on online social networks. This means that P2P lending is a relatively new field of research [1]. Peer-to-peer lending is becoming more and more popular around Europe because lenders are seeking to achieve higher returns than offered by banks and borrowers are applying for loans in market sectors where banks refuse to lend. Another reason why peer-to-peer lending is becoming more and more popular is that lenders offer borrowers lower interest rates than banks. In an article in the Economist [19] it states that lenders usually lend money to borrowers via online internet auctions. The loans issued often consist of smaller sums from different lenders who provide money to borrowers with a preset interest rate, and all without the intervention of banks [19]. Although this market is growing rapidly, the current volume of P2P loans is equivalent to only a small proportion of the loans provided by traditional financial institutions, which suggests that P2P platforms cannot compete with financial institutions. Tristan Hugo-Webb [15] says that P2P lending platforms may be considered an interesting phenomenon, but that they do not pose a threat to traditional loan providers. However, the origin of P2P lending provides a new, valuable source of funding in the global financial world after 2008, and has the potential to fill the loan gap in mature as well as developing markets. In future, the global economy may benefit from consumers and businesses being allowed to obtain funds through P2P lending.

Couch surfing involves offering free accommodation anywhere in the world. It began in 2004. Within weeks it became clear that the service was much sought after and popular among people. The statistics show that 2 million registered members use the service in more than 230 countries worldwide [9]. Couch surfing combines social network technologies with cultural exchange to create inspirational experiences and meaningful connections that improve global awareness and understanding. Since its establishment in

2004, the CouchSurfing Company has switched from being a non-profit organisation making connections between travellers and local communities to a global community of people with the same travelling mindset and who participate in thousands of connections [8].

2 Data and Methods

The purpose of this paper is to determine the size of the population in the Czech Republic that participates in the sharing economy through the use of social networks and other channels with the purpose of identifying and describing individual factors (variables) affecting this market. The presented data were collected via an online questionnaire posted on social networks from June 1 to September 5, 2017 on Facebook. The authors used the aforementioned social network in order to distribute the questionnaires, as they believed they would be able to gain adequate number of randomly selected respondents in order to achieve the above mentioned objective of the contribution. This hypothesis was not confirmed; therefore, in the following part of the contribution, the authors decided only to deal with the description of the results in the sample and the results of the statistical tests will not be extrapolated to the whole population of the Czech Republic. In total, 88 randomly selected respondents completed the questionnaire via the aforementioned social network, with 78.41% stating that they participate in the sharing economy and 21.59% stating they do not. Since these were randomly selected respondents using Facebook social network, it was not possible to get basic information on the given respondents. The data obtained through the survey were analysed using Excel 2010 and R statistical software. When taking into consideration the small number of respondents (N=88), in particular due to the break down thereof according to multi-category variables (region and forms of participation in the sharing economy), some of the poorly represented categories were combined and the dependencies not generalised for the whole population across the social networks involved. As referred to above, only the description of the situation for the selected sample is only presented in the form of contingency tables and/or diagrams was involved.

For verification of the hypotheses, Pearson´s chi-square test, which is one of the most widely used statistical tests of independence in the contingency table, will be primarily used. The measurement for accepting or declining the hypotheses is so-called p-value with the Pearson´s chi-square test, which enables to verify whether the random variable has a pre-determined probability distribution. The hypotheses are tested on the level of significance "α"as follows: H_0 = null hypothesis p-value > α is accepted, alternative hypothesis H_a is declined. The following formula serves for calculation of the Pearson´s chi-square test:

$$X^2 = \sum_{i=1}^{k} \frac{(X_i - N_{pi})^2}{N_{pi}} \quad (1)$$

X^2 chi-square
X_i real count
N_{pi} expected count of type i

3 Results and Discussion

The first question in the survey sought to determine the age of the population in the sample set that participate in the sharing economy in the Czech Republic across the selected social networks. The arithmetic mean age of the men in the sample set was 31.73 years and of the women 29.22 years. The variable ratio of "male and female age" is illustrated in Figure 1.

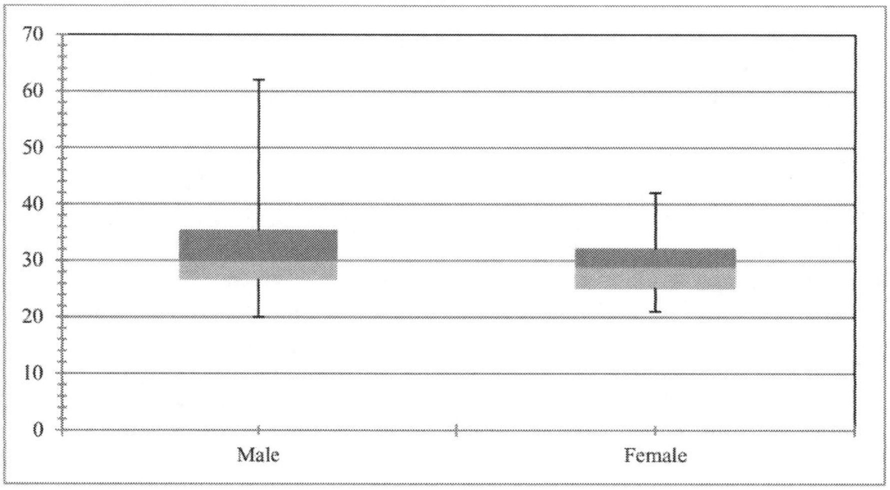

Fig. 1. Age structure of participants in the sharing economy
Source: Authors.

Figure 1 shows that the median age is 30 years for the men and 29 years for the women. The maximum age of the men in the sample set who participate in the sharing economy across the selected social networks was 62 years whilst that of the women was 42 years. Conversely, the minimum age among the men was 20 years and among the women 21 years. Figure 1 also clearly shows that the 25% age quantile is 26.75 years for the men and 25.25 years for the women, and that the 75% age quantile for the men is 35.25 years and 32.06 years for the women. The standard deviation of age is 8.01 years for the men and 6.82 years for the women. The coefficient of variation is 25.2% for the men and 23.3% for the women. In summary, the age level in the sample is only slightly higher among the men than the women, but the absolute and relative variability of age is substantially higher among the men. The value of the variation coefficient shows that both data files are homogenous.

Another objective was to identify whether the median age of the respondents in the sample who do and do not participate in the sharing economy is significantly different. This result is typically obtained from a two-sample t-test of median age values in the considered population. However, this test requires data normality of the "age" variable in both samples, or at least sufficient data in both samples. With only 19 measurements available in the set of respondents who do not participate in the sharing economy, it was necessary to first test the normality of the "age" variable in both samples. R statistical software was used to apply the Shapiro-Wilk test of normality for this purpose.

The computer outputs were $W = 0.9241$, p-value $= 0.000445$ for the respondents who participate in the sharing economy, and $W = 0.8013$, p-value $= 0.001195$ for the respondents who do not participate in the sharing economy. This means that the null hypothesis of age normality in both sets was rejected at the 5% significance level.

A non-parametric two-sample Wilcoxon test was applied using R statistical software to identify whether the age was significantly different in both considered population samples. The null hypothesis H_0: both population median ages are equal, was tested against the alternative hypothesis H_1: both population median ages are not equal. The computer outputs were $W = 733.5$, p-value $= 0.4283$. The null hypothesis was therefore not rejected. This means that at the 5% significance level it was not proved that the median age of the populations (in the sample) participating and not participating in the sharing economy across the selected social networks is significantly different.

The next question in the survey sought to determine whether participation in the sharing economy is affected by the factor "Permanent Address". Due to the lack of data for some regions, the results for the sample set are presented in the form of a contingency table. Regions with the least data are summarised in the category Other Regions. Table 2 shows the frequency (in percent) of the population in the sample set that participate in the sharing economy according to region and gender.

Table 2. Structure of participants in the sharing economy according to permanent address (region) and gender

Region	Male	Female	Total
Prague	28.99%	4.35%	33.33%
Central Bohemian Region	14.49%	4.35%	18.84%
Moravian-Silesian Region	7.25%	1.45%	8.70%
Zlín Region	7.25%	1.45%	8.70%
Other regions	28.99%	1.45%	30.43%
Total	86.96%	13.04%	100.00%

Source: Authors.

The table shows that most participants in the sharing economy in the sample set are located in Prague and the Central Bohemian Region. The responses to this question may be highly biased due to the low number of respondents and/or the structure of the respondents, both of which must be taken into consideration. In the tested sample, it has been revealed there is also a significant difference between the men and women. There are 2.5 times more male than female participants.

Participation in the sharing economy (within the tested sample) was subsequently tested against the other identified variables (gender, employment and education of respondents). For this purpose, contingency tables were used and Pearson's chi-squared test of independence applied accordingly.

Figure 2 shows the participation of the population tested in the sharing economy across the selected social networks according to gender. The results reveal that 68.18% of men and only 10.23% of women participate in the sharing economy.

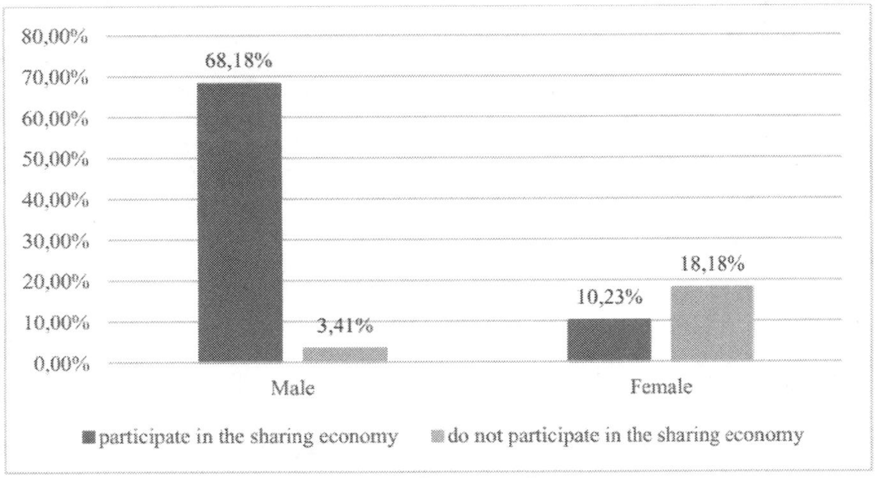

Fig. 2. Participation in the sharing economy according to gender
Source: Authors.

 https://doi.org/10.1051/shsconf/20196101025

Pearson's chi-squared test was applied in the relevant contingency table to test the null hypothesis H_0: participation in the sharing economy does not depend on respondents´ gender, against the alternative hypothesis. The outputs from the R statistical software were as follows: X-squared = 37.099, df = 1, p-value = $1.123.10^{-9}$. The null hypothesis was therefore rejected. It was demonstrated that at the 5% significance level the participation of the population from the tested sample in the sharing economy across the selected social networks is significantly dependent on respondents´ gender.

The dependency on the other identified variables was tested likewise. Figure 3 shows the participation of the population from the tested sample in the sharing economy across the selected social networks according to employment. The results reveal within the tested sample, the sharing economy is used by employees (45.45%), self-employed (19.32%), students (7.95%) and others (5.68 %).

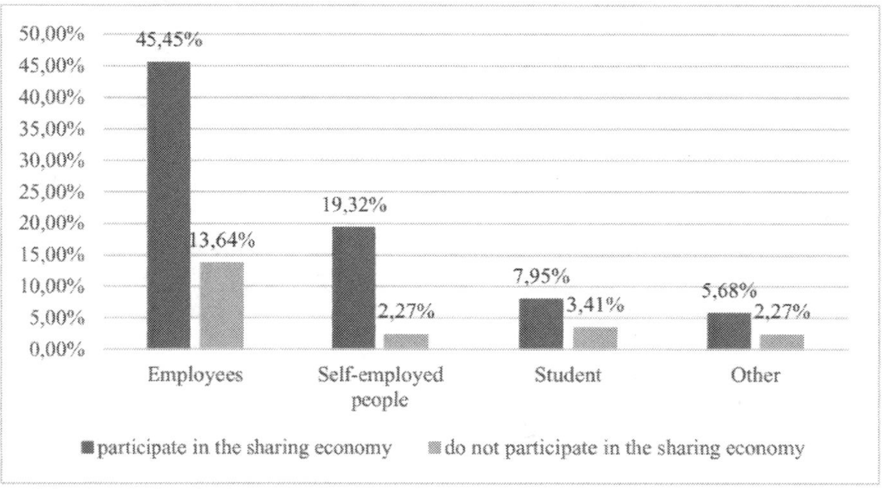

Fig. 3. Participation in the sharing economy according to employment
Source: Authors.

Pearson's chi-squared test was applied in the relevant contingency table to test the null hypothesis H_0: participation in the sharing economy does not depend on the employment of respondents involved in the tested sample, against the alternative hypothesis. The outputs from the R statistical software were as follows: X-squared = 2.061, df = 3, p-value = 0.5598. The null hypothesis was therefore confirmed. It was demonstrated that at the 5% significance level the participation of the population tested in the sharing economy across the selected social networks is not significantly dependent on respondents' employment.

Another variable that was possibly deemed to play an important role with regards to participation in the sharing economy was the education of respondents. Figure 4 shows that the majority of tested participants in the sharing economy across the selected social networks are tertiary-educated.

https://doi.org/10.1051/shsconf/20196101025

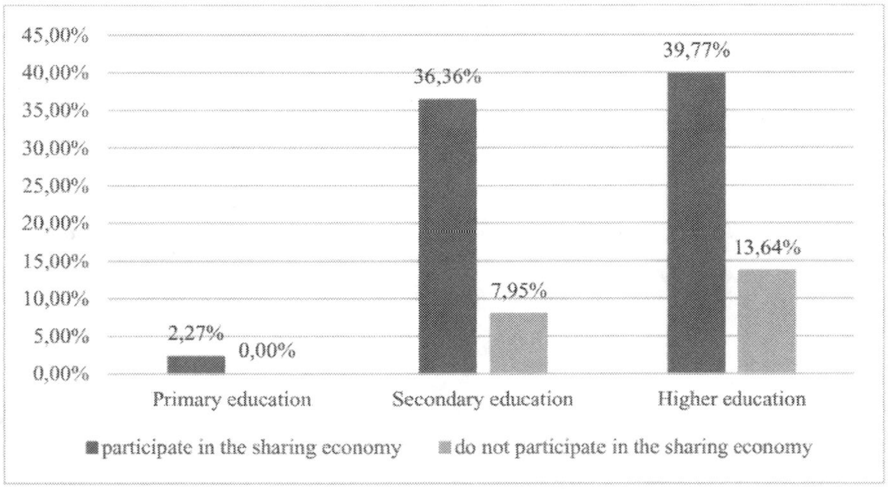

Fig. 4. Participation in the sharing economy according to education
Source: Authors.

The results were similar to those of the previous Pearson's chi-squared test of independence (p-value = 0.5253). This means that the test did not demonstrate that participation in the sharing economy was significantly dependent on the education of the respondents from the tested sample.

Similar results were obtained for the other identified variables. It can therefore be concluded that, on the basis of the identified and tested variables, participation of the tested population in the sharing economy across the selected social networks is only affected by respondents' gender.

Figure 5 shows the segments of the sharing economy that are utilised the most by respondents in the population of the sample set.

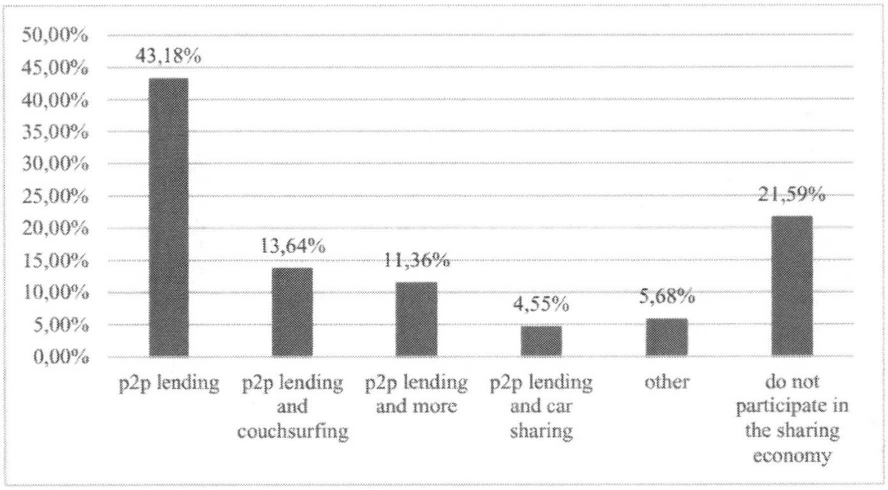

Fig. 5. Participation in the sharing economy according to segment
Source: Authors.

P2P lending is by far the biggest segment, followed by P2P lending and couch surfing. The chart shows that P2P lending (55.07%) and P2P lending couchsurfing (17.39%) are the mostly widely utilised services by respondents within the sharing economy.

4 Conclusion

The objective of this paper was to examine the age gap between male and female respondents participating in the sharing economy across selected social networks. It was determined that the gap between the arithmetic mean for male and female respondents was 2.51 years, that the gap between the upper age limit of the male and female participants in the sharing economy was 20 years, and that the lower age limit gap was negligible (only one year). The age breakdown of the male and female participants in the sharing economy was illustrated using a box diagram. A contingency table was used to describe the structure of participation in the sharing economy across the social networks according to the region of the permanent address of the respondents and their gender. Another objective of this paper was to identify other variables that may affect participation in the sharing economy of the respondents involved in the tested sample. The respondents' gender was the first variable studied by the authors. The data obtained from a questionnaire survey were entered into a contingency table and subjected to a Pearson's chi-squared test of independence, the results of which demonstrated the importance of gender for participation in the sharing economy. This finding did not apply to the other identified variables in the analysis. The authors also determined that among the population of the sample set P2P lending was the most utilised service within this market segment.

References

1. A. Bachmann et al., Online peer-to-peer lending - A literature review. *Journal of Internet Banking and Commerce*, **16**(2), 1-18, (2011)
2. I. B. Armet, et al., Peer-to-peer carsharing: Exploring public perception and market characteristics in the San Francisco Bay area. *Transportation Research Record: Journal of the Transportation Research Board*, **2**(2416), 27-36, (2014)

3. R. Belk, Why Not Share Rather Than Own? *The ANNALS of the American Academy of Political and Social Science,* **611**(1), 126-140, (2007)
4. R. Belk, You are what you can access: Sharing and collaborative consumption online. *Journal of Business Research*, **67**(8), 1595-1600, (2014)
5. F. Bonciu, Impact of the sparing economy on the labour market. *Romanian Economic and Business Review*, 43-51, (2016)
6. R. Botsman, R. Rogers, Harvard Business Review, *Beyond Zipcar: Collaborative Consumption* [online], Available at: https://hbr.org/2010/10/beyond-zipcar-collaborative-consumption (2010)
7. R. Botsman, Fast company, *The Sharing economy lacks a shared definition* [online], Available at: https://www.fastcompany.com/3022028/the-sharing-economy-lacks-a-shared-definition (2013)
8. CouchSurfing International, *CouchSurfing Raises $7.6 Million in Series A From Benchmark Capital and Omidyar Network to Expand Services and Fuel Growth in Cultural Exchange Network* (2011)
9. Couchsurfing.cz, *Co je Couchsurfing*. [online] Available at: www.Couchsurfing.cz (2018)
10. A. Erceg, Carsharing situation in Croatia. *Ekonomski vjesnik: Review of Contemporary Entrepreneurship, Business, and Economic Issues*, **27**(1), 183-195, (2014)
11. M. Felson, L. Spaeth, Community Structure and Collaborative Consumption: "A Routine Activity Approach". *American Behavioral Scientist*, **21**(4), 614-624, (1978)
12. L. Gansky, *The mesh: why the future of business is sharing,* (2010)
13. A. Gregory, G. Halff, Understanding public relations in the 'sharing economy'. *Public Relations, Review*, **43**(1), 4-13, (2017)
14. R. Hampshire, C. Gaites, Peer-to-Peer Carsharing Market Analysis and Potential Growth. *Transportation Research Record Journal of the Transportation Research Board*, **2217**(1), 119-126, (2011)
15. T. H. Webb, Peer-to-Peer Lending Market 2015, Platforms Revolutionizing Traditional Practices, *M2 Presswire* [online] Available at: https://www.m2.com/m2/web/publication.php/m2presswire (2015)
16. Pricewaterhousecopers, Consumer Intelligence Series: The Sharing Economy, *PricewaterhouseCoopers* [online] Available at: https://www.pwc.com/cz/cs.html (2015)
17. S. Shaheen, A. Cohen, *Innovative mobility carsharing outlook carsharing market overview, analysis, and trends,* (2016)
18. P. Teffer, *The 'sharing economy' lacks a common definition,* (2017)
19. The Economist, Banking without banks. *The Economist Newspaper Limited* [online] Available at: https://www.economist.com/finance-and-economics/2014/02/28/banking-without-banks (2014)

Management technologies of ensuring environmental protection as the territory development strategic priority

Olena Sushchenko[1], *Iryna* Trunina[2], *Oleksii* Klok[3] and *Olha* Loseva[4,*]

[1]D.Sc. (Economics), Professor, Simon Kuznets Kharkiv National University of Economics, Ukraine
[2]D.Sc. (Economics), Professor, Kremenchuk Mykhailo Ostrohradskyi National University, Ukraine
[3]PhD (Economics), Simon Kuznets Kharkiv National University of Economics, Ukraine
[4]PhD candidate, Simon Kuznets Kharkiv National University of Economics, Ukraine

Abstract. The issues of changing the paradigm of strategic management of sustainable development and management technologies for environmental protection are considered on the basis of analysis of sustainable development strategies. The Eastern European countries such as Belarus, Bulgaria, Czech Republic, Hungary, Moldova, Poland, Romania, Slovakia and Ukraine were used as the focus group. The country profiles were made based on the management technologies used to protect the environment. The analysis of existing models of the quantitative estimation of efficiency of sustainable development with the purpose of revealing the indexes capable to assess the efficiency of management technologies were carried out. Based on the indexes of the Economic Performance Index, the strengths and weaknesses of each country were identified (based on 24 EPI indexes in general and EPI sub-indexes in particular). Proceeding from the results of the content analysis and the components of the economic efficiency index, the practice of strategic environmental protection management as a strategic priority of sustainable development is based on increasing resource efficiency and environmental awareness, development of eco-technologies and instruments of environmental management.
Key words: Management technologies, Environmental protection, Development strategic, Strategic priority.

1 Introduction

The effectiveness of the development of the country is influenced, among other things, by the decision-making model at the planning and management level. With the adoption of the program plan "Agenda 21", the prevailing decision-making systems proved to be inconsistent in the context of the transition to sustainable development, since they divided social, economic and political factors, while sustainable development implies their full integration [1]. Thus, the program plan, adopted in 1992, made significant adjustments of

* Corresponding author: olha.loseva@hneu.net

the approaches to the strategic management of the territory development. As a result, the governments began to make significant institutional changes in order to ensure a systematic approach to the issue of sustainable development of the country. Also, the activities planned within the framework of «Integration of environment and development in decision-making» (Chapter 8), which describes the integration of environment and development of the policy, planning and management, assumed the adoption of national strategies for sustainable development.

The strategic management of the development of the administrative territory is a special kind of management that focuses on accomplishing the tasks set, registers the necessary steps to achieve an objective in the future. At the same time, the strategic management is the process that determines the sequence of actions to develop and implement a strategy. The sequence includes setting objectives, developing a strategy, identifying the necessary resources and maintaining communication with the external environment - all this al-lows you to achieve objectives. The strategy, as a key element in the strategic management in the context of this study, is considered as a set of measures for achieving sustainable development of the territory, which was developed on the basis of assessment and analysis of current development trends.

The ecological perspectives of sustainable development are based on ensuring proportionality between meeting the current and future needs of the society and natural conditions, taking into account the possibilities of using natural resources and landscapes. The awareness of the role of the environment on an equal footing with the economic and social component in ensuring sustainable development of the territory is an important stage on the way to reducing the negative human impact on the environment, overcoming global crises and poverty [2]. The key tasks to be resolved in the near future in this context are as follows: improvement of the quality of the environment, effective use of natural resources and waste management, maintenance of environmental stability, mitigation of climate change and adaptation to environmental changes caused by climate change.

At the present stage, it is impossible to ensure the sustainable development of each country without taking into account the environ-mental component. The global human impact on the environment leads to the realization of the need to attract new management technologies to ensure environmental protection. The transformation of approaches to strategic management of the territory is caused by an ever-increasing range of management technologies and improvement of territory management as a whole.

The objectives of this study are to identify common management technologies of protecting the environment and assess their effectiveness on the basis of the analysis of the national strategy for sustainable development.

2 Data and Methods

The protection of the environment as an element of sustainable development of the modern world is in the sphere of interests of the largest international organizations: the United Nations, the World Economic Forum, the European Commission, the Organization for Economic Co-operation and Development, the World Bank Group and so on.

According to the United Nations definition, the environmental protection is the activity of maintaining or restoring the quality of the environment aimed at preventing pollutant emissions or reducing the presence of pollutants in the environment by changing the characteristics of goods and services, consumption patterns, production methods, by recycling, preventing the degradation of the landscape and ecosystems, etc. [3].

Thus, the object of the research is the National Strategy for Sustainable Development as the fundamental document in the issues of strategic development of the country. The focus group of countries whose national sustainable development strategies were under study, was limited to a number of countries in Eastern Europe, such as Belarus, Bulgaria, Czech Republic, Hungary, Moldova, Poland, Romania, Slovakia and Ukraine.

The existing strategies, developed between 2001 and 2014, were analyzed at the first stage of the research. Table 1 contains a description of their main characteristics.

Table 1. National strategies for sustainable development in Eastern Europe

Country	Name of the Strategy	Time Bound	Developer
Belarus	National Strategy for Sustainable Development of the Republic of Belarus	2004-2020	the research Institute of Economy of the Ministry of Economy
Bulgaria	The National Development Programme: Bulgaria 2020	2012-2020	the Council of Ministers
Czech Republic	The Czech Republic Strategy for Sustainable Development	2004-2014 (2030)	the Government Council for Sustainable Development
Hungary	National Framework Strategy on Sustainable Development of Hungary	2012-2024	National Council for Sustainable Development
Moldova	the National Development Strategy 2020	2012-2020	"Expert-Grup" Analytical Center
Poland	Poland 2025 - a long-term strategy for sustainable development	2000-2025	Ministry of Investment and Development
Romania	National sustainable development Strategy Romania 2013-2020-2030	2008-2030	Ministry of Environment and Sustainable Development, National Centre for Sustainable Development in Bucharest
Slovakia	The National Strategy for Sustainable Development for the Slovak Republic	2001-2020	Ministry of the Environment of the Slovak Republic, United Nations Development Program
Ukraine	The Strategy for Sustainable Development "Ukraine 2020"	2015-2020	Administration of the President, Executive Committee of the Reforms and other

Source: Authors.

3 Results

3.1 The profiles of the countries

Belarus. In Belarus, the current status and changes in the environment are monitored and assessed on an ongoing basis in the context of environmental protection as a strategic priority of the country's development. The strategy implies the implementation of a set of measures aimed at gradual reduction of the anthropogenic impact on the environment, creation of conditions for building up and restoring the natural ecosystems of Belarus to an acceptable level. The country's environmental policy is based on the key ecological imperative, such as a man in the center of attention; environmental protection should become an integral part of the overall socio-economic development and cannot be considered in isolation from it; improvement of people's living standards. The features of management technologies of ensuring environmental protection include anti-crisis control, structural and technological adjustment [4].

Bulgaria. According to The National Development Program: Bulgaria 2020, the key aspect of the management of the territory on the path to sustainable development of the country is improving and integrating the national system of planning and management of regional development and the spatial planning and implementation of innovative solutions in the management practice. The conscious nature management is considered as one of the key components of sustainable development of the country, along with overcoming such problems as corruption, and structural and functional optimization of the government. The sustainable development strategy aims at combining environmental quality protection with ongoing economic growth in a way that is sustainable in the long term. On this basis, human capital plays a vital role in achieving the goals of sustainable development [5].

Czech Republic. Adaptive and crisis management, emergency planning formed the basis of "The Czech Republic Strategy for Sustainable Development" as main management technologies of environmental protection. The improvement of the quality of life of the population involves the introduction of technologies of active planting of greenery, reduction of the level of pollution and measures to improve the quality of spatial planning, systematic support of recycling. The ecological component in the system of sustainable development of the republic is expected to search for technologies for resolving the conflict between environmental protection (conservation of natural landscapes) and economic development (construction of new infrastructure facilities). The strategy implies the intensification of economic incentives for the transition to sustainable patterns of production and consumption and promotion of sustainable development of municipalities and regions [6].

Hungary. According to [7], sustainable development is a management method that emphasizes the long term, value preserving utilization of natural resources, enhancing prosperity without destroying diversity, complexity and ecosystem services. Prudent resources management and forward-looking management are a basis of development. The strategy implies the use of technologies aimed at achieving environmental awareness and sustainable lifestyle through the system of currently running programs assuming that in the future the ecological way of life will be perceived as a natural component. The development of knowledge and innovation-based nation is considered as an incentive to sustainability, while much attention is paid to improving the model of adoption of environmental solutions in the field of environmental protection.

Moldova. In view of the adoption of The National Development Strategy 2020, the paradigm for the development of the country's economy was substantially changed. The optimization of managerial policy in the matters of healthcare, culture, social security and environmental protection, are crucial for the country's sustainable development. As part of the strategy, the transition to green economic development is planned, including more effective management in the field of sustainable development by integrating and strengthening environmental aspects in all areas of the country's socio-economic development. The management technologies in this context imply significant institutional changes through reforms and modernization [8].

Poland. A long-term strategy for sustainable development of Poland is the structural basis of the country's development and a weighty basis for the implementation of specialized programs and plans for the development of human resources, regional development, including environmental protection, energy and climate policies, etc. The strategy is based on the principles of openness, unhindered access to information, active social communication and citizens' participation in the process of making managerial decisions. In view of the tendency of intensification of anthropogenic human impact on the environment, the strategy involves targeted technological measures to improve the quality of its components. By applying restrictive strategies towards sustainable development in the country, the assessment of the impact of adopted policy documents on the environment is made [9].

Romania. "National Sustainable Development Strategy Romania 2013-2020-2030" was developed on the basis of the program documents common to the member states of the

European Union and covers the implementation of a number of tasks, such as: improvement of energy efficiency and extended use of renewable energy resources in order to mitigate the effects of climate change. At the current stage, the reduction of energy consumption is expected to be realized through active promotion of the Green House Program. Thus, the analysis of the country's sustainable development strategy in the context of environmental protection is testimony to the perception of sustainable development as the only true and rational perspective for the development of the society. The key management technologies of environmental protection within the framework of sustainable development are the application of anti-crisis management practices, the ability to anticipate the effects of climate change, to develop cross-sector contingency plans comprising portfolios of alternative crisis-management solutions in case of natural or man-made disasters [10].

Slovakia. The tasks formulated in [11] on the basis of the analysis of the world experience of strategic management of sustainable development are based primarily on ethical principles, global liability and capacity of a small country to provide assistance in solution of international problems. The tasks to be addressed within the frame-work of the environmental protection strategy are as follows: to overcome the consequences of prolonged unregulated use of limited natural resources, the use of management technologies that promote reasonable consumption, the improvement of long-term planning practice, the management of natural resources (special management system, in order to properly use, protect or avoid threatening of agricultural soil, waste management. The active application of the possibilities of environmental management and technologies of environmental legislation in Slovakia is directed at minimizing the risk factors of the environment, reducing the anthropogenic impact. The forecasting is considered as an opportunity to regulate economic and urbanizing activities for the rational use of resources and contribute to making the landscape management ecologically friendly.

Ukraine. It was found in the course of the retrospective analysis of the legal and regulatory framework of Ukraine in the context of the transition to sustainable development that the strategic management became a practice only in 2015 (The Strategy for Sustainable Development Ukraine 2020 [12]). Prior to this period, the unified strategy of Ukraine's sustainable development in various legal wordings was in the draft phase. The modern policy of sustainable development of Ukraine reflects the principle of parity of the economic, social and environmental components thus recognizing that long-term progressive development is not possible in conditions of environmental degradation. Consequently, there is an objective need to create a unified state system of management, regulation and control over compliance with resource-ecological standards, restrictions and requirements for environmental management and environmental quality. The management technologies are based on the complex approach, that is, the interests of the state are considered in the con-text of the paradigm of sustainable development of mankind. The strategy provides for strengthening of cross-sectoral integration, continuous interaction, coordination and cooperation of the authorities and society; mechanisms of mutual responsibility and assessment, a clear monitoring system that ensures the coordination of all government programs and activities with the objectives written in the Strategy.

3.2 Indexes

At the second stage, the analysis of existing models of quantitative assessment of the effectiveness of sustainable development is made to identify indexes that can assess the management technologies of environmental protection as the territory development strategic priority.

The study defines [13] the sustainable development strategy as a coordinated set of participatory and continuously improving processes of analysis, debate, capacity-strengthening, planning and investment, which integrates the economic, social and environmental objectives of society, seeking trade-offs where this is not possible.

Table 2 gives the main integral indexes which are used in the world practice of assessing the sustainability of development [14-23].

Table 2. Indicators of Sustainable Development

Abbreviation	Name of the Index	Sustainable Development		
		Social	Environment	Economic
ANS	Adjusted Net Saving	1	1	1
EDP	Environmentally Adjusted Net Domestic Product	0	1	1
EF	Ecological Footprint	0	1	0
EPI	Environmental Performance Index	0	1	0
ESI	Environmental Sustainability Index	0	1	0
GCI	Global Competitiveness Index	1	0	1
GGDP	Green gross domestic product	0	1	1
GS	Genuine saving indicator	1	1	1
HDI	Human Development Index	1	0	1
ISEW	Index of Sustainable Economic Welfare	1	1	1
SSI	Sustainable Society Index	1	1	1

Source: According to [14-23], authors.

The column "Sustainable Development" of Table 2 illustrates the inclusion of one or another component of sustainable development (social, environment, economic) in the index, where: rating 1 implies a direct relation of the indexes, included in the integral assessment, to this area of sustainable development; rating 0 implies the indirect relation of the indexes, included in the integral assessment, to this area of sustainable development.

The obtained data indicate that 90% of the studied indexes in one way or another assessing the sustainable development of the country take into account the state of the environment. At the same time, such indexes as Ecological Footprint, Environmental Performance Index, and Environmental Sustainability Index illustrate only the state of the environment as an element of sustainable development.

On the basis of the analysis of the resources that calculate the relevant indexes, we can claim that the most relevant data includes such indexes as the Environmental Performance Index (2018 indicators available) [17], the Global Competitiveness Index (2017) [23] and the Human Development Index (2016) [24], the Sustainable Society Index (2016) [22]. Thus, the comparative analysis of the countries under study is rational to make on the basis of the Environmental Performance Index.

According to the EPI calculation methodology, the key categories in which all 24 indexes are grouped are: air quality, water and sanitation, heavy metals, biodiversity and habitat, forests, fisheries, climate and energy, air pollution, water resources, agriculture [17]. Fig. 1 represents the specific weight of each indicator in the structure of the index.

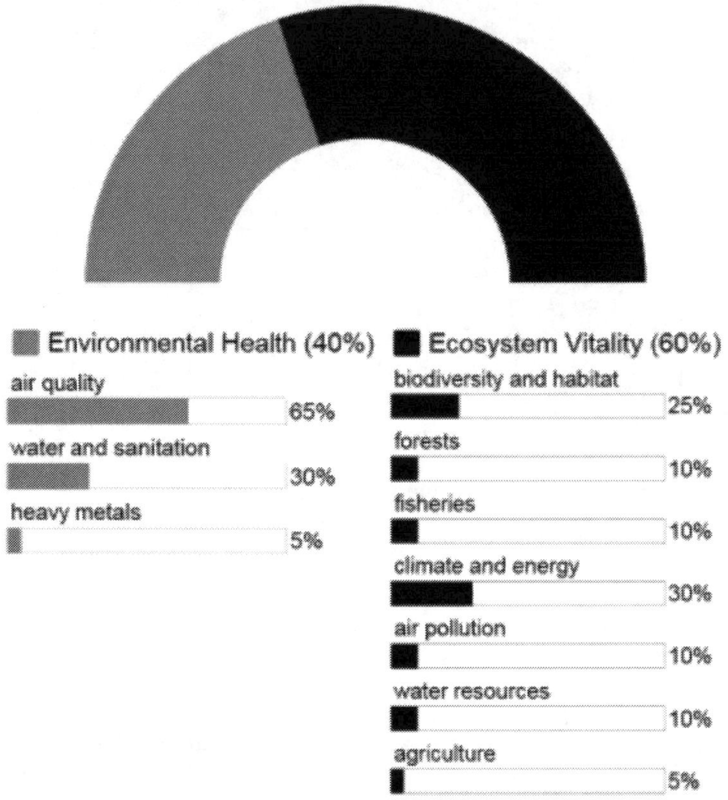

Fig. 1. Structure of indicators
Source: Authors.

The key sub-indexes for the calculation of EPI-2018 are Environmental Health (EH), Ecosystem Vitality (EV). The former is used to assess the degree of environmental impact on human health, the latter, in turn, focuses on indexes that reflect the biodiversity losses and climate change. Whereas the policy objective of Environmental Health has an empirical basis for deriving weights, the selection of weights in Ecosystem Vitality, is more subjective [17]. The sub-index values of EH, EV are shown in Table 3.

Table 3. EPI sub-index values

Country	EPI	EH	EV
Slovakia	70,60	63,87	75,08
Bulgaria	67,85	69,60	66,68
Czech Republic	67,68	68,69	67,01
Hungary	65,01	57,67	69,90
Belarus	64,98	69,55	61,94
Romania	64,78	58,67	68,85
Poland	64,11	58,71	67,72
Ukraine	52,87	64,44	45,16
Moldova	51,97	60,29	46,42

Source: According to [17], authors.

The results obtained indicate that Slovakia is the absolute leader among the countries under study, according to the index (EPI = 70.60). The average index for the sample of countries was EPI = 63.32 in view of the rather low indexes of such countries as Ukraine and Moldova. In the global ranking, Slovakia takes the 28th place in the world (from 180 countries), while Ukraine and Moldova in 2018 took 109th and 112th places respectively.

Since the task of the study is to assess the effectiveness of management technologies directed at protecting the environment, we consider it expedient to identify the strengths and weaknesses of each country (based on 24 EPI indexes in general and EPI sub-indexes in particular) for the purpose of subsequent search of the interconnection of the environmental support technologies and quantitative indexes of environmental efficiency provided by the strategy. The range of problems of environmental efficiency lies in the sphere of reducing pressure on the environment and, as a result, on human health, stimulating the viability of ecological systems and stable management of natural resources, including counteraction to the climate change, the burden of economic activity on the environment, the effectiveness of state policy in the sphere of ecology.

Table 4 includes the assessment of the strengths and weaknesses of activities for the protection of the environment in Eastern Europe. Based on the data received, such global problem as air pollution is the most vulnerable place in the country's ecological efficiency (in particular, Romania, Poland, Slovakia, Bulgaria and Czech Republic are the countries affected by this problem most of all). The solution of the problems connected with biodiversity conservation and habitat conservation requires finding new management approaches for such countries as Belarus and Moldova. For Ukraine, the most urgent problem is overcoming rapid negative climate changes. Opposite from the point of view of Ukraine's positions in the rating is the experience of Slovakia and Romania, where technologies for overcoming climate pollution and reducing greenhouse gas emissions are effectively working.

Table 4. Countries Rank

Country	2018 EPI	EH	EV	The best index position	The worst index position
Slovakia	28	89	3	Climate & Energy (9)	Air Quality (133)
Bulgaria	30	66	24	Biodiversity & Habitat (23)	Air Quality(89)
Czech Republic	33	69	22	Heavy Metals (5)	Air Quality (97)
Hungary	43	107	12	Agriculture (5)	Air Quality (146)
Belarus	44	67	48	Water Resources(18)	Biodiversity & Habitat (114)
Romania	45	105	15	Climate & Energy (15)	Air Quality (136)
Poland	50	104	18	Biodiversity & Habitat (6)	Air Quality (144)
Ukraine	109	87	139	Heavy Metals (15)	Climate & Energy(143)
Moldova	112	102	127	Agriculture (35)	Biodiversity & Habitat (160)

Source: Authors.

Air pollution is a global problem affecting people in all countries and socio-economic groups. The unsatisfactory condition of the environment, primarily air pollution, causes an increase in morbidity and mortality among the population. On the basis of the experience of Western European countries, the most common management technologies in this respect can be considered:

– air management,

– air protection,

– monitoring and evaluation of climatic parameters and air quality,

– implementing clean and affordable modern technologies inside homes,

– furthering public awareness on air pollution,

– encouraging scientific progress in regards to air pollution and its subsequent effects on health and climate change.

The most common practice in the issues of energy and climate is the adaptation of the country to changing climatic conditions, which is due to anthropogenic influence. The strategies for sustainable development of the countries from the focus group may imply the following:

– improvement of energy efficiency and expanded use of renewable energy resources,
– investments into the energy sector,
– use of potential of renewable climatic resources (solar and wind energy),
– regulation of economic and urbanizing activities,
– ecologisation of landscape management.

In order to preserve the existing biodiversity, which is affected by such negative influence factor as urbanization, the presence of heavy and energy industries in the country, the intensification of agriculture and forestry, mass tourism, the countries strive to level out the impact of noise, emissions and other factors of impact on the biodiversity of the territory. This aspiration is manifested in sustainable development strategies in the following way:

– conservation and strengthening of life-supporting, self-regulatory, self-purifying and self-supporting natural mechanisms,
– adaptive management, for a reduction in the number of endangered varieties and species of plants and animals and the preservation of biodiversity,
– elaborating and implementing action plans with regards to endangered animal and plant species,
– forest fire prevention and sustainable utilization management of the forest resources.

4 Discussion and Conclusion

Thus, based on the content analysis of the strategies, we may state that the main objective of biodiversity conservation is to ensure the reasonable use, reproduction and protection of animal and plant species, their genetic and biological diversity and renewal of natural ecosystems, while maintaining regional and global environmental balance and production possibilities of the biosphere.

The characteristic feature of sustainable development strategies in the context of environmental protection as a strategic priority at the level of separate territories is the continuous adjustment and cyclical renewal of the strategy in the dialogue with stakeholders to ensure comprehensive sustainable development. At the present stage, the key importance is given to the following technologies:

– rise of resource efficiency (effective management of resources throughout their life cycle), which in turn is closely dependent on eco-innovations, changes in the structure of production and consumption),
– rise of awareness (conscious nature management is considered as one of the key components of protection of environment), while the practice of creating a competitive environment in the form of ratings is widespread;
– development of ecological technologies (green technologies); along with the growing popularity of the practice of the use of renewable energy sources, there is an active support of funding research, subsidies, eco-labeling and government "green" procurements;
– effective ecological management, which includes new methods of recording, monitoring and reporting on environmental pollutants and wastes, using the best available information and technology.

References

1. Agenda 21, Earth Summit, *The United Nations Program of Action from Rio, 1992* [online]. Available at: https://sustainabledevelopment.un.org/content/documents/Agenda21.pdf (1992)
2. Environmental protection expenditure, European Commission [online]. Available at: http://ec.europa.eu/eurostat/web/environment/environmental-protection-expenditure (2018)
3. *Glossary of Environment Statistics, Studies in Methods*, Series F, No. 67, United Nations, New York (1997).
4. *National Strategy for Sustainable Development for the Period to 2020 of the Republic of Belarus*, The National Sustainable Development Commission of Belarus. Minsk, p. 76 [online]. Available at: http://www.by.undp.org/content/dam/belarus/docs/OON_sMall%20-%201.pdf (2004)
5. *The National Development Programme*: Bulgaria 2020 [online]. Available at: https://www.eufunds.bg/archive/documents/1357828564.pdf (2018)
6. *The Czech Republic Strategy for Sustainable Development*, Prague [online]. Available at: https://www.mzp.cz/C125750E003B698B/en/czech_republic_strategy_sd/$FILE/KM-CR_SDS_eng-20041208.pdf (2004)
7. *National Framework Strategy on Sustainable Development of Hungary* [online]. Available at: https://www.stakeholderforum.org/fileadmin/files/National%20Framework%20Strategy%20on%20Sustainable%20Development.pdf (2018)
8. *National Development Strategy 2020*. Moldova 2020 [online]. Available at: http://particip.gov.md/public/files/Moldova_2020_ENG1.pdf (2018)
9. *Strategia zrównoważonego rozwoju polski do roku do roku 2025* [online]. Available at: http://snep.edu.pl/sms/materialy/strategia%20zrownowazonego%20rozwoju%20polski%20do%20roku%202025.pdf (2018)
10. *Romanian National Strategy for Sustainable Development – Horizons 2013-2020-2030*. Ministry of Environment and Sustainable Development, Bucharest [online]. Available at: http://sdgtoolkit.org/wp-content/uploads/2017/02/National-Sustainable-Development-Strategy-of-Romania-2013-2020-2030.pdf (2008)
11. *National Strategy For Sustainable Development For The Slovak Republic* [online]. Available at: https://www.thegef.org/sites/default/files/ncsa-documents/Slovakia_NSSD_Final.pdf (2001)
12. *On the Strategy of sustainable development "Ukraine – 2020": Decree of President of Ukraine from 01.12.2015 № 5/2015* [online]. Available at: http://zakon3.rada.gov.ua/laws/show/5/2015?nreg=5%2F2015&find=1&text=%F1%F2%E0%EB&x=0&y=0 (2015)
13. Dalal-Clayton B., Bass S. *Sustainable Development Strategies: a resource book*. OECD & UNDP [online]. Available at: https://www.sd-network.eu/pdf/resources/Dalal-Clayton,%20Bass%20(2002)%20-%20Sustainable%20Development%20Strategies%20-%20A%20Resource%20Book.pdf (2001)
14. *World development indicators 2007* [online]. Available at: http://documents.worldbank.org/curated/en/336241468138277212/World-development-indicators-2007 (2007)

15. *Environmentally Adjusted Net Domestic Product*, The United Nations Statistics Division [online]. Available at: https://unstats.un.org/unsd/environmentgl/gesform.asp?getitem=467 (2018)
16. *Ecological Footprint,* Global Footprint Network, Advancing the Science of Sustainability [online]. Available at: https://www.footprintnetwork.org/our-work/ecological-footprint (2018)
17. *Environmental Performance Index* [online]. Available at: https://epi.envirocenter.yale.edu (2018)
18. Socioeconomic Data and Applications Center, *Environmental Sustainability Index* [online]. Available at: http://sedac.ciesin.columbia.edu/data/collection/esi (2018)
19. *Green gross domestic product*, Green Growth Knowledge Platform [online]. Available at: http://www.greengrowthknowledge.org (2018)
20. *Genuine saving as a sustainability indicator*, The World Bank [online]. Available at: http://documents.worldbank.org/curated/en/908161468740713285/Genuine-saving-as-a-sustainability-indicator (2000)
21. F. M. Chelli, M. Ciommi, G. Chiara, Index of Sustainable Economic Welfare: A comparison of two Italian regions. *Procedia – Social and Behavioral Sciences*, **81**, 443-448, (2013)
22. *Sustainable Society Index* [online]. Available at: http://www.ssfindex.com (2017)
23. *The Global Competitiveness Report 2017-2018*, World Economic Forum Annual Meeting [online]. Available at: https://www.weforum.org/reports/the-global-competitiveness-report-2017-2018 (2017)
24. *Human Development Reports*, United Nations Development Programme [online]. Available at: http://hdr.undp.org/en (2018)
25. N. V. Nenkov, M. M. Petrova, Y. Dyachenko, Intelligence Technologies in Management and Administration of Justice. *3rd International Multidisciplinary Scientific Conference on Social Sciences and Arts SGEM 2016 Conference Proceedings*, 385-392, (2016)
26. N. Nenkov, O. Sushchenko, Y. Dyachenko, Role of chief information officer within the system of human resource development in the service organizations (tourism). *Economic Annals-XXI*, **165**(5-6), 97-103, (2017).

An imitation model of the financial-economic mechanism of taking strategic decisions at the enterprise

Iryna Trunina[1,*], *Iryna* Kushal[2], and *Denys* Zagirniak[3]

[1]Doctor of Economics, professor, Kremenchuk Mykhailo Ostrohradskyi National University, Ukraine
[2]PhD candidate, associate professor, Volodymyr Dahl East Ukrainian National University, Ukraine
[3]PhD candidate, associate professor, Kremenchuk Mykhailo Ostrohradskyi National University, Ukraine

Abstract. The crisis conditions of the enterprise functioning restrain the possibilities for the innovation activity and lay down higher requirements to the determination of the tasks of the financial strategy of development. To enhance the adaptive possibilities the authors propose to use an imitation-class mathematical model, indicate the basic stages of its creation, determine the structure: a control system; a functional system consisting of functional financial and economic blocks, an information system including a database. We distinguished the basic indices for the analysis of the ability of the enterprise to take strategic decisions of both financial and economic components of the mechanism of their realization. The practical realization of the proposed imitation model of the financial-economic mechanism of taking strategic decisions based on the calculation of the integral index provides the required information basis for the grounded strategic decision and further development of the enterprise.
Key words: model, mechanism, management, decision, enterprise

1 Introduction

The transformation condition of the economy development requires from the top managers not only certain skills and knowledge of the management practice, but also taking into account all the factors of influence on the enterprise activity and, first of all, taking into account the risk factor inherent in the nature of strategic decisions. One can solve this problem by creation of models, which will make it possible not only to facilitate the stages of the routine analysis but also to automate them. All this enables the manager to have additional time for the creative approach to taking strategic decisions, which will improve their quality in the future and enhance the level of the management work.

In practice the situation when the enterprise takes strategic decisions under the conditions of risk occurs most often as risk is an inseparable part of economic life. One can choose a decision reducing the risk but it is impossible to avoid it. Most often an enterprise does not try to avoid risk when taking strategic decisions but it tries to minimize it using the instruments of insurance and financial market. At the stage of the development of a

* Corresponding author: deniszagirniak@gmail.com

managerial decision it is expedient to use modeling methods to reduce the cost of further insurance against financial losses after their realization.

Well-known economists of the world have performed fundamental research in this field. So, J.K. Lafta made an essential contribution into the research of the category of "mechanism" concerning taking strategic decisions. He considered it an "apparatus" or a "device" [1]. W. Cherchmen, R. Ackoff, L. Arnof stated that the presence of a mechanism of taking strategic decisions is the key to success, the efficiency in realization and modeling their basic parameters [2]. A. Kulman had the same opinion [3].

Modeling does not cancel the use of experience and intuition of the person who takes the decision; on the contrary, their significance grows. Modeling is a certain formalized type of experience and makes it possible to control the intuitive representations of the leaders of the enterprise. The person who takes a decision can supplement the model with the required qualitative analysis taking into account his/her own experience. Thus, the model supports the manager in the process of making strategic decisions but in no way replaces him/her [4].

2 Data and Methods

Top managers should pay special attention to the creation of the models of taking strategic decisions because: the solutions are of a non-changeable character; the termination of the realization of the initiated long-term projects or even essential changes in them are sometimes impossible due to economic criteria; the success of strategic decisions, first of all, exclusively depends on top managers; the effect of the taken strategic decisions is measured by thousands and millions monetary units. Under such conditions the support of strategic decisions by means of modeling is necessary and topical, certifies the expediency of the use of this method during taking decisions and justifies material and labor costs of its use.

The use of modeling in the process of taking decisions makes it possible to support such advanced managerial principles as system, multivariate analysis, taking into account the uncertainty of the activity conditions, financial-economic risk, the complexity of taking into account the financial, economic social and other consequences of decision realization.

3 Results

3.1 The purpose of the research

One of the most important conditions of taking high-quality strategic decisions in the field of finance consists in the availability of the financial-economic mechanism of the process of their taking and the improvement of this mechanism. So, the creation of a model of financial-economic mechanism of taking strategic decisions is the purpose of the research. It will promote the decisions efficiency, the increase of financial profit, the simplification and automation of the decision-taking process, the improvement of their organizational and informational support. Besides it is the key factor of management success.

The use of modeling in the process of taking strategic decisions does not reduce the significance and the intuition of the manager; on the contrary, when used correctly, it promotes the development of the initiative of the employees who are to put these decisions into practice.

3.2 The analysis of the basic parameters of the imitation model of the financial-economic mechanism of taking strategic decisions at the enterprise

Having generalized the information from economic literature and having performed substantial research, the authors can propose the following definition of a "model of a financial-economic mechanism of taking strategic decisions". It is a dynamic object reflecting the interdependences and relations of its components and serving to identify the alternatives of financial and economic development under the influence of the external and internal factors of the enterprise activity environment.

Under the conditions of transformation economy the need for efficient scientifically substantiated economic forecasts does not decrease, on the contrary, it essentially increases. That is why it is necessary to look for the ways for the correlation of the content and organization of the development of financial-economic forecasting with the changes taking place in all the fields of activity. A complex program-oriented forecast of the financial and economic development of the enterprise under the conditions of the transformation economy is necessary. It causes the use of mathematical models whose creation requires taking into account certain elements. They include a problem situation; the term of taking a strategic decision; the necessary resources; a system of factors subject to management; unmanageable factors; the state of the environment (by the unmanageable factors); the system of restrictions (resource, time, legal, social etc.); a system of manageable factors relations; a system of unmanageable factors relations; the result of the actions; a system of the criteria for the determination of the actions result correspondence to the posed purpose; a system of the usefulness functions for the estimation of the result by the criteria system; a system of functions for the estimation of the expected effect of the actions realization; a complex estimation of the alternatives; an optimal strategic decision.

In the course of the research, it was found that the quality of a strategic decision would increase if senior executives would take into account the manifestation of the laws of the organization, namely the law of synergy. In making strategic decisions, synergy means the strategic advantages of the enterprise, in which synergies between the components of the financial and economic decision-making mechanism take the form of strategic business development. Synergetic relationships between the components of the financial and economic mechanism for making strategic decisions play the major role, since they are the building forces of the constituent parts, and vice versa, their weakness leads to the breakdown of communications and its disintegration as a holistic creation. Therefore, the practical and organizational problem is the choice of methods, approaches and system of indicators for assessing synergy effects when making strategic decisions, the solution of which requires the introduction of a certain evaluation mechanism, which will allow to quantify the results of the manifestation of synergy and determine the components of the financial and economic mechanism of adoption solutions that need to be strengthened to get synergistic benefits in the future. One of the main conditions for the emergence of synergistic effects when making strategic decisions is the presence of an effective financial and economic mechanism in the enterprise.

The concrete forms of manifestation of synergy in enterprises when making strategic decisions are: team synergy and improvement of corporate culture; cost reduction; managerial competence (managerial synergy); trade synergy; operational synergy; investment synergy; accelerating the movement of information; cash flow growth (financial synergy), etc.

By the positive values of all the parameters of the financial and economic component, there is certainly a synergistic effect and the question of its quantification is raised. Justification by means of an appropriate system of calculations of the most economically advantageous strategic alternative solution based on multicriteria optimization is one of the possible ways of practical realization of the synergetic methodology, namely: the sources of the synergistic effect, the classification of synergy, the consideration of the principles and principles of synergetics, the determination of the type of the system under study and nature of the manifestation of synergy depending on it.

Open systems are characterized by the presence of more powerful sources of development, metabolism and energy with the environment. All this is also relevant for economic systems and the enterprise is the most open system, which is inherent in measures of strategic management. For a company as an open system in a market environment, the manifestation of synergy will be fundamentally different, depending on the level of competition.

At high intensity of competition, the main losses of synergy of the enterprise are changing the direction of development, that is, enterprises are forced to lose additional synergy reserves for structural or assortment policy, etc. But intense competition forces most of companies to start balanced strategic and tactical measures. As the intensity of competition decreases, there is a general decrease in the "energy tone" of the enterprise. That is, because of the low intensity of enterprise competition, the lower value of final synergy. At zero intensity of competition, synergy effects are zero. This situation is dangerous in nature.

It should be emphasized that the synergy effect is a somewhat complex phenomenon that needs to be planned and used. The management of an enterprise should create the necessary conditions for the realization of the potential synergy effect. Otherwise, there may be a negative synergy effect.

In the practice of strategic management, the calculation of a positive synergistic effect ensures the verification of strategic decisions at the development stage for the purpose of meeting the strategic goals of the objective conditions of functioning in the market, using the integral indicator, since the financial and economic components are not homogeneous. As a result of the calculation of the integral indicator, one can judge the effectiveness of the whole team, and above all, the effectiveness of senior management, those who take strategic decisions at the enterprise.

Calculation data of integral indicators of estimation of economic and financial components of making strategic decisions will be the starting point for building an effective model of the mechanism of making strategic decisions of enterprises and an indicator of determining the success of their activity at the appropriate stage of development.

The basic stages of the creation of the model of financial-economic mechanism of taking strategic decisions at an enterprise include: the determination of financial purposes; the determination of the manageable components and the structure of the model; the choice of the model type in accordance with the financial purpose; the estimation, testing and modification of the manageable components of the model; the verification of the model adequacy to the conditions of financial-economic development; economic interpretation of the obtained results.

The formation of adequate requirements to the creation of the model of financial-economic mechanism of taking strategic decisions rests on the difference between the strategic decisions concerning the field of activity (the object of management on the whole) and the enterprise. So, for their realization in practice the managers must know if the acting financial-economic mechanism meets the requirements of the decision (capital, manufacturing capacity, cost policy, property status, etc.), if it is possible to approve of the decision under the existing activity conditions.

The solution to this problem consists in the use of an imitation-class economic-mathematical model with the following advantages: it allows the description of the process simulated with higher adequacy than others; it is characterized by certain flexibility of the structure, algorithms and parameters of the system; it provides the possibility for the application of computer hardware, which essentially reduces the time and cost of the task solution.

The proposed structure of the imitation model of the financial-economic mechanism of taking strategic decisions consists of three parts: a control system; a functional system including three blocks; an information system including a database.

In its turn, the control system contains a modeling control block; a dialog block, a block for processing the modeling results; an event calendar.

Due to the proposed imitation model of the financial-economic mechanism of taking strategic decisions at the enterprise it is possible not only to calculate its basic parameters but also to estimate its correspondence to the requirements of the managerial decision, i.e. to determine the degree and the ability of the enterprise to take a certain decision.

In the authors' opinion, the readiness of the enterprise to take strategic decisions is characterized by a financial component (the efficiency of the use of the capital, the efficiency of the activity, the structure of expenditure, the production program, the taxation policy) and an economic component (the characteristic of the managerial structure, industrial engineering, management organization), which reflects on the proposed imitation model of financial-economic mechanism of taking strategic decisions by the composition of its functional system (the financial and economic block).

This analysis is performed in the imitation model of the financial-economic mechanism of taking strategic decisions at the enterprise in the functional system, namely, in the block of calculation of the financial indices concerning taking strategic decisions and in the block of calculation of the economic indices concerning taking strategic decisions. The indices are to be calculated in dynamics before and after taking strategic decisions to determine the lines of improvement.

According to the presented structural-logical scheme of the analysis of the ability of the enterprise to take strategic decisions and the determination of the indices of the estimation of the financial-economic mechanism of taking strategic decisions included into the functional system of the imitation model, this process is to be performed in the following order: the formation of the system of indices that by their nature reflect the features and regularities of each component of the enterprise ability (financial, economic) to take strategic decisions; the calculation of the indices values, the determination of their significance and reference values; the calculation of the integral index of the ability of the enterprise to financial-economic development and taking strategic decisions; making conclusions by the results of the analysis.

The analysis of the theoretical developments and the diagnostics of the financial-economic mechanism of taking strategic decisions at the enterprise made it possible to identify the following indices for the analysis of the ability of the enterprise to take strategic decisions:

I. The indices of the analysis of the financial component of the ability of the enterprise to take strategic decisions: a) the efficiency of the use of the capital: the coefficient of autonomy, financial leverage, long-term financial independence, relation of the long- and short-term liabilities, the period of capital turnover, of equity turnover, financial dependence, the concentration of loan capital, financial stability; b) the efficiency of the activity: of the own working capital turnover, provision of current assets by own funds, current liquidity, that characterizes the component of short-term liabilities in the capital of the enterprise, profitability of sales on net profit, sales for gross income; c) the production program: the dynamics of the change of the proceeds from sales, net profit; d) the cost structure: the structural dynamics of the cost by the elements, constant and variable cost; e) the property status: the share of current assets in the working capital, the share of fixed means in assets, the coefficient of the fixed assets wear, renewal of the fixed assets, mobility of the assets; f) the taxation policy: the index of the general taxation level, the coefficient of taxation efficiency, tax-capacity of the sold product, taxation of the enterprise income, taxation of the cost, taxation of profits, preferential taxation.

II. The indices of the analysis of the economic component of the enterprise ability to take strategic decisions: a) industrial engineering: the coefficient of the automation of the production process, the coefficient of uninterrupted process, the coefficient of renewal; b) the organization of management: the coefficient of position regulation, the coefficient of the completeness of the coverage of management functions, the coefficient of the economy of the management system, the coefficient of the reliability of the operation of the management personnel; c) the structure of management: the coefficient of the structure flexibility, the coefficient of the structure specialization, the coefficient of the structural subdivisions regulation, the coefficient of the organizational structure orientation to the market.

The presented indices vary within the interval from 0 to 1 and are stimulators as to their nature. One should determine the total amount of the financial and economic components of the readiness to taking strategic decisions based on the integral index by formula (1):

$$Ig\sum_{i=1}^{n} V_{ij} \frac{X_{ij}}{Xe_{ij}}, \tag{1}$$

where V_{ij} – the significance of the i-th index of the j-th component; X_{ij} – the factual value of the i-th index of the j-th component; Xe_{ij} – the reference value of the i-th index of the j-th component.

The higher the value of the integral index is the higher the level of the ability to take strategic decisions is.

3.3 The comparative analysis of the results of modeling the financial-economic mechanism of taking strategic decisions

The practical realization of the proposed imitation model of the financial-economic mechanism of taking strategic decisions based on the calculation of the integral index of the financial and economic components will provide the necessary database for the substantiation of the managerial decision and further financial-economic development of the enterprise.

We will represent the financial-economic mechanism of taking strategic decisions at the enterprise, which corresponds to the chosen financial strategy of development, by means of a production function enabling the estimation of the resource return, working out a variant of the development plan and the optimization of enterprise functioning and it will be of the form (2):

$$Y = f\left(R_1, R_2, \ldots, R_i, \ldots R_n\right), \tag{2}$$

where Y – the desired parameters of the financial-economic mechanism of taking strategic decisions at the enterprise in accordance with the chosen financial strategy of development; R – the factor indices of the i-th resources included into the components (financial and economic) of the mechanism of taking managerial decisions; n – the number of the factor indices of financial-economic mechanism of taking strategic decisions.

The following groups of assumptions determine the function of the corresponding financial-economic mechanism of taking strategic decisions: mathematical assumptions – the function of the readiness of the financial-economic mechanism of taking managerial decision is continuous and twice differentiated; economic assumptions:

1) should at least one resource be absent, it is impossible to take strategic decisions (3) [5]:

$$\begin{aligned} &Y(0, R_1, \ldots, R_i, \ldots R_n) = \ldots = \\ &= Y = f(R_1, R_2, \ldots, 0, \ldots R_n) = \ldots = \\ &= f(R_1, R_2, \ldots, R_i, \ldots 0) = 0 \end{aligned} \quad (3)$$

2) the growth of factor indices results in the increase of the efficiency (ability) of the financial-economic mechanism of taking strategic decisions (4):

$$\frac{\partial Y}{\partial R_i} > 0 \text{, at } R_i >, i = 1, 2, \ldots n. \quad (4)$$

3) the cost increase does not result in the decrease of the efficiency of the financial-economic mechanism of taking strategic decisions (5):

$$\frac{\partial^2 Y}{\partial R_i^2} < 0 \text{, } i = 1, 2, \ldots n. \quad (5)$$

Based on the above said, the character of the factor indices influence on the efficiency of the financial-economic mechanism of taking strategic decisions is variable. So, it is necessary to determine their phase coordinates in the imitation model and represent them as $r_1, r_2, \ldots r_n$, and the ability to take a particular strategic decision – as functions $y_1(t)$, $y_2(t)$, … , $y_n(t)$. We will take the initial moment of functioning of the financial-economic mechanism of taking strategic decisions in time as t_0. Then we can consider that at the enterprise there is a financial-economic mechanism of taking strategic decisions, corresponding to the financial strategy of development, if for every $t > t_0$ it is possible to calculate its position exactly and unambiguously, i.e. to find functions $r_1(t)$, $r_2(t)$, …, $r_n(t)$, characterizing the change of the phase coordinates in time. The dynamics of values r and $y = (y_1, y_2, \ldots, y_n)$ will represent the process of realization of the strategic decision in time by the corresponding financial-economic mechanism. Based on the above said, we provide the process of realization of the financial strategy of enterprise development with the indices of readiness to taking strategic decisions such as profitability of sales, accumulation, labor productivity, etc.

The main task of the financial-economic development of the enterprise based on the above listed assumptions is formed in the following way: at the initial moment of time t_0 the enterprise is in the phase condition y_0; it is necessary to choose such financial-economic mechanism of taking strategic decisions $y(t)$, that in time t_1 will take the enterprise in the phase condition y_1. Values t_1 and y_1 are set by the chosen financial strategy of development and characterize the ability of the enterprise to take a particular strategic decision. If the financial strategy of the enterprise development implies the provision of the financial-economic mechanism of taking strategic decisions with the determined factor indices of the *i*-th resources (6):

$$y(t) = (r_1(t)), (r_2(t)), \ldots, (r_n(t)) = const\ . \quad (6)$$

Negative influence of the external or internal environment may put the financial-economic mechanism of taking strategic decisions out of the stable condition, which makes the determination of the future state (y_1) impossible under such conditions [6]. For efficient realization of the financial-economic mechanism of taking strategic decisions at the enterprise, we should have a set of its determined functions $y(t)$, which allow returning it into the working condition y_0 from any point y_1. Under such conditions we can represent the mathematical model of the financial-economic mechanism of taking strategic decisions at the enterprise by the following system of differential equations (7):

$$\begin{aligned} y_1 &= f_1(r_1,\dots r_n, y_1(t),\dots y_n(t)) \\ y_2 &= f_2(r_1,\dots r_n, y_1(t),\dots y_n(t)) \\ &\dots\dots\dots\dots \\ y_n &= f_n(r_1,\dots r_n, y_1(t),\dots y_n(t)) \end{aligned} \quad (7)$$

where $f(r,y)$ – functions of the ability of the financial-economic mechanism to take strategic decisions.

This system with known $t > t_0$, y_0 and the parameters of the financial and economic components of the mechanism of taking strategic decisions allows the unambiguous determination of the scheme of financial development of the enterprise.

In accordance with the proposed simulation model, we represent by means of the system of differential equations the ability of the financial and economic mechanism to make strategic decisions for the enterprise. The choice of indicators included in it was carried out using the expert estimation method. Experts were specialists in the number of 10 people working in the economic posts of the higher saber research enterprises. Practitioners have determined by the ball system from 1 to 10 the main factors influencing the area of research enterprises and the main indicators by which these factors can be estimated.

The main factors of the financial component of the experts were: efficiency of capital use, efficiency of activities, production program, cost structure, property status, tax policy. Accordingly, the indicators selected by experts were the following: the coefficient of autonomy, current liquidity, return on capital, the share of variable costs, mobility of assets, tax capacity of sales. The main economic factors according to experts: organization of production, management organization, organizational structure and indicators, by means of which they can be estimated - accumulation coefficient, profitability of sales, expenses per 1 monetary unit of revenues.

These indicators were calculated according to the financial statements of 5 large enterprises of one type of activity (industry) and one region in dynamics for 6 years and the results are presented in digital terms. For a comprehensive presentation of the enterprise ability to take strategic decisions it is reasonable to create a model according to the normative and average values of indices of the activity types (8):

*the factual value for the enterprise	$Y_f = f_f(0.58; 1.82; 0.15; 1.04; 2.63; 0.20; 0.44; 0.29; 11.57)$	
*the normative value	$Y_n = f_n(0.70; 2.00; 0.15; 0.67; 1.00; 0.18; 0.30; 0.50; 0.09)$	(8)
* the value, average for the activity type	$Y_s = f_s(0.14; 0.50; 0.08; 0.67; 4.45; 0.18; 2.31; 0.08; 0.09)$	

If the selected indicators were not normative, the authors were guided by the average type of activity (industry).

3.4. The problems and prospects of modeling the financial-economic mechanism of taking strategic decisions

For clarity we present the obtained results of the imitation modeling of the financial-economic mechanism of taking strategic decisions at the enterprise in a 3D space, with which purpose we grouped the indices according to three features: those that characterize the sources of financing (the coefficient of autonomy, current liquidity and return on equity) and the assets control (assets mobility, the tax capacity of the sold product, the profitability of sales), and those that present the cost policy (the share of variable costs, the coefficient of accumulation, managerial costs per 1 monetary unit of profit. Applying the extrapolation method to each of the indices groups, we find the coordinates for the enterprises, for the activity type and the normative value of the indices (if there was no determined normative for the index, an average value in the type of activity was used). We plot the first group of indices on axis x in the 3D space, the second and the third group respectively on axes y and z (Fig. 1) [5].

The obtained graphic model according to calculations formula 8 makes it possible to determine the line of change of the parameters of the financial-economic mechanism of taking strategic decisions according to the chosen financial strategy of the development and realization of the particular managerial decision in its scope. The graphic model is constructed according to the calculations of formula 8 and reflects the functions of actual, medium-sized and regulatory capabilities. With its help, we can instantly see what indicators need to be changed in our mechanism to achieve the goals of both financial and economic goals. Ideally, these three pieces should be as close as possible to each other.

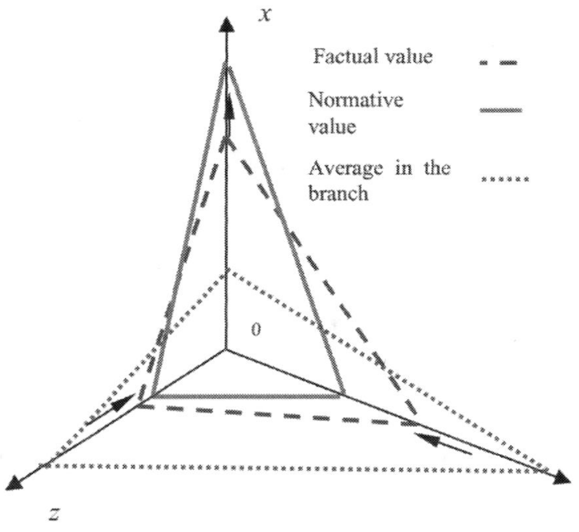

Fig. 1. A graphic model of the financial-economic mechanism of taking strategic decisions at the enterprise
Source: [5].

We would like to emphasize the fact that the parameters of the financial-economic mechanism of taking strategic decisions cannot take absolutely arbitrary values and are subject to various restrictions: by the amount of the required capital, technological requirements, the personnel potential, etc. [6].

4 Discussion and Conclusion

Thus, the main result of our research includes: the interpretation of the "model of the financial-economic mechanism of taking strategic decisions" and creation of an imitation model, providing the enterprise with the properties of competitive advantage and market flexibility. The introduction of the imitation model of the financial-economic mechanism of taking strategic decisions at the enterprise will make it possible not only to calculate its basic parameters but also to estimate its correspondence to the requirements of the strategic decision and to determine the degree of the enterprise ability to realize them within the scope of the financial strategy of development.

References

1. I. Trunina, Development of entrepreneurship entity competitive strategy using competence-based approach. *Actual problems of economics*, **11**(173), 206-213, (2015)
2. O. Sushchenko, I. Trunina, D. Zagirniak, Determination of electrical engineering business entity's core fields of competence as sources of its competitive advantages. *Proceedings of the 2017 IEEE International Conference on Modern Electrical and Energy System (MEES)*, 344-347, (2017)
3. A. Kulman, *Economic mechanisms.* Moscow: Publishing Group "Progress", (1993)
4. R. Buhner, *Personal management*. Landsberg: Verlag Moderne Industrie, (1997)
5. T. Kalinescu, O. Antipov and I. Kushal, *Strategy of social development: mechanism of decision-making process*, 360, (2016)
6. W. Cherchmen, R. Ackoff, L. Arnof, *Introduction to the study of operations.* Moscow, 488, (1968)

Corporate Governance Mechanisms: Their Strengths, Weaknesses and Complementarity

Jan Urban[1,*]

[1] Institute of Technology and Business in České Budějovice, Faculty of Corporate Strategy, Okružní 517/10, 37001 České Budějovice, Czech Republic

Abstract. By corporate governance mechanisms we understand both market and non-market processes, including corporate rules and measures that tackle, internally and externally, two corporate governance problems, i.e. the vertical governance problem that arises between shareholders and managers, and the horizontal governance problem occurring between shareholders themselves. Efficient corporate governance mechanisms, aligning various stakeholder's interests, help to put the right managers in the right place, providing them at the same time with the right set of incentives and constraints. While some corporate governance mechanisms serving these goals, both internal and external from the corporation perspective, are more efficient in dealing with vertical corporate governance, other solve better horizontal governance issues. The main emphasis of the article is laid on the relative role and limits of market and non-market corporate governance mechanisms solving vertical governance problems as well their complementarity and substitution possibilities. Its method is mainly based on secondary analysis of case studies in corporate governance.

Key words: corporate governance, agency costs, public corporation, corporate performance

1 Introduction

Corporate governance refers to mechanisms of interests′ alignment through which corporations are controlled and directed by their shareholders, other stakeholders (employees, creditors, investors, customers, suppliers etc.), as well as the public as a whole [1, 2]. Without effective governance mechanisms, the principal-agent problem may arise, enabling the managements ("agents") to engage in "opportunistic behavior"[2], detrimental to the enterprise‘s owners, stakeholders as well the economy as a whole [3, 4, 5, 6].

* Corresponding author: urban@mail.vstecb.cz

[2] Opportunistic behavior occurs where one party takes advantage of its superior knowledge, in order to further his/her interests, e.g. by failing to disclose such information to the other party [5, 7, 8, 9, 10].

The typical corporate governance problem has a both vertical and horizontal aspect [9, 11]. The vertical aspect of the problem is especially relevant in corporations with dispersed shares ownership. It arises between senior managers and shareholders and its focus is on keeping the senior management loyal to shareholders as well as competent enough for the goals of the firm´s management.

The horizontal aspect of the corporate governance problem occurs between dominant and dispersed shareholders; its focus is on preventing the shifts in value from dispersed to dominant or controlling shareholders. It appears in countries/corporations, including the Czech Republic [10, 12, 13], that fail to develop mechanisms that would effectively stop dominant stockholders' self-dealing. i.e. attempting to benefit from transactions that are being executed on the behalf of the corporation.

The two aspects of the corporate governance problem share one similar characteristic: in each the controlling agent may extract "rent" (private benefit) at the cost of remaining shareholders [9, 11, 12]. The way through which they can do so, however, differs and so do the mechanisms to minimize the scope of the problem. This article **is mainly looking at the problem of vertical corporate governance (the shareholder-manager interests´ alignment problem) and the relative efficiency of market and non-market mechanisms of its prevention.**

The shareholder-manager interests´ alignment problem may appear in two main forms which usually differ in terms of their visibility or transparency. The first concerns the possibility of diversion of firm´s assets, the second managers´ (in)competence [6, 14, 15].

As for assets´ diversion, managers can divert value from the firm into their own hands either less transparently, e.g. by having the firm sell goods at low prices, buy at high prices (to and from entities that the managers control etc.) or, more transparently, e.g. by paying themselves (excessively) high salaries.

The other form of shareholder-manager interests´ alignment problem, i.e. incompetence, concerns the fact that managers may not be up to running the firm in the first place. This may happen either because their selection was a mistake (which may have only appear with a delay), or because changed circumstances made the manager no longer suitable for the job [14, 16].

Mechanisms aligning the interests of shareholders and senior managers can be – from the perspective of the corporation - both market and non-market. The former cover mainly competitive markets of good, capital and labor (including managers), the latter are predominantly based on performance based management incentives and external corporate governance.

The firm's product, capital, managerial and labor markets, are often the most important mechanisms of corporate governance. Their function is based on their ability to punish low-performing and reward well-performing organizations. If the firm, due to its bad management, cannot profitably sell its products, it sooner or later disappears from the market; if it, due to the same reason, cannot raise capital, it cannot grow, and if it cannot get and retain good managers and employees, it will waste its resources. Also, if its executives manage the firm poorly, they lose their market reputation; competitive markets distinguishing between performing and non-performing firms may thus also distinguish between performing and non-performing managers [1, 4, 15].

The importance of markets as corporate governance mechanisms leads sometimes economists to underrate the importance of non-market governance instruments. The role of market mechanisms in corporate governance, though crucial, has, however, its limits. It is diminished mainly in markets with lower competition as well as in cases when they tend to act only from a longer term perspective.

2 Data and Methods

The analysis of corporate government mechanisms rests on economic, management and organization theory [2, 3, 6, 8, 17, 18] as well as data sources provided by case studies of specific corporate governance problems and issues [4, 5, 6, 14, 15].

The basic assumption of this analysis, supported by literature [1, 16, 17] is that though efficient corporate governance helps in both putting the right manager in the right place and providing him with the right set of incentives and constraints, specific corporate governance mechanisms may, however, not be in the position to reduce both diversion and incompetence issues simultaneously. Some may affect one, some the other. Thus the issue of complementarity of market and non-market governance mechanisms arises.

2.1 Market Mechanisms of Corporate Governance

Markets may be good at some corporate governance tasks, but weak at others. They can, for example, be efficient at limiting the most alarming types of management incompetence, but less good at limiting diversion, especially if this represents a small part of the firm's total value. Also, markets tend to be imperfect mainly in terms of not obeying the principles of perfect competition, sometimes substantially so. If this is the case, instead of acting as a tight constraint, they often have their limits in terms of governance mechanisms.

2.1.1 The case of imperfect managerial and labor markets

The issue of market imperfection can be illustrated on the case of managerial and labor markets. It is true that managers usually want to be compensated and promoted, and thus try to act accordingly. If they do badly, they'll jeopardize their compensation and their future career prospects. But many senior managers are at the end of their careers: they aren't usually moving anywhere, except into retirement. Labor markets thus may be important in selecting the "right" people for these final positions, but less efficient thereafter: once "selected" for their final jobs, managers face labor market constraints that are weak. Furthermore, most of these senior managers at large firms are already very rich and more money may only motivate them weakly.

Employee labor markets can play their role within the labor-management dimension of corporate governance. A firm that treats employees badly, fails to motivate them or doesn't induce them to be productive will usually face labor market problems. Good and/or ambitious employees will leave, those less productive will stay. But even this labor market has its limits. The point is that in many countries external labor markets are not fully flexible.

Concerning the managerial and labor markets in general, the point here is not that these markets utterly fail to constrain managers, but that they are not perfect constraints. Eventually, if the firm applies poor personnel policy it will hit a limit in one or another of these markets. But it can, depending on its market setting, take some time.

2.1.2 Performance based executive compensation schemes

One or two decades ago, performance compensation, especially incentives based on shares or share options, seemed a promising way to align senior manager's interests with that of shareholders, and thereby reduce the vertical corporate governance problem [19]. The theory was simple: managers who become equity owners have a higher motivation to raise the company financial performance. If managers get options to buy the company's stock and if their management of the firm induces stock price to rise, the manager's cash in options and make money. If their management of the firm doesn't induce its stock price to rise, then the managers' options are without value.

Incentive compensation certainly is an important instrument of corporate governance, but it too has its limits. First, for incentive compensation to be effective, the shares on which the options are written must be accurately priced. Second, with shareholders scattered and not themselves writing the incentive contract, the options may not fully motivate managers while they can successfully enrich them. Since the common options contracts seem to fall short of an ideal contract, there's good reason to think that the options actually used don't resolve the vertical governance problem as well as they could. They have typically been based on a general rise in the company's share price and not its performance compared to that of other (competing) firms.

Many analyses also conclude that incentive compensation works best in the long run, and its efficiency depends on the time-span. Over approximately a decade, managers at strongly performing companies do indeed earn much more than managers at weakly performing companies.

Another problem of the incentive compensation institution, not noted in the economic literature, is that even if the compensation is well attuned to firm′s performance, senior managers, once they become rich, do not turn the firm over to a new crew of "hungry" managers. They instead continue, while wealthy, running the firm. For these people surely more money remains better than less money, but one can question how strongly incentive compensation motivates them. Its role thus might rather be to motivate managers seeking to be truly wealthy (those lower down in the firm's hierarchy), not those already there.

2.2 Non-market corporate governance mechanisms

Non-market corporate governance mechanisms cover mainly lawsuits, corporate information disclosure rules, intrinsic management motivation and external, i.e. social corporate governance instruments. Their role lies in their ability to work where markets cannot reach. In other words, while markets deal sometimes only with gross deviations, non-market corporate governance mechanisms can deal with subtler governance failures.

In those cases, non-market institutions can be more important than markets. That is, sometimes the non-market institutions are better, cheaper, and faster at governing the firm than the market constraints.

Second, the non-market corporate governance institutions can strengthen the market mechanisms. Good internal decision-making can make sure good managers come, stay, and perform and can thus facilitate the firm's adaptation to the markets' pressures. Good non-market governance institutions can thus become one of the bases for competition in all three markets.

2.2.1 Corporate lawsuits and corporate information disclosure rules

Threat of corporate lawsuits, i.e. the possibility to sue company managers, represents often the main non-market corporate governance instrument. In most countries there are two broad ways for shareholders to sue managers: the breach of fiduciary duties (the duty of loyalty and the duty of care) and breach of securities law obligations [4].

Executives who divert company assets from the firm to themselves or their relatives (i.e. steal) violate typically the first of these duties, while managers who make the firm to sell its products at a lower price (compared with the market price) will typically violate the second of these duties. The latter case, breach of securities law obligations relates most basically to the company's quality of information disclosure, and can involve public enforcement actions as well as private lawsuits.

Unlike market restraints which are not likely to reduce insiders' opportunistic behavior strongly (since the value diverted from the company can be relatively low in relation to the size of the firm´s assets and markets cannot work well if they don't know about the wrong-doing), law-based institutions, serving mainly as a deterrent, can be in these cases more efficient. They are, however, usually less efficient as a tool for solving the problem of managerial incompetency. The reason is that corporate courts often won't listen to complaints about managerial decision mistakes.

Corporate information disclosure rules, on the other hand, are usually better at tackling management failures since shareholders can see the bottom line results that managers produce. At the same time, however, they are often less efficient at handling diversion because, mainly in large firms, personal wealth can often be obtained without destroying the corporate bottom line.

2.2.2 Intrinsic motivation of managers

The economic theory usually does not cover the issue of managers´ "internal" motivation. The fact, however, is that managers do not act solely for remuneration, but also for the satisfaction of "doing a good job", no matter how circumstantial this motivation can be. As professionals they want the firm to do well. They are often used to work for shareholders, and often do so even when the other institutional constraints are not tight. Or, managers who feel good about themselves if they introduce a new product or technology innovation, and make shareholders profit.

These notions of professionalism are softer and less well understood than the other institutions, but that doesn't mean that they are absent, or unimportant.

As far as intrinsic motivation theory is concerned, several scientifically proven as well as and some alternative motivation theories were created. Miner, 2005 summarizes that managers motivation can be among other aspects characterized by attitude towards authority, need for competition, need for power, need for affirmation, need for different and differential status, wish for doing routine managerial duties etc. From these theories is clear that only material motivation is not quite appropriate and not enough for managers (researcher also found weak, neutral or even negative correlation between managerial compensations and managerial work efficiency, other also found that in dominantly intellectual duties/work the best motivation is psychological (intrinsic) one, and for dominantly physical duties/work the best motivation is material (extrinsic) one. Therefore, according to those results dominant physical work should be rewarded by material strategies and dominant intellectual work should use more psychological/intrinsic strategies - what is quite opposite the practice in corporate world.

Furthermore, more general human motivation theories have some important managerial implication e.g. Maslow's need hierarchy, Alferder theory of three stages, theories of achievement motivation (McClelland and Atkins), two factor theories (Herzberg), theory of role motivation (Miner), theory of justice in social exchange (Adams), integral process models of motivation, self-determination motivation theories, flow theories, also old well-known and (mis)used punishment-reward theory. Though these theories can give important insights into management intrinsic motivation and are therefore important for the application of practical management motivation measures, their analysis goes beyond the scope of this article focused on alternative corporate governance institutions.

2.2.3. External corporate governance: issues of corporate social legitimacy

Corporate governance can have another, external dimension [16]. Being neither horizontal, nor vertical, nor necessarily governing the relationships among owners and managers, this dimension is of governing the firm in a way that it is legitimate in its society.

This legitimacy dimension is mainly determined by the outsiders: the society demands that the firm structure its governing institutions in a way that leads to implementing into its objectives some of the goals of public policy. Issues of legitimacy can be, however, extended in further directions. For example, one might view the problem of corporate governance in organizational terms, as how to motivate a team, how to keep it flexible but effective, etc.

This dimension is less pronounced in some countries and more in others. It is usually absent in academic economic literature though analyzed in management literature. An example is the role of employees in corporate governance. They are typically seen as inputs or suppliers that managers transform into sales and profits. Even though employees have to be paid well and have to think well of their work/corporate environment, this is not usually seen as a corporate governance issue (in terms of aligning the interests of company stakeholders). In some countries, however, where social considerations and the role of employees have much more impact, their role tends to be formalized, e.g. in the form of Germany inspired codetermination, in which employees end up inside the boardroom. Generally speaking, the role of employees in decision making can be rather technical one-where to employee is given the freedom/duty to take part in technical improvements or strategic one, when the employees have the authority to take part in decisions concerning important corporate management topics including investment directions and distribution of earned profit.

The external dimension of corporate governance focuses less on relationships between shareholders and managers or among shareholders. It could, however, have a "legitimacy" angle, and it could have an "efficiency" angle, as such a focus might yield greater productivity and more utility for those inside the organization.

This issue of corporate governance and legitimacy can also be seen more abstractly: how, can one organize a business entity that survives politically and socially and how have corporate institutions adapted to be legitimate in society? This kind of inquiry gets so far little attention in the academic literature, although it covers important issues like which institutions are legitimate, which ones don't attract negative attention from the society, or which ones manage to resist outside incursions better than others even if they turned out to be less effective in another (economic) dimensions.

The external dimension of corporate governance reflects the fact that business entity is also social entity which can also improve the life of employees by integrating them in different levels of decision making (not only in technical/operational sense, but also in strategical/political sense where it is possible) in order to improve the vitality of the company (efficacy and making money not only for „company" but also for employees and their interests what can have huge positive results in overall motivation and corporate social legitimacy sense).

3 Recommendations and Suggestions for Practice

As we have seen, various (imperfect) corporate governance institutions can play the role of complements and/or substitutes. Though each of them can affect the alignment of managers and shareholders' interests (some primarily in decreasing the probability of insider diversions of value, and others in increasing the quality of management), in most cases they can substitute and complement each other. In these cases, their overall efficiency increases [1, 11].

Substitution is mainly effective if one institution becomes too costly, and the demand for the other corporate governance institutions rises. For example, competitive product markets confine managerial discretion (fierce competition can sharply lower corporate profitability if the managers' fail to take optimal decisions), and in that sense serve as a substitute for the other vertical governance institutions. Furthermore, lower profits usually lower the firm's stock price and make the incentive compensation less remunerative, and signal takeover entrepreneurs that a new opportunity. Thus, in markets and/or countries with historically strong product market competition firms can prosper without really strong governance constraints on managers because product market competition keeps managers working effectively.

Efficient corporate governance has also a positive spill-over effect. If markets are competitive, but all incumbent managers not performing properly, improving one corporate governance institution or another in one of the competitors, e.g. introducing incentive compensation or nominating a more dynamic board, can lead to reaction of the other competing firms as well. They can react by innovating but also by improving their own internal corporate governance systems, with better boards, better compensation, etc.

Similar principles concern the complementarity of institutions. Better information, for example, leads investors to price securities more accurately. It can, however, also motivate managers who feel ill-at-ease to see their shares price decline to make the value of those shares rise. Boards that got better feed-back (of declining shares prices) can be more inclined to take the action that might be needed, e.g. design better compensation plans or foster the professionalism of managers. Takeover entrepreneurs might also be inclined to act.

The topic of corporate governance is relevant. Stockholders are demanding improved governance of companies by strengthening the responsibilities and competencies of directors and managers and more responsive shareholder's relations. Companies with diffused ownership often suffer from the fact that the stockholders' agenda - maximizing profits/shareholder value - can be at odds with agendas of their managers. In this setting, aligning both parties' interests become the goal of institutions of corporate governance. The aim of these institutions, i.e. rules and mechanisms, is to tie managers to the interests of stockholders.

4 Conclusion

The stockholder-manager alignment problem appears in two main forms. The first concerns diversion, the second competence. Managers can divert value from the firm into their own hands: they can have the firm transfer funds to their own bank accounts (or a relative's), or have the firm sell goods at low prices to (or buy at high prices from) entities that the managers control. More transparently, they can pay themselves (excessively) high salaries. The other form concerns the fact that managers may not be up to running the firm, either because one or another manager never was up to the task (their selection was a mistake) or more plausibly because changed circumstances made the manager no longer right for his company.

Well-functioning corporate governance institutions put the right manager in the right place and give that manager the right set of incentives and constraints. This goal, however, hides a couple of issues. The first is that each specific corporate governance institution does not uniformly reduce both. Some affect one, some affect the other. Both problems, however, bring costs to shareholders.

Market and non-market corporate governance mechanisms, can thus be regarded as **specialized:** they can play the role of complements and/or substitutes. Lawsuits, for example, compared with markets, are more effective in tackling conflicts of interest while markets and other close to market mechanisms, e.g. managements′ incentive instruments, are better suited to lead managers toward pro-shareholder decision-making and/or reduce the problems of management′s incompetence.

Though each of these mechanisms can affect the alignment of managers and shareholders' interests (some primarily in decreasing the probability of insider diversions of value, and others in increasing the quality of management), in most cases they can substitute and complement each other. In these cases, their overall efficiency increases.

References

1. N. Kang, Analysing 'System Change': The Role of Institutional Complementarity in Corporate Governance. *International Centre for Corporate Social Responsibility Nottingham University Business School Nottingham University, Research Paper Series,* (43-2006), (2006)
2. R. Adams, B. E. Hermalin, M. S. Weisbach, The Role of Boards of Directors in Corporate Governance: A conceptual Framework and Survey. *Journal of Economic Literature,* **48**(1), 58-107, (2008)
3. R. V. Aguilera, G. Jackson, Comparative and International Corporate Governance. *The Academy of Management Annals*, **4**(1), 485-556, (2010)
4. B. J. Bushee, M. E. Carter, J. J. Gerakos, Institutional Investor Preferences for Corporate Governance Mechanisms. *Journal of Management Accounting Research: Fall 2014*, **26**(2), 123-149, (2014)
5. G. Ferrarini, N. Moloney, Executive Remuneration in the EU: The Context for Reform. *Law Working Paper* [online], (32/2005), Available at:

https://ecgi.global/sites/default/files/working_papers/documents/SSRN-id715862.pdf (2005)

6. J. P. Hawley, S. J. Kamath, A. T. Williams, *Corporate Governance Failures: The Role of Institutional Investors in the Global Financial Crisis*. University of Pennsylvania Press, (2011)
7. A. L. Friedman, S. Miles, *Stakeholders Theory and Practice*. OUP Oxford, (2006)
8. M. G. Gentile, Corporate Governance and Accountability: What Do We Know and What Do We Teach Future Business Leaders? *The Aspen Institute Business & Society Program*, (2011)
9. N. Ozkan, Do corporate governance mechanisms influence CEO compensation? An empirical investigation of UK companies. *Journal of Multinational Financial Management*, **17**(5), 349-364, (2007)
10. M. J. Roe, *Political Determinants of Corporate Governance: Political Context, Corporate Impact*. New York: Oxford University Press, (2006)
11. J. B. Miner, *Organizational Behavior 1: Essential Theories of Motivation and Leadership*. Routledge, (2015)
12. M. J. Roe, The institutions of corporate governance. *Harvard Law and Economics Discussion Paper*, (488), (2004)
13. M. Singh et al., Agency costs, ownership structure and corporate governance mechanisms. *Journal of Banking & Finance*, **27**(5), 793-816, (2003)
14. Z. Caha, J. Urban, *Etické řízení organizace: Od etického kodexu k etickému řízení*. RAM-Verlag, (2017)
15. P. C. Fiss, *Institutions and Corporate Governance* [online]. Available at: https://poseidon01.ssrn.com/delivery.php?ID=470002118005101069118030096091105092096075037082045007068102004085117001006095014029013055021115109098101084093025011086081021055058036076048083127031111031090107065006017091023089026012112086014089066080072078125005026118091103116089067100064081098&EXT=pdf (2007)
16. C. M. Daily et al, Corporate Governance: Decades of Dialogue and Data. *Academy of Management Review*, **28**(3), (2013)
17. Z. Caha, M. Vokoun, The Impact of the Presence of a Code Of Ethics on the Economic Prosperity of a Company. *Innovation Management, Entrepreneurship and Sustainability 2017*, 72-78, (2017)
18. M. Hilb, *New Corporate Governance: Successful Board Management Tools*. Springer-Verlag berlin Heidelberg, (2012)
19. Z. Caha, Central Europe: Ethical Overlaps of Environmental and Economic Interests in Coming Years. *Science and Engineering Ethics*, **24**(6), 1801-1807, (2017)

Enterprise Architecture within the Process of Permanent Change

Jan Váchal[1] and *Jarmila* Straková[1]

[1]Institute of Technology and Business in České Budějovice, Faculty of Corporate Strategy, Department of Management, Okružní 517/10, 37001 České Budějovice, Czech Republic

Abstract. Permanent changes that happen within the external surroundings of an enterprise, respectively within the macro-environment, significantly influence its efficiency and behaviour. The framework for these changes is made up by Enterprise Architecture (EA). Impacts of these changes are distinct in both, the internal environment of an enterprise as well as in its mid-environment, being of dynamic nature, and their elimination is a basic assumption of keeping the enterprise´s competitiveness. The article focuses on analysing the macro-environment factors´ influence on the changes in enterprise EA from the perspective of their sector differentiation. Tests have been carried out both ways – for chosen macro-environment factors, as well as macro-environment areas as a whole. The research survey was carried out on a file of 456 enterprises from the whole of the Czech Republic, out of which the production and industry sectors included 187 enterprises, and the service sector 255 enterprises. Pearson´s chi-quadrate test was used as a statistical method. At the same moment, macro-environment factor prediction was tested, respectively the importance attached by top managers to the individual macro-environment factors in the future. Research results have proven a significant sector differentiation from the perspective of macro-environment factors´ impact on EA, while a greater influence of these factors has been indicated mainly in the service sector. It may be assumed that top managers begin to prove a risk symptom caused by a change in the economic cycle.

Key words: uncertainty, probability, fuzziness, expected value, fuzzy number

1 Introduction

Enterprise changes may be of different dimension, intensity, and range. One of the most significant and decisive changes within the enterprise is the change in enterprise architecture (EA), expressing all basic enterprise dimensions within its system conception. Mainly we are talking about enterprise vision, strategy, aims, enterprise functions, processes, and organization structure, database, software, hardware, and space localization. The term of EA has been used sing the end of the past century [25], but its definition as well as content is non-uniform, and greatly differentiated. Some authors define it as a complex enterprise definition matching the given time period [3, 17, 22, 23]. Other authors define it as a strategical quantity within enterprise activity [2, 3, 13-15]. According to the Gartner company [7], enterprise architecture is a process of efficient transformation of enterprise vision and mission into enterprise change via creation, communication, and improvement of key requirements, principles and models that describe the enterprise´s future situation, and enable its development. Some authors [6, 12, 24] integrate enterprise architecture within information enterprise systems to reach enterprise aims while managing IT sources more efficiently, flexibly, and transparently. Recent works include professional interviews, case studies and surveys that are prepared with a broader focus on the EA management in organizations both in the commercial and IT context [8, 20].

Changes happening in enterprise architecture result mainly from a globalizing external enterprise environment. They reach both, the enterprise mid-environment and internal enterprise resources; in total they reach the complete enterprise architecture. In real enterprise practice dynamic process [4] in EA change set happens aiming to keep competitiveness and sustainability [15, 18]. These changes are marked by a range of authors as strategic changes [2, 3, 15, 23], as decisive changes by other authors [13, 18]. Classification of these changes is introduced by Veber [23], and the term of ´a great change´ is being introduced too, Christensen [10, 11], on the contrary, understands it as a ´breakthrough´ change. According to Dan Schendel and Charles W. Hofer [21] a process dealing with enterprise strategy, such as a change in enterprise focus, in resource use, competition advantage, market position, suppliers and buyers may be understood the strategic change.

The aim of enterprise architecture and its innovation is to react to constantly changing competition environment. Enterprise architecture transformation process is dependent on targeted implementation projects including all basic process areas of the given enterprise [9, 19]. To manage organization changes, Lapalme [16] perceive enterprise architecture and project portfolio management as the most important pro-growth and value-creating enterprise tool. The architecture framework TOGAF describes the methods of creation, management and application of framework and information systems used for creating a model of developing business architecture [1]. TOGAF (The Open Group Architecture Framework) is owned and managed by The Open Group consortium. This non-profit consortium consists of a number of experts from the whole world. TOGAF has many key components, one of them being the ADM (Architecture Development Method) as a process-based model for creating and managing the development and changes of business architecture [5]. Enterprise architecture is a managers´ tool for an efficient and flexible decision-making within changes in an enterprise so that the enterprise is able to exist in a long-term period. The submitted article focuses on the external environment´s influence on the EA nowadays even predicting future development.

2 Methodology

The testing set included 456 enterprises from the whole of the Czech Republic. Enterprises were broken down from the perspective of size categorization, sector differentiation, and scope. The set from the perspective of enterprise size categorization included 109 micro-enterprises (up to 10 employees), 140 small enterprises (up to 50 employees), 124 medium enterprises (up to 250 employees), and 83 big enterprises (more than 250 employees). From the perspective of sector differentiation, 187 enterprises came from the sector of production and industry, 255 enterprises from the service sector. The research survey was done through a vast questionnaire survey focused on the issue of enterprise environment, respectively its individual components (internal enterprise resources, mid-environment, and macro-environment), and strategy. Using Pearson´s chi-quadrate test, macro-environment factor dependence/independence has been determined (individually as well as for the macro-environment area as a whole) on the profitability of enterprises, respectively their influence on EA from the perspective of sector differentiation. Tests have been carried out on a standard level of significance of 0.05, i.e. with reliability of 95%. Research hypothesis relied in the assumption that factors analysed as significant from the perspective of enterprise profitability reflect the need of their stabilization or change within enterprise architecture, at the same time.

Prediction, respectively what significance enterprise top managers assign to individual macro-enterprise factors in the future, was tested as well. For each manager, their total deviation from the optimal model via quadrate loss function may be determined. Based on this, 60 of the best top managers, 30 from the service sector and 30 from production enterprise sector were chosen. These managers were trusted in future development predictions. Within prediction determination, average values of individual variables from a subset were calculated, used further as a prediction of the optimal macro-environment situation in the future.

3 Results and Discussion

The following tables (table 1 and 2) state the results of macro-environment dependence factors on enterprise profitability, respectively on necessary changes in EA for industry and production sector (table 1) and Service Sector (table 2). Tab. 1 shows that none of the analysed factors has proven statistically significant value. In the rate of obsolescence of means of production and technology transfer, a higher influence on EA changes may be assumed. These factors have been also mentioned frequently within structured interviews with representatives of top management. Results of related internal enterprise resource analyses and mid-environment, which are not the subject of this contribution, have indicated a significantly higher importance in internal enterprise resources in comparison to macro-environment factors.

In service sector (table 2), the following factors influencing the change of EA have been proven: enterprise lifecycle, tax policy, the rate of obsolescence of services provided, inflation, and technology transfer.

Table 1. Dependence of macro-environment factors on profitability, respectively changes in EA enterprises operating in production and industry sector.

Macro-environment Factors	p-Value	Chi-Quadrate
Demographic Development	0.8687	3.8694
Income Distribution	0.419	8.1499
Lifestyle	0.4067	8.2793
Level of Education	0.7105	5.4322
Citizen Mobility	0.7181	5.3632
Leisure Time Approaches	0.8195	4.3982
Governmental Expenditures on Research, Development	0.9651	2.4259
New Patents, Discoveries	0.6297	6.1569
Technology Transfer	0.2526	10.18
Rate of obsolescence of the means of production	0.108	13.1142
GDP Development	0.4945	7.3971
Interest Rates	0.864	3.9235
Inflation	0.7114	5.4248
Enterprise Lifecycle	0.3301	9.1464
Unemployment	0.9358	2.9769

Government Stability	0.7162	5.3809
Foreign Trade Regulation	0.8432	4.1514
Tax Policy	0.3843	8.5207
Legislation	0.4224	8.1139

Source: Authors.

Table 2. Macro-environment factor dependence on profitability, respectively EA changes in enterprises operating in service sector

Macro-environment Factors	p-Value	Chi-Quadrate
Demographic Development	0.6355	4.3049
Income Distribution	0.5733	6.6636
Lifestyle	0.7021	5.5089
Level of Education	0.05759	12.2029
Citizen Mobility	0.2911	7.334
Leisure Time Approaches	0.4484	5.7791
Governmental Expenditures on Research, Development	0.3557	8.8434
New Patents, Discoveries	0.05365	15.2954
Technology Transfer	**0.01552**	18.8799
Rate of Service Obsolescence	**0.01499**	18.9758
GDP Development	0.2726	7.5549
Interest Rates	0.1738	8.9965
Inflation	**0.01514**	15.7527
Enterprise Life Cycle	**0.003161**	8.1819
Unemployment	0.4159	8.1819
Government Stability	0.1104	13.0415
Foreign Trade Regulation	0.02244	14.7326
Tax Policy	**0.0148**	15.8122
Legislation	0.05753	12.2059

Source: Authors.

Tested factors were organized into 4 basic areas: socio-cultural, technological, economical, and political-legal. Table 3 states the dependence of the above-mentioned areas as a total on profitability (EA change) of enterprises.

Table 3. Dependence of macro-enterprise areas on profitability, respectively changes in EA enterprises.

Factors	p-Value	Chi-Quadrate
Production and Industry Sector		
Socio- Cultural	0.5791	4.7287
Technological	0.04612	12.812
Economical	0.06491	11.8708
Political-Legal	0.03092	13.887
Service Sector		
Socio-Cultural	0.02671	14.2751
Technological	0.00000003647	45.5452
Economical	0.000001417	37.4832
Political	0.000003536	35.4432

Source: Authors.

As opposed to results within an individual evaluation of each factor independently, significant values were proved in technological and political-legal area, in all areas for service sector, in a group factor rating (see table 3). In production sector an assumption may be made that mainly the technological area including the innovation process, and political-legal area are significant ones, in service sector, a higher level of interdependence between this sector and macro-environment, and the straightforward effect of its factors on EA changes.

Another part of the research was focused on predicting optimal situation of macro-environment from the perspective of opinion level to managers (table 4). As it has been suggested by prediction outcomes, in both sectors tested significant changes may be expected, different in individual sectors, in both individual factor significance as well as the intensity of their action.

In production and industry sector, technology transfer and intellectual property may be understood as future core factors. Follow-up important factors include also macro-economic indicators (interest rates, inflation, GDP development), factors from the socio-cultural area are relatively significant as well. In service sector, enterprise life cycle, income distribution, technology transfer and level of education belong among core factors.

Despite a great sector differentiation, a range of common factors that will have to be solved within EA, may be found. As the results suggest, some factors will be necessary to be strengthened via EA changes, some stabilized, in some factors their current meaning will be lowered and rather harmonized with other follow-up factors.

During a detailed analysis of macro-environment factors, fears of economic cycle change and succession of economic recession begin to appear in managers′ approaches throughout both sectors. A growing influence of macro-indicators, anti-crisis and pro-growth factors, such as technology transfer, patent or enterprise life cycle may be set as examples. The importance of these factors, however, should be of core character, these factors should play a significant role in the change of enterprise architecture.

 https://doi.org/10.1051/shsconf/20196101029

Table 4. Macro-environment Optimal Situation Prediction from the Perspective of Opinion Levels of Managers from the Perspective of Sector Differentiation

Macro-environment Factors	Variable Average Values for Production and Industry Sector	Variable Average Values for Service Sector
Demographic Development	2.566667	2.666667
Income Distribution	2.666667	**2.766667**
Lifestyle	2.666667	2.6
Level of Education	2.533333	**2.7**
Citizen Mobility	2.666667	2.566667
Leisure Time Approaches	2.666667	2.533333
Governmental Expenditures on Research,	2.633333	2.666667
New Patents	**2.733333**	2.666667
Degree of Obsolescence of Means of Production	**2.733333**	**2.766667**
Degree of Obsolescence of Means of Production	2.433333	2.666667
GDP Development	2.633333	2.633333
Interest Rates	**2.733333**	2.666667
Inflation	2.633333	2.466667
Enterprise Life Cycle	2.666667	**2.9**
Unemployment	2.466667	2.5
Government Stability	2.566667	2.6
Foreign Market Regulation	2.5	2.633333
Tax Policy	2.5	2.433333
Legislation	2.466667	2.4

Source: Authors.

4 Conclusion

Results from the external factor research within the context of changes in enterprise architecture may be perceived as input and orientational, for the research survey was happening at a period of a significant economic growth, and this fact did influence the value of outputs obtained in real terms. Despite this statement, the following conclusions may be made:

- Based on the prediction of future situation, significant changes in enterprise architecture may be expected, relying mainly in the change of significance of individual factors, but also in the intensity of these changes and their effect. Sector enterprise differentiation is being strictly projected within these changes.
- In production and industry sector, significant factors of EA changes may be understood from the current perspective as well as prediction mainly by factors from the technological area (technology transfer and the degree of obsolescence of means of production).
- EA changes in service sector will be caused, in the future, by factors blending through all macro-environment areas tested, mainly within the enterprise life cycle, income distribution, and technology transfer. Changes will be increased by demographic development, governmental expenditures and research expenditures,

the area of intellectual property, the degree of obsolescence of means of production, and the level of interest rates.

Theses with similar focus do not commonly occur in expert printing, authors are, however, aware of the heaviness, and the up-to-date characteristics of this issue. They do believe that once finished, the outputs may be used while ensuring the competitiveness of Czech enterprises, perhaps their stabilization, and while preparing for a period of time when economic growth is less intense and pro-growth.

References

1. E. Amalia, H. Supriadi, Development of enterprise architecture in university using TOGAF as framework. *AIP Conference Proceedings* [online], **1855**, 1, Available at: https://aip.scitation.org/doi/abs/10.1063/1.4985527 (2017)
2. R. Bevan, *Change Making: Tactics and Resources for Managing Organizational Change.* CreateSpace Independent Publishing Platform, (2011)
3. R. L. Daft, *Organization Theory and Design.* Cengage Learning, (2007)
4. Y. Doz, M. Kosonen, *Fast Strategy. How Strategic Agility Will Help You Stay Ahead of the Game.* FT Press, (2011)
5. A. Ferrugento, A. Rocha, Evolution of Methodological Proposals for the Development of Enterprise Architecture. *New Contributions in Information Systems and Technologies,* 351-359, (2018)
6. F. Gampfer, A. Jürgens, M. Müller, R. Buchkremer, Past, current and future trends in enterprise architecture: A view beyond the horizon. *Computers in Industry*, **100**, 70-84, (2018)
7. I. Gartner, *Enterprise Architecture (EA) | Gartner IT Glossary. IT Dictionary, IT Terms | Gartner IT Glossary* [online], Available at: http://www.gartner.com/it-glossary/enterprise-architecture-ea/ (2012)
8. B. T. Hazen, R. V. Bradley, J. E. Bell, J. In, T. A. Byrd, Enterprise architecture: A competence-based approach to achieving agility and firm performance. *International Journal of Production Economics,* **193**, 556-577, (2017)
9. M. Hugoson, et al., Enterprise architecture design principles and business-driven IT management. *International Conference on Business Information Systems,* **57**, 144-155, (2010)
10. C. M. Christensen, *The Innovator's Dilemma: When New Technologies Cause Great Firms to Fail.* HighBridge Audio, Abridged edition, (2006)
11. C. M. Christensen, J. Grossman, J. Hwang, *The Innovator's Prescription: A Disruptive Solution for Health Care.* McGraw-Hill Education, (2009)
12. E. Kaoutar, B. Bounabat, Strategic Alignment Assessment Based on Enterprise Architecture. *International Conference on Information Management and Evaluation*, 179-187, (2010)
13. J. P Kotter, Leading Change: Why Transformation Efforts Fail. *Harward Business Review*, (1995)
14. J. P Kotter, *Srdce změny: Skutečné příběhy o tom, jak lidé mění své organizace.* Management Press, (2003)
15. J. P. Kotter, *Vedení procesu změny.* Management Press, (2004)
16. J. Lapalme, *Three Schools of Thought on Enterprise Architecture.* IEEE Computer Society, (2012)

17. M. Lankhorst, *Enterprise Architecture at Work: Modelling, Communicationand Analysis.* Springer-Verlag Berlin Heidelberg, (2009)
18. R. Luecke, *Managing Change and Transition: 7 Practical Strategies to Help You Lead During Turbulent Times.* Harvard Business School Press, (2003)
19. J. W. Ross, et al., *Enterprise architecture as strategy: creating a foundation for business execution.* Harvard Business School Press, (2006)
20. G. Shanks, M. Gloet, I. A. Someh, K. Frampton, T. Tamm, Achieving benefits with enterprise architecture. *The Journal of Strategic Information Systems,* **27**(2) 139-156, (2018)
21. D.E. Schendel, Ch.W. Hofer, *Strategic management. A new view of business policy and planning.* Little, Brown & Company, (1979)
22. M. Van den Berg, M. Van Steenbergen, *Building an enterprise architecture practice.* Springer, (2006)
23. J. Veber, et al., *Management: Základy, moderní manažerské přístupy, výkonnost a prosperita.* Management Press, (2011)
24. R. Winter, J. Schelp, Enterprise architecture governance: the need for a business-to-IT approach. *Proceedings of the 2008 ACM symposium on Applied computing,* 548-552, (2008)
25. J. A. Zachman, A framework for information systems architecture. *The IBM systems Journal*, **26**(3), 276-292, (1987)

Estimation of the development of the Euro to Chinese Yuan exchange rate using artificial neural networks

Marek Vochozka[1,*] and *Jaromír* Vrbka[1]

[1]Institute of Technology and Business in České Budějovice, School of Expertness and Valuation, Okružní 517/10, 37001 České Budějovice, Czech Republic

Abstract. The exchange rate is one of the most monitored economic variables, from the position of individual citizens or economists, financial institutions or entrepreneurs. In the long run, it is a reflection of the condition of the economy, and in the short and medium term it has a significant impact on the economy. The time series of currency development maps past developments, current status, and is also able to predict future developments. This article analyzes the time series of the development of EUR to Yuan exchange rate using artificial intelligence. It aims to evaluate this development and to indicate the prediction of the future development of EUR to Yuan.
Key words: time series, analysis, exchange rate, EUR, Yuan

1 Introduction

When talking about the development of currency rates, it is necessary to realize that numerical series are an essential part and an important aspect of the whole process. So why are the numerical series important? Numerical series map past developments, current status, and also allow you to generate predictions.

We can label the time series as observing data that is organized, from the perspective of time, from the past to the present. Time series analysis is mainly used for the aforementioned prediction of the future, which is created using different methods. An example of the analysis is the study of the modular structure of the global exchange network for which Mai [1] used correlation matrices. She found that the East Asian currency module is more closely correlated than the European currency module. The strong correlation is the result of strong joint currency movements in the region.

According to Kenneth [2], the link between the exchange rate and the economic fundamentals provides a new basis for empirical studies. Based on the situational analysis of the data and the empirical analysis of the commercial econometric model, we can see that increasing Chinese exports affect the Euro currency market [3]. The exchange rate is one of the most watched economic variables, whether from the position of individual citizens or economists, or financial institutions and entrepreneurs. While in the long run it is a reflection of the condition of the economy, in the short and medium term, it has a

* Corresponding author: vochozka@mail.vstecb.cz

significant impact on it [4]. The exchange rate, as well as a number of other economic variables (eg gross domestic product, interest rates, etc.), can be tracked both in nominal terms and in real terms [5].

The economical view of the exchange rate often uses the real exchange rate optics, the product of the nominal exchange rate and the ratio between price levels abroad and the domestic economy [6]. Fundamentally, the currency crisis has been hit by the financial crisis, which has caused severe volatility in the currency areas of the fixed exchange rate regimes [7]. There are constant changes in the rates that are based on central banks around the world. Changes in rates can be influenced by both institutional tools and their trading on international foreign exchange markets. By buying and selling them, the country tries to get the currency into its target level [8]. Conversely, currencies tend to react to these activities by appreciating or depreciating the currency. Appreciation means the currency gains value. This situation is advantageous for importers, since they will buy more foreign currencies for the same amount of Czech currency. Depreciation is the opposite, and it is a devaluation of the currency itself, which will strengthen another currency. This situation concerns floating rates. Floating rates are determined by foreign exchange markets and are dependent on demand and supply [9].

This article will measure the development of the EUR compared to the Yuan, China's national currency. Yuan was added to the Special Drawing Rights (SDR) by the International Monetary Fund in 2016, which confirmed the success of China's economic development. This merge in practice means that the central banks of individual member states of the International Monetary Fund included Yuan, which raised the value of Chinese bonds on international markets [10]. Bond yields are estimated at CZK 24 trillion over the next five years. Some critics have spoken out sharply against the adoption of the Chinese national currency in the SDR, mainly because of the devaluation of the official Yuan exchange rate in 2015. The Chinese central bank weakened its currency by almost three and a half percent. Since 2005, it has been the biggest exchange rate change, and China has described it as a "one-off depreciation" [11]. The value of the currency is determined by the Chinese central bank, which sets the average exchange rate around which the market rate, irrespective of supply or demand, may deviate on a daily basis [12]. China, of course, is striving to make the most of international trade and uses Yuan as an alternative to the dollar and the euro [13].

As in China, the shared currency of the European Union is managed by one institution, the European Central Bank. The Central Bank was established in 1998 by the Treaty on European Union and its seat is in Frankfurt. Its task is to manage the EU's single currency and to ensure price stability. Of the 27 member states, only 16 of them adopted a common Euro currency [14]. The development of the Euro currency is related to the development of other currencies with political and economic changes. The exchange rate mainly affects the two largest economic sectors, which are enterprises and households [15]. Long-term efficiency of the existence of the Euro requires wage and price flexibility and, above all, high labor mobility, in which the Chinese market stands out [16]. An essential part of the exchange rate is stable political integration, which is also dependent on a number of factors.

2 Data and methods

Data for analysis is available on the website of the World Bank [17] etc. The information on the exchange rate of the Euro and Yuan will be used for the analysis. The timeframe for which the data will be available begins on October 6, 2009 and ends on October 21, 2018. We always record the daily exchange rate of both currencies. This is a total of 3,303 records of input data. The unit is the number of Yuans per 1 Euro. Descriptive characteristics of the data are given in Table 1.

Table 1. Characteristics of the datafile

Samples	Data statistics	
	Case (Input variab.)	Euro to Chinese Yuan (output) (goal)
Minimum (Training)	1.000	6.56070
Maximum (Training)	3303.000	10.29420
Průměr (Training)	1643.786	8.08101
Standard deviation (Training)	939.260	0.75919
Minimum (Testing)	11.000	6.56980
Maximum (Testing)	3302.000	10.25190
Average (Testing)	1664.707	8.06546
Standard deviation (Testing)	957.966	0.76986
Minimum (Validation)	20.000	6.58290
Maximum (Validation)	3297.000	10.30830
Average (Validation)	1677.677	8.09958
Standard deviation (Validation)	1438.254	1.20789
Minimum (Overall)	1.000	6.56070
Maximum (Overall)	3303.000	10.30830
Average (Overall)	1652.000	8.08147
Standard deviation (Overall)	953.638	0.76206

Source: Authors.

For data processing, DELL's Statistica version 12 will be used. For calculating neural structures, we use the Data Mining tool, specifically Neural Networks (ANS – Automated Neural Networks). We will generate multilayer perceptron networks and neural networks of basic radial functions. The independent variable will be time (or measuring designation - case). We will determine the Euro to Yuan exchange rate as the dependent variable. We divide the time series into three sets - training, testing, and validation. The first group will be 70% of input data. Based on the training set of data, we generate neural structures. In the remaining two sets of data, we always leave 15% of the input information. Both groups will serve us to verify the reliability of the found neural structure, or the found model. The delay of the time series will be 1. We will generate 100,000 neural networks. We will úreserve 5 of them with the best characteristics[†]. In the hidden layer, we will have at least two neurons, at most 50. In the case of the radial basic function, there will be at least 21 neurons in the hidden layer, at most 30. For the multiple perceptron network we will consider these distribution functions in the hidden layer and in the output layer:

- Linear,
- Logistic,
- Atanh,
- Exponential,
- Sinus.

Other settings are left by default (ANS – automated neural networks).

[†] We will be orientated using the smallest square method. We will terminate network generation if there is no improvement, ie to reduce the sum of squares. So we will preserve those neural structures whose sum of squares of residuals to the actual development of the Euro to Yuan exchange rate will be as low as possible (ideally zero).

3 Results

Based on the established procedure, 100,000 neural networks were generated. Five networks have been preserved, showing the best parameters. Their overview is given in Table 2.

Table 2. Overview of preserved neural networks

Network name	Training performance	Testing perf.	Validation perf.	Training error	Testing error	Validation error	Training algorithm	Error function	Activation of hidd. lyr.	Output activ. funct.
RBF 1-24-1	0.983554	0.984770	0.984081	0.009370	0.009020	0.009390	RBFT	Sum of sq.	Gauss	Identity
RBF 1-29-1	0.981894	0.981738	0.983707	0.010306	0.010730	0.009532	RBFT	Sum of sq.	Gauss	Identity
RBF 1-30-1	0.984826	0.985312	0.984546	0.008650	0.008742	0.009129	RBFT	Sum of sq.	Gauss	Identity
RBF 1-28-1	0.984362	0.984673	0.983238	0.008913	0.009007	0.009832	RBFT	Sum of sq.	Gauss	Identity
RBF 1-26-1	0.983486	0.984695	0.984107	0.009407	0.009014	0.009311	RBFT	Sum of sq.	Gauss	Identity

Source: Authors.

These are only the neural networks of the basic radial function. The input layer has only one variable - time. In the hidden layer, neural networks contain 24 to 30 neurons. In the output layer, we have logically a single neuron, and the only output variable is the Euro to Yuan exchange rate. For all networks, the RBFT training algorithm was applied. All neural structures used the same function to activate the hidden neuron layer, namely the Gaussian curve. They also use the same function to activate the outer layer of the neurons, and this function is identity (see Table 2).

Training, testing and validation performance is also interesting. In general, we are looking for a network that has the same performance in all sets of data (we recall that the distribution of data into sets was random). The error should be as minimal as possible.

The performance of individual sets of data is expressed as a correlation coefficient. The values of the individual data sets according to specific neural networks are presented in Table 3.

Table 3. Correlation coefficients of individual data sets

	Correlation coefficients		
	Balance (Training)	Balance (Testing)	Balance (Validation)
1.RBF 1-24-1	0.983554	0.984770	0.984081
2.RBF 1-29-1	0.981894	0.981738	0.983707
3.RBF 1-30-1	0.984826	0.985312	0.984546
4.RBF 1-28-1	0.984362	0.984673	0.983238
5.RBF 1-26-1	0.983486	0.984695	0.984107

Source: Authors.

The table shows that the performance of all preserved neural structures is approximately identical. Slight differences do not affect the performance of individual networks. The correlation coefficient of all training data sets ranges from more than 0.981 to over 0.984. The value of the correlation coefficient of testing data sets reaches the same values as the training set of data. The correlation coefficient of the validation data set of all neural networks is above 0.983. At the same time, let's that the error rate in all sets of data is in the range of from nearly 0.009 to more than 0.01. Differences in the error of aligned time series in the individual datasets are almost negligible.

In order to select the most suitable neural structure, we need to analyze the results obtained. Table 4 provides the basic statistical characteristics of each set of data for all neural structures.

Table 4. Statistics of individual sets of data according to preserved neural structures

	Prediction statistics Output (goal): Euro to Chinese Yuan				
Statistics	1.RBF 1-24-1	2.RBF 1-29-1	3.RBF 1-30-1	4.RBF 1-28-1	5.RBF 1-26-1
Minimum prediction (Training)	6.85480	6.70498	6.85756	6.86930	6.79547
Maximum prediction (Training)	10.21765	9.92194	10.20493	10.14413	10.17991
Minimum prediction (Testing)	6.85511	6.70617	6.85749	6.86959	6.79648
Maximum prediction (Testing)	10.21718	9.92191	10.20358	10.14384	10.17832
Minimum prediction (Validation)	6.85528	6.70502	6.85825	6.87002	6.79581
Maximum prediction (Validation)	10.21006	9.92164	10.19128	10.14073	10.18000
Minimum residues (Training)	-0.45897	-0.70049	-0.40403	-0.39654	-0.45631
Maximum residues (Training)	0.56285	0.41491	0.47297	0.49684	0.58830
Minimum residues (Testing)	-0.47026	-0.69190	-0.38432	-0.38399	-0.43711
Maximum residues (Testing)	0.47346	0.40493	0.43641	0.45880	0.41700
Minimum residues (Validation)	-0.45710	-0.56530	-0.35886	-0.38681	-0.42207
Maximum residues (Validation)	0.57571	0.42783	0.46780	0.50829	0.58714
Minimum standard residues (Training)	-4.74158	-6.90008	-4.34425	-4.20036	-4.70472
Maximum standard residues (Training)	5.81469	4.08702	5.08551	5.26279	6.06549
Minimum standard residues (Testing)	-4.95139	-6.67959	-4.11032	-4.04606	-4.60389
Maximum standard residues (Testing)	4.98511	3.90916	4.66740	4.83434	4.39212
Minimum standard residues (Validation)	-4.71717	-5.79008	-3.75590	-3.90102	-4.37405
Maximum standard residues (Validation)	5.94113	4.38203	4.89604	5.12617	6.08474

Source: Authors.

Ideally, the individual statistics of the neural network are cross-sectionally same in all sets (minimum, maximum, residue, etc.). In the case of preserved neural networks, the differences in time series are minimal, both in absolute terms and in the case of residues. However, we are not able to clearly identify which of the preserved neural networks has the most appropriate results. All neural structures appear to be usable in practice.

Figure 1 is a graph showing the actual development of Euro and Yuan as well as the development of predictions using individual generated and preserved networks (or even time series).

It is clear from the graph that all neural networks predict the evolution of the Euro to Yuan exchange rate at individual intervals slightly differently. However, what is important is no the the similarity between the predictions of individual networks, but the similarity (or the degree of consistency) with the actual development of the exchange rate of both currencies. In this respect, all preserved neural networks seem very interesting at first glance. They respect the basic guidelines of the curve assessing the evolution of the Euro to Yuan exchange rate, and at the same time tend to perceive the extremes of this curve.

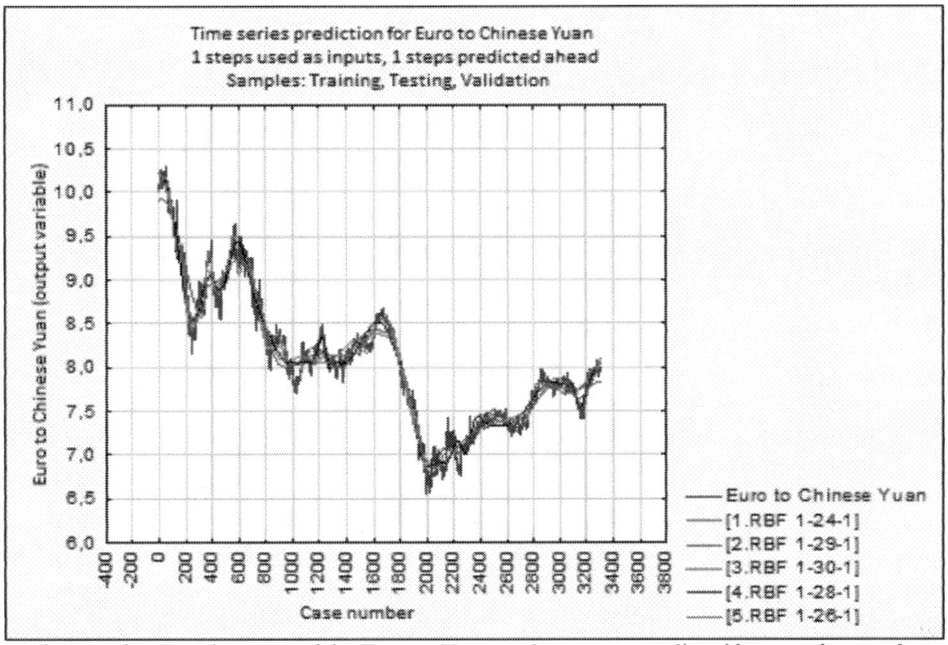

Fig. 1. Link graph – Development of the Euro to Yuan exchange rate predicted by neural networks compared to the actual exchange rate in the monitored period
Source: Authors.

Given that the chart in Figure 1 includes 3,303 records of data on the Euro to Yuan exchange rate, it may appear to be unclear. Therefore, it is appropriate to demonstrate the situation at the selected data interval. The graph in Figure 2 provides a comparison of the actual development of the Euro to Yuan exchange rate at the interval of the last 100 days of the reference period, ie from July 14 to October 21, 2018.

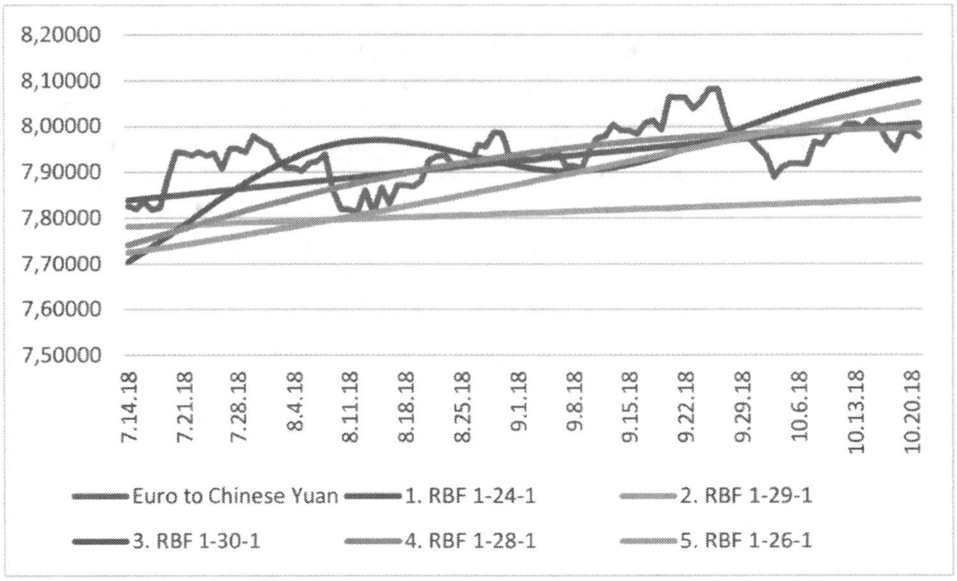

Fig. 2. Link graph – Development of the Euro to Yuan exchange rate predicted by neural networks compared to actual exchange rates from July 14 to October 21, 2018
Source: Authors.

It can be seen from the graph that none of the preserved neural networks can fully copy the actual exchange rate of Euro to Yuan in the observed interval. The truth is, however, that the perceived closest to reality is the network 1.RBF 1-24-1. At the beginning of the period under review, it is almost the same as the actual value of Yuan, and at the end of the period under review, it is not too different from the target value. The difference in both cases is in the hundredths of a Yuan. But even from this perspective, the least accurate network, 3.RBF 1-30-1 differ from the actual exchange rate by less than 0.15 Yuan. The examination of residues also appears interesting. The development of residues in the period from 14 July to 21 October 2018 is the subject of Figure 3.

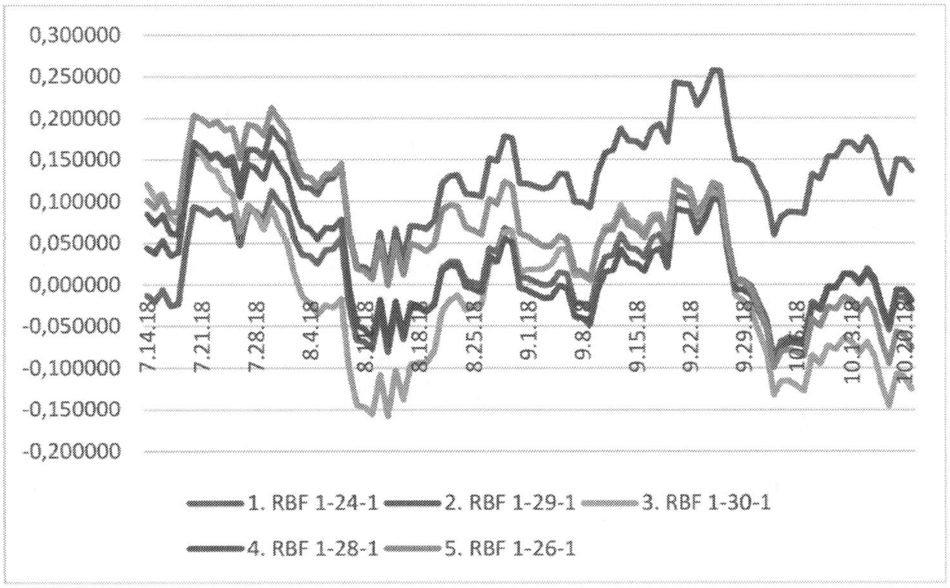

Fig. 3. Development of residues of aligned time series from 14 July to 21 October 2018
Source: Authors.

The graph shows that the sum of the residues approaching zero can be recorded in the observed period for all neural networks except the 2.RBF 1-29-1 network. In this case, residues produce relatively high positive values. To illustrate the situation, Table 5 offers the sum of the residuals of all aligned time series.

Table 5. Sum of the residues of the individual aligned time series

	1. RBF 1-24-1	2. RBF 1-29-1	3. RBF 1-30-1	4. RBF 1-28-1	5. RBF 1-26-1
Sum of residues	10.753689	1.369654	10.598169	4.290912	0.676559

Source: Authors.

When we ignore the individual fluctuations of residues throughout the whole monitored period, the sum of the residues will ideally be zero. The closest to zero is the sum of the fifth neural network, for which it is almost 0.676. On the contrary, the highest sums are shown by 1. RBF 1-24-1 and 3. RBF 1-30-1. Always above the value of 10. The last analysis would then show that the most successful time series is 5. RBF 1-26-1. However, we must realize that in the case of 3303 measurements, the sum of residues at the value of 10 is absolutely minimal. Therefore, we maintain that the most successful neural structure is the network 1.RBF 1-24-1.

4 Conclusion

The aim of the paper was to align the time series mapping the Euro to Chinese Yuan exchange rate using artificial neural networks to find an artificial neural structure with a performance greater than 0.95.

In general, each prediction is given by a certain degree of probability with which it is to be fulfilled. As we predict the future development of any variable, we try to estimate the future development of this variable on the basis of previous years' data. Although we can include most of the factors influencing the target variable in the model, we always simplify reality, and we always work with a certain degree of probability that some of the predicted scenarios will be fulfilled.

In the case of our contribution, we aimed to predict the relationship between two currencies. We have not taken into account any other variables than time. We have not addressed growth, possibly a fall in gross domestic product, the volume of mutual transactions, the relationship of other countries to the currencies analyzed, the volume of mutual tourism and many others. We only dealt with the machine prediction of the future development of the two-currency exchange rate. Therefore, we performed the time series alignment using neural networks. All neuron structures show above-average performance (higher than the desired 0.95) and present a minimal rate of error. Still, we analyzed five preserved neural networks. We cared about which time series is more successful than the others. We concluded that the most successful network is the neural structure of 1.RBF 1-24-1. Although having the second highest sum of residues, in the case of basic statistics, it keeps up with other preserved networks. At the same time, we must state that it is the most similar to the actual development of the Euro and Yuan exchange rate. This network offers performance in training, test, and validation sets of data always above the 0.98 correlation coefficient. It also shows minimal error. So we can conclude that the aim of the paper has been met. We have generated a high-performance neural network with the ability to predict the exchange rates of Euro to Yuan in days and weeks.

It is often a major mistake for neural networks to overtrain. This means that networks have excellent characteristics – minimum error and very high performance. Yet their results tend to be pointless. In this case, they were not. All preserved networks are usable with a certain degree of accuracy, and as previously noted, the 1.RBF 1-24-1 network gives even more accurate results than other networks.

References

1. Y. Mai, H. Chen, J. Z. Zou, S. P. Li, Currency co-movement and network correlation structure of foreign Exchange market. *Physica A: Statistical Mechanics and its Applications*, **42**, 65-74, (2018)
2. K. Clements, Y. Lan, J. Si, Uncertainty in currency mispricing. *Applied Economics*, **50**(20), 2297-2312, (2017)
3. Y. Wang, Y. Yang, Analysis on the Impact of RMB Exchange Rates on Manufacturing Trade between China and EU Zone. *Proceedings of 2016 2nd International konference on humanities and social Science Research,* **70**, 42-45, (2016)
4. J. Budík, L. Smolíková, Advanced investment strategy for trading major currency pairs. *Littera Scripta*, **6**(1), 18-29, (2013)
5. K. Jebran, I. Amjad, Dynamics of volatility spillover between stock market and foreign Exchange market: evidence from Asian Countries. *Finnancial Innovation*, **2**(1), (2016)
6. L. Komárek, M. Motl, Jak si vede náš měnový kurz. *ČNB* [online]. Available at: https://www.cnb.cz/cs/o_cnb/blog_cnb/prispevky/komarek_motl_20180222.html (2018)
7. J. Aizenman, Optimal Currency Area: A twentieth Century Idea for the twenty-first Century? *4th International Symposium in Computational Economics and Finance (ISCEF)*, **29**(2), (2016)
8. S. Wang, X. Wei, Relationships between exchange rates, economic growth and FDI in China: An empirical study based on the TVP-VAR model. *Littera Scripta*, **10**(1), 166-179, (2017)

9. R. Šmach, Jak se mění kurzy. *Kurzy měn ČNB* [online]. Available at: http://www.kurzymencnb.cz/Jak-se-meni-kurzy.php (2013)
10. J. C. Cuaresma, I. Fortin, J. Hlouskova, Exchange rate forecasting and the performance of currency portfolios. *Journal of Forecasting*, **37**(5), 519-540, (2018)
11. T. Suzuki, The renminbi Exchange rate reform and its implications for Asian markets. *China quarterly of international strategic studies*, **2**(4), 485-506, (2016)
12. L. Shi, P. Sheng, M. Vochozka, The reduction cost of nonperforming loan: evidence from China´s commercial bank. *Applied Economic Letters*, **24**(7), 456-459, (2016)
13. N. Kadlčáková, L. Komárek, Foreign Exchange Market Contagion in Central Europe from the Viewpoint of Extreme Value Theory. *Prague Economic Papers*, **26**(6), (2018)
14. J. T. Jalles, C. Mulas-Granados, J. Tavares, C. C. Caro, Single currency and fiscal performance: the early Euro Area in perspective. *Applied Economics letters*, **25**(6), 415-419, (2018)
15. K. Glinka, Adaptations within the Financial Market in China after Global Financial Crisis. Oeconomia Copernicana. *Oeconomia Copernicana*, **7**(4), 565-591, (2016)
16. J. A. Frankel, New Estimation of China´s Exchange rate régime. *Pacific economic review*, **14**(3), 346-360, (2009)
17. World Bank. *The World Bank*. [online]. Available at: http://www.worldbank.org/ (2018)

Comparison of neural networks and regression time series in estimating the development of the EU and the PRC trade balance

Jaromír Vrbka[1,*], *Zuzana* Rowland[1] and *Petr* Šuleř[1]

[1]Institute of Technology and Business in České Budějovice, School of Expertness and Valuation, Okružní 517/10, 37001 České Budějovice, Czech Republic

Abstract. China, by GDP, is the second largest economic power, and hence also a key player in the field of international relations. As far as the EU is concerned, it is China's largest trading partner. From this point of view, it is clear that monitoring export and import development between these partners is essential. This paper therefore aims to compare two useful methods, namely the accuracy of time series alignment through regression analysis and artificial neural networks, to assess the evolution of the EU and the People's Republic of China trade balance. Data on the export and import trends of these two partners since 2000 have been used, and it is clear that the trade balance was completely different that year than it is now. The development over time is interesting. The most appropriate curve is selected from the linear regression, and from the neural networks three useful neural structures are selected. We also look at the prediction of future developments while taking into account seasonal fluctuations.
Key words: trade balance, export and import, linear regression, neural networks

1 Introduction

Balance of payments is the payment value of all economic transactions between the country and other countries over a period, usually a calendar year. The balance of payments is made up of export and import transactions of goods and services, the movement of investment resources, gifts and transfers of money or gold [1]. The balance of payments is in the form of an account where credit and debit items are recorded. In the case of surplus exports, the trade balance is an asset, otherwise it is a liability. This is related to the balance of payments balance, which tracks the difference between export and import on the current account [2].

China is with its GDP the second largest economic power and the key player in the field of international relations [3]. On the development of the Chinese economy we can see progress in transforming the low-cost economy into innovation and high-tech products [4]. The first trade relations of the European Union with China were established in 1975. The Trade and Economic Cooperation Agreement, which includes three basic principles of

* Corresponding author: vrbka@mail.vstecb.cz

cooperation – business, economic and the tasks of the Joint Committee became the fundamental document of trade relations between the EU and China in 1985 [5]. International trade is also subject to a political dialogue that seeks to correct the Long-Term Policy for China-European Relations [6]. Until 2020, the EU-China Strategic Partnership Agenda applies, covering the area of security, exchange and sustainable development [7].

Successful business ties between China and the EU can be expressed by a figure from 2015 when trade exchanges amounted to 520.897 billion EUR. Looking at the European Union's trade balance, we must say that it is unfavorable [8]. The difference between export and import is almost doubled in favor of imports from China [9]. In the years 2014-2015, exports grew by almost 16 percent. The European Union is China's biggest trading partner, which was confirmed by the Chinese National Statistics Office, which stated that imports from China in 2017 amounted to 400 billion USD. The table below shows the development of exports and imports over a five-year period [10].

Table 1. The development of exports and imports over a five-year period

Export (USD trillions)	Import (USD trillions)	Return (USD trillions)	Balance (USD billions)	
2012	2.05	1.82	3.87	230.31
2013	2.21	1.95	4.16	259.01
2014	2.34	1.96	4.30	382.46
2015	2.27	1.68	3.95	593
2016	2.09	1.58	3.68	510.7
2017	2.21	1.73	3.94	476

Source: [10].

Taking into account the diversity of the sector in which the European Union and China are linked, we can expect a positive development of cooperation, but we should not forget the dangerous factors [11]. These factors can be included, for example, in cybernetics, and in particular their security. Cyber security is today a concept that employs all three of the largest trading partners, China, the European Union and the United States. International trade in the future will also affect relations between China and the United States, whose relations are still not politically stable. Another risk, which could be negatively affected, is the problem of dumping. The European Union is now addressing sixteen cases of likely imports of goods below the market price. China is facing a problem of large distribution at the expense of efficiency of one industry. China's economic growth carries various risks. In addition to the internal ones, it is mainly the fact that the economies of the EU countries that have failed to adapt to the effects of the financial crisis can only keep exports, of which a significant proportion goes to China [12]. Given that Chinese quality and technical level of production are rising, commitments on the EU's current account balance of payments are deepened by mutual commercial relationships, which negatively affects the unemployment rate and government debt. Thanks to acquisitions, China is buying the know-how of EU companies, thus gradually taking over its own export markets. In this way, businesses are losing their competitiveness, and economic risk is rising [13]. The European-Chinese market can certainly offer us much more, the question is whether China will realize its priorities and effectively transform them into satisfaction for both parties.

2 Data and methods

Data for analysis is available on the website of the World Bank etc. For the analysis, trade balance data between the EU and the People's Republic of China (hereinafter referred to as the PRC) will be used. This will be the difference between total exports and imports between the two trading partners from the EU perspective. The time period for which the data will be available is the monthly balance starting from January 2000 and ending in July 2018. This is about 223 input data. The unit is the Euro.

The descriptive characteristics of the data are given in Table 2.

Table 2. Characteristics of the data file

Samples	Month (Input variable)	Balance (Output (goal))
Minimum (Training)	36526.00	-2.066917E+10
Maximum (Training)	43252.00	-3.102813E+09
Average (Training)	39916.52	-1.066424E+10
Standard deviation (Training)	1949.82	4.128087E+09
Minimum (Testing)	36586.00	-1.888250E+10
Maximum (Testing)	43282.00	-3.118169E+09
Average (Testing)	39702.30	-1.022066E+10
Standard deviation (Testing)	2174.13	4.848833E+09
Minimum (Validation)	36951.00	-1.937546E+10
Maximum (Validation)	43040.00	-3.497008E+09
Average (Validation)	40047.88	-1.137918E+10
Standard deviation (Validation)	3096.16	7.279221E+09
Minimum (Overall)	36526.00	-2.066917E+10
Maximum (Overall)	43282.00	-3.102813E+09
Average (Overall)	39904.26	-1.070440E+10
Standard deviation (Overall)	1963.77	4.274381E+09

Source: Authors.

Interesting, of course, is price development over time. Figure 1 therefore shows the selected statistical characteristics in graphical form, including the histogram of the input data.

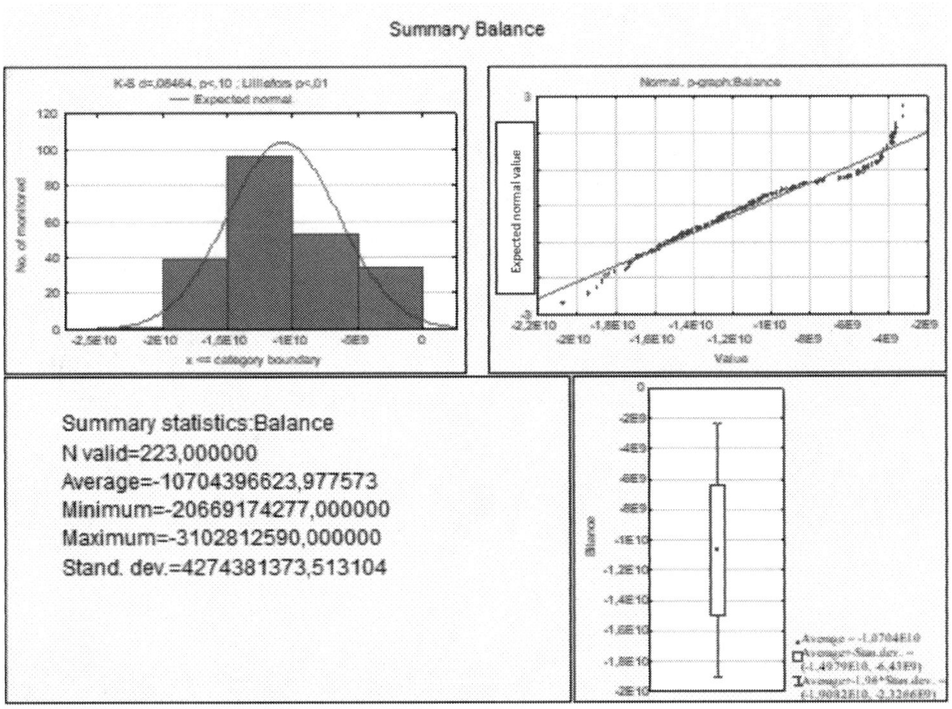

Fig. 1. Graph of basic statistical characteristics
Source: Authors.

For data processing, DELL's Statistica version 12 will be used. First, a linear regression will be performed. Afterwards, neural networks will be used for regression.

Linear regression will be performed on the monitored data sample for the following functions:

- Linear,
- Polynomial,
- Logarithmic,
- Exponential,
- Multiparameter of weighed distances,
- Multiparameter negative-exponential smoothing.

First, the correlation coefficient, i.e. the dependence of trade balance on time, will be calculated. Next, we will work with a level of significance of 0.95.

Then regression will be performed using neural structures. We will generate multilayer perceptron networks and neural networks of basic radial functions. Time will be the independent variable. We will determine the trade balance between the EU and the PRC as the dependent variable. We divide the time series into three sets – training, testing, and validation. The first group will have 70% of input data. Based on the training set of data, we generate neural structures. In the remaining two sets of data, we always leave 15% of the input information. Both groups will serve us to verify the reliability of the found neural structure, or the found model. The delay of the time series will be 1. We will generate 10,000 neural networks. We will preserve 5 of them with the best characteristics[†]. In the

[†] We will be orientated using the smallest square method. We will terminate network generation if there is no improvement, i.e. to reduce the sum of squares. Thus, we will preserve those neural

hidden layer, we will have at least two neurons, at most 50. In the case of the radial basic function, there will be at least 21 neurons in the hidden layer, at most 30. For the multiple perceptron network we will consider these distribution functions in the hidden layer and in the output layer:

- Linear,
- Logistic,
- Atanh,
- Exponential,
- Sinus.

Other settings are left by default (using the ANS tool – automated neural network).

In conclusion, we compare the results of linear regression and regression using neural networks. Comparison will not take place in the form of residue analysis (minimum, maximum values, spread of residues, etc.), but at expert level and experience of the assessor, economist.

3 Results

Linear regression

The correlation coefficient is -0.7668, which represents a significant statistical indirect dependence of the trade balance on the development over time. The determination coefficient reaches 0.9001.

A point graph was constructed (see Figure 2), where the points were intersected by a regression curve, in this case linear. The line parameters are shown in the figure.

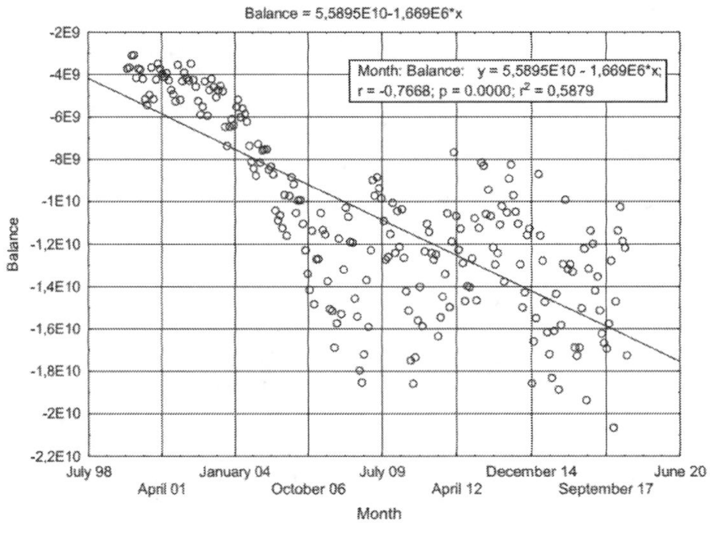

Fig. 2. The Eu and the PRC point graph intersected by the regression curve – linear curve
Source: Authors.

The solid line represents the regression function

Figure 3 represents the intersection of the point graph with a polynomial function.

structures whose sum of squares of residue to the actual EU and PRC trade balance will be as low as possible (ideally zero).

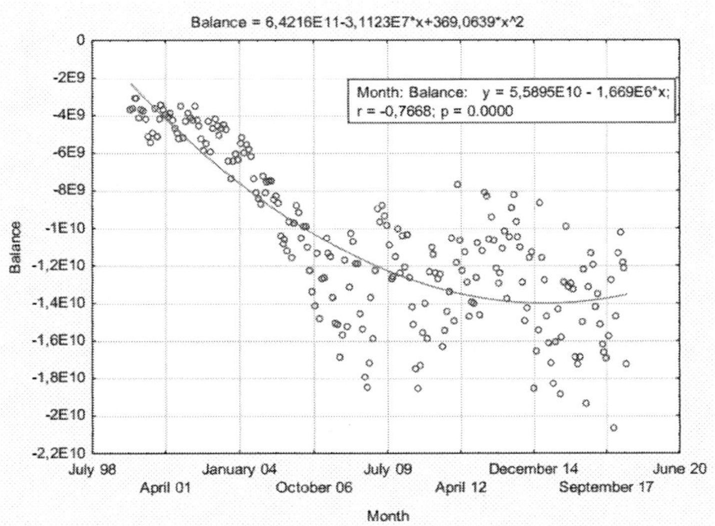

Fig. 3. The EU and the PRC trade balance point graph intersected by a regression curve – polynomial function
Source: Authors.

The polynomial function better outlines the trade balance developments between the EU and the PRC than is the case with the linear function.

Figure 4 shows a point graph intersected by a logarithmic function.

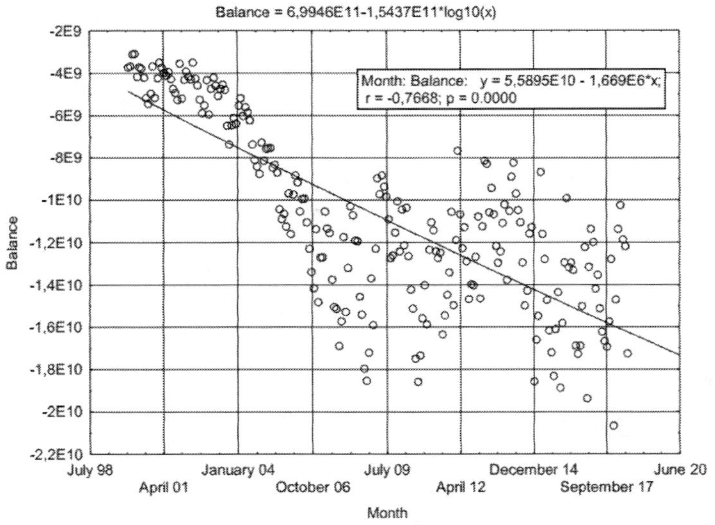

Fig. 4. EU and PRC trade balance point graph intersected by a regression curve – logarithmic function
Source: Authors.

Due to the shape of the curve and the location of individual points in the graph, it is obvious that the logarithmic function rather copies the linear function and is not suitable for the regression and will not estimate the evolution of the EU and the PRC trade balance optimally.

Figure 5 represents the LOWESS (locally weighted scatterplot smoothing) or LOES (locally estimated scatterplot smoothing) function.

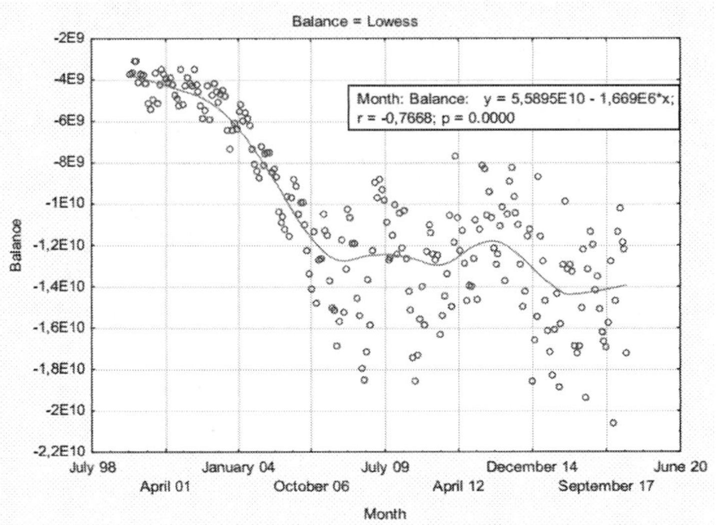

Fig. 5. The EU and the PRC trade balance point graph intersected by a regression curve – LOWESS
Source: Authors.

Regression is calculated for such a curve at partial intervals. The LOWESS curve aligns the time series and monitors its trend appropriately. LOWESS curve intersection looks very interesting. It can fairly reveal the trend of the trade balance of the partner countries.

Figure 6 provides a point graph of the EU and PRC trade balance, intersected by smallest squares of weighed distances function.

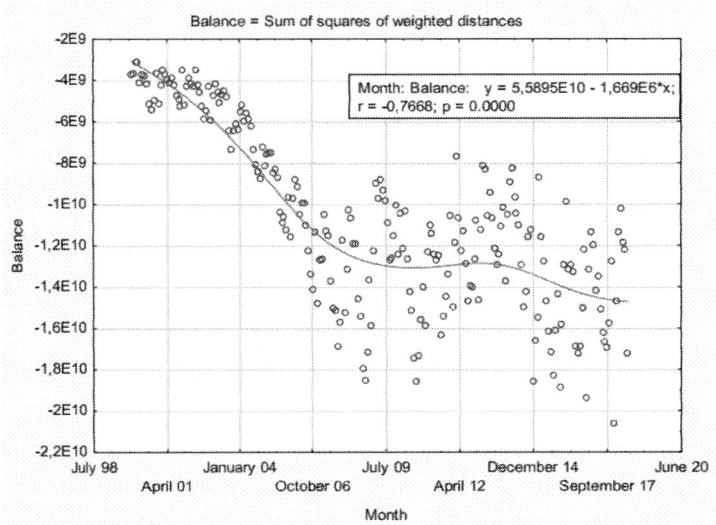

Fig. 6. The EU and PRC trade balance point graph intersected by the regression curve – Sm. sq. of weighted distances
Source: Authors.

The curve fairly well copies the development of the trade balance over its entire interval. However, we must state that in detail, the curve is less accurate than the previous LOWESS curve.

Figure 7 provides the intersection of the function obtained by smallest squares of negative-exponential smoothing.

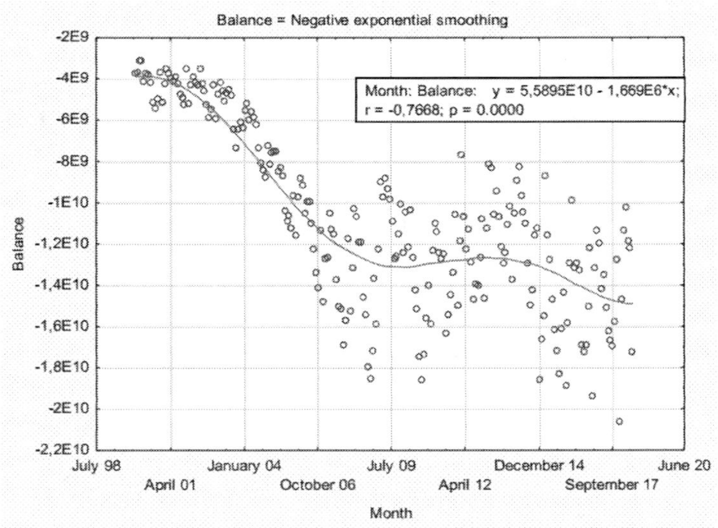

Figure 7. EU and PRC trade balance point graph intersected by a regression curve – Sm. sq. negative exponential smoothing
Source: Authors.

Even this curve appears to be interesting and appropriate for a possible prediction.

As noted above, the correlation coefficient indicates a significant statistical indirect dependence of the target variable on development over time. Nevertheless, it is very distant from the value of -1. If we only evaluated the results by comparing the development of the trade balance between the EU and the PRC and the shape of the regression curve while taking into account a simple linear regression, we could certainly say that the LOWESS curve is closest to the development. This follows curves obtained by the smallest squares method, namely negative-exponential smoothing and weighed distances. All three copy the basic development of trade balance between the EU and the PRC.

Neural structures

Based on the established procedure, 10,000 neural networks were generated. Five networks have been preserved, the ones showing the best parameters. Their overview is given in Table 3.

Table 3. Overview of preserved neural networks

Network name	Training perf.	Testing perf.	Valid. perf.	Training error	Testing error	Validation error	Training algorithm	Error function	Activ. of hidd. lyr.	Output activ. funct.
RBF 1-24-1	0.905183	0.905623	0.926697	1.500796E+18	2.368994E+18	1.609908E+18	RBFT	Sum of sq.	Gauss	Identity
RBF 1-29-1	0.908078	0.881347	0.926506	1.459978E+18	2.571730E+18	1.559369E+18	RBFT	Sum of sq.	Gauss	Identity
RBF 1-26-1	0.908992	0.894534	0.933502	1.443607E+18	2.334177E+18	1.522580E+18	RBFT	Sum of sq.	Gauss	Identity
RBF 1-29-1	0.887327	0.887465	0.928457	1.767475E+18	2.534595E+18	1.365003E+18	RBFT	Sum of sq.	Gauss	Identity
RBF 1-29-1	0.912620	0.881659	0.927402	1.388482E+18	2.542374E+18	1.485960E+18	RBFT	Sum of sq.	Gauss	Identity

Source: Authors.

These are only the neural networks of the basic radial function. The input layer has only one variable - time. In the hidden layer, neural networks contain from 24 to 29 neurons. In the output layer, we have a single neuron and the only output variable is the export balance of the EU and the PRC. For all networks, the RBFT training algorithm was applied. In addition, all neural structures used the same function to activate the hidden layer of neurons, namely the Gaussian curve. They also use the same function to activate the outer layer of neurons, and this function is identity (see Table 3).

Training, testing and validation performance is also interesting. In general, we are looking for a network that has the same performance in all sets of data (we remind that the data was randomly distributed). The error should be as small as possible.

The performance of individual sets of data is expressed as a correlation coefficient. The values of the individual data sets according to specific neural networks are presented in Table 4.

Table 4. Correlation coefficients of individual data sets

	Balance (Training)	Balance (Testing)	Balance (Validation)
1.RBF 1-24-1	0.905183	0.905623	0.926697
2.RBF 1-29-1	0.908078	0.881347	0.926506
3.RBF 1-26-1	0.908992	0.894534	0.933502
4.RBF 1-29-1	0.887327	0.887465	0.928457
5.RBF 1-29-1	0.912620	0.881659	0.927402

Source: Authors.

The table shows that the performance of all preserved neural structures is approximately identical. Slight differences do not affect the performance of individual networks. The correlation coefficient of all training data sets ranges from almost 0.89 to more than 0.91. The value of the correlation coefficient of test data sets reaches more than 0.887. The correlation coefficient of the validation set of data for all neural networks is above 0.926. In order to select the most suitable neural structure, we need to analyse the results obtained. Table 5 provides the basic statistical characteristics of each set of data for all neural structures.

Table 5. Statistics of individual data sets according to preserved neural structures

Statistics	1.RBF 1-24-1	2.RBF 1-29-1	3.RBF 1-26-1	4.RBF 1-29-1	5.RBF 1-29-1
Minimum prediction (Training)	-1,871511E+10	-1,947388E+10	-1,766263E+10	-1,873705E+10	-1,711803E+10
Maximum prediction (Training)	-3,194390E+09	-2,393646E+09	-3,512380E+09	-2,187089E+09	-3,099840E+09
Minimum prediction (Testing)	-1,817011E+10	-1,535797E+10	-1,487201E+10	-1,584944E+10	-1,566790E+10
Maximum prediction (Testing)	-3,152334E+09	-2,445794E+09	-3,543154E+09	-3,310599E+09	-3,302471E+09
Minimum prediction (Validation)	-1,594921E+10	-1,545983E+10	-1,583557E+10	-1,677209E+10	-1,606970E+10
Maximum prediction (Validation)	-3,210110E+09	-3,745279E+09	-4,477256E+09	-2,682479E+09	-4,243404E+09
Minimum residues (Training)	-6,487865E+09	-6,484048E+09	-7,099190E+09	-6,816668E+09	-6,486602E+09
Maximum residues (Training)	4,486943E+09	3,800359E+09	4,809439E+09	5,740114E+09	4,622375E+09
Minimum residues (Testing)	-4,556967E+09	-4,745440E+09	-4,659938E+09	-5,208299E+09	-4,529743E+09
Maximum residues (Testing)	5,079201E+09	6,510952E+09	5,428001E+09	5,388652E+09	6,731699E+09
Minimum residues (Validation)	-4,903684E+09	-4,318523E+09	-3,827757E+09	-4,820044E+09	-4,821168E+09
Maximum residues (Validation)	3,076065E+09	3,131742E+09	2,597524E+09	3,247646E+09	2,224139E+09
Minimum standard residues (Training)	-5,295914E+00	-5,366276E+00	-5,908597E+00	-5,127380E+00	-5,504870E+00
Maximum standard residues (Training)	3,662602E+00	3,145223E+00	4,002856E+00	4,317614E+00	3,922790E+00
Minimum standard residues (Testing)	-2,960697E+00	-2,959129E+00	-3,050094E+00	-3,271460E+00	-2,840887E+00
Maximum standard residues (Testing)	3,299997E+00	4,060054E+00	3,552819E+00	3,384744E+00	4,221871E+00
Minimum standard residues (Valid.)	-3,864755E+00	-3,458285E+00	-3,102089E+00	-4,125572E+00	-3,955020E+00
Maximum standard residues (Valid.)	2,424349E+00	2,507907E+00	2,105085E+00	2,779725E+00	1,824561E+00

Source: Authors.

Ideally, the individual statistics of the neural network are cross-sectional in all sets (minimum, maximum, residue, etc.). However, in the case of aligned time series, the differences are minimal. Therefore, we are not able to clearly identify which of the preserved neural networks has the most appropriate results.

Figure 8 is a graph showing the real trade balance between the EU and the PRC, as well as the development of predictions using individual generated and preserved networks.

The graph shows that all neural networks predict the trade balance at different intervals slightly differently. However, the similarity between predictions of individual networks is not important, but the similarity (or degree of consistency) with actual trade balance developments is. Based on the optical comparison of the curves, we can say that the RBR 1-24-1 and RBF 1-29-1 networks show at certain intervals weaker results than the remaining three preserved neural networks. The networks 2. RBF 1-29-1, 3. RBF 1-29-1, and 5. RBF 1-29-1 respects the curve guidelines that assess the development of the trade balance and at the same time tend to perceive the extremes of this curve.

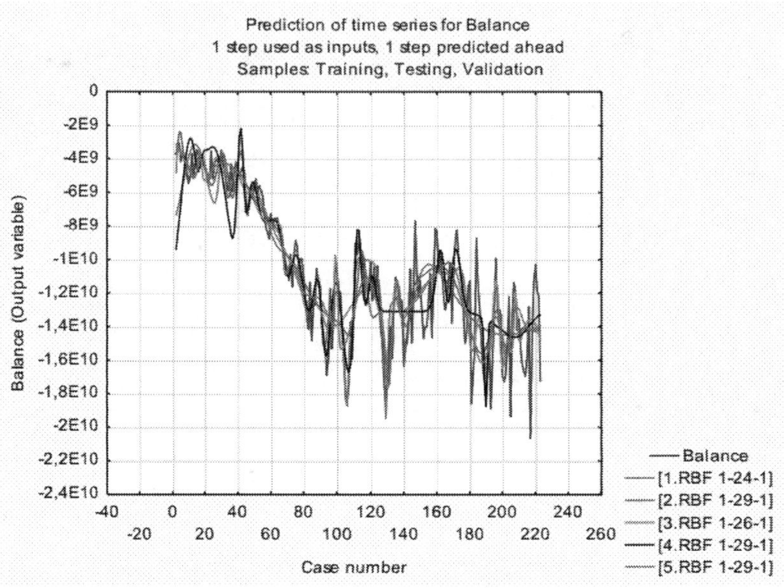

Fig. 8. Line graph – The EU and the PRC trade balance predicted by neural networks compared to the real trade balance in the monitored period
Source: Authors.

4 Conclusion

The aim of the paper was to compare the accuracy of time series alignment using regression analysis and neural networks on the example of the trade balance between the EU and the People's Republic of China.

In general, each prediction is given by a certain degree of probability with which it is to be fulfilled. As we predict the future development of any variable, we try to estimate the future development of this variable on the basis of previous years' data. Although we can include most of the factors influencing the target variable in the model, we always simplify reality, and we always work with a certain degree of probability that some of the predicted scenarios will be fulfilled. It is only a mathematical variable that evaluates the relationship of two or more variables that were generated in the past. And we are trying to predict future developments in accordance with past developments. As with linear regression, as well as regression using neural networks, there is simplification – and quite substantial. We only work with two variables – input (time) and output (trade balance EU X PRC). So, we completely ignore other input variables, which often have a significant impact on the trade balance between the EU and the PRC (international political situation, taxation in both countries, production factor prices, state export support and many others). However it is still true that we estimate aggregated variables better than the partial ones. For some analyses, we can use regression results using neural networks that show a higher degree of accuracy. This may be trend-setting for estimating industry development for the need for individual business valuation, etc.

At the same time, we can state that due to the great simplification of reality, it is not possible to predict the emergence of extraordinary situations and their impact on the EU and PRC trade balance (perhaps in the short term, yes, but certainly not long term). Ideal would be a prediction for the next few days, but for such a short prediction it is currently not possible to obtain data.

The trade balance of both countries can be determined on the basis of statistical methods, causal methods and intuitive methods. In this case, we have been comparing statistical methods. However, they only gave us a possible framework for the development of the monitored variable. It is important to work with information on the possible future development of the economic, political or legal environment. If we can predict their development, we can then project it into the monitored trade balance. At the same time, however, the personality of the evaluator comes into play - an economist who, on the basis of his knowledge and experience, corrects the price determined by the statistical methods and specified on the basis of causal links.

Optically, the LOWESS curve was the best out of linear regression, followed by the smallest squares curve by negative-exponential smoothing, and the curve obtained by the smallest square-weighed distance. Of the neuronal networks, three of the five preserved neural structures proved to be usable in practice: 2. RBF 1-29-1, 3. RBF 1-29-1 and 5. RBF 1-29-1. If we look at the performance from a correlation coefficient point of view, only three of the neural networks listed here are left to be used, of which there is practically no difference.

The aim of the paper was fulfilled.

Given the real trade balance between the EU and the PRC, it is advisable to predict future developments while taking into account seasonal fluctuations.

References

1. C. Z. Liu, F. Wang, Empirical Study on elasticity approach to balance of payments in China – Synchronously discussing the influence of RMB apreciation to payments balance of current account. *International Conference on Management Science and Engineering,* (2004)
2. J. Stefaniak, J. Kuczewska., Export Specialization in Services of the Visegrad Countries. *Equilibrium: Quarterly Journal of Economics and Economic Policy,* **11**(2), 265-284, (2016)
3. J. L. Zamora Ortiz, Balance of Payments Constrained Growth in China: An Application of the Autoregressive Distributed-Lag Modelling Approach. *Revista Ecorfan*, **6**(15), (2015)
4. L. Yan, Does foreign direct investment provide desirable development finance? The case of China. *China and World Economy*, **15**(2), (2007)
5. I. L. Rocha, Manfacturing as driver of economic growth. *PSL Quartely Review,* **71**(285), (2018)
6. P. Balaz, *Globalization of the World Economy and the Impact of Chinas Expansion on the Competitivness of EU*, 14-25, (2014)
7. E. Soukiazis, Are the Transition Economies – Balance-of-Payments Constrained? An Aggregate and Multisector Approach Applied to Central and Eastern Europe. *Eastern European Economics*, **55**(5) (2017)
8. C. Angelache, P. Anghel, Main Developments and Perspektives of the European Union. *Romanian Statistical Review*, 2, 57-79, (2017)
9. J. Mayer, Towards a new normal growth strategy: China in comparative perspective. *Journal of Chinese Economic and Business Studies*, **14**(2), (2016)
10. Embassy of the Czech Republic in Beijing, *China: Foreign Trade and Investment* [online]. Available at: https://www.businessinfo.cz/cs/clanky/cina-zahranicni-obchod-a-investice-19052.html (2018)

11. P. Balaz, J. Pasztorova, *Changes in Development of Interantional Trade in the Period of Financial Crisis – Consequences for Economic Interests of the Slovak Republic*, (2014)
12. R. Higgs, Against the Whole Concept and Construction of the Balance of International Payments. *Independent Review*, **23**(2), (2018)
13. J. Schymik, Globalization and the evolution of corporate governance. *European Economic Review,* **102**(48), (2018)

Research on development of China E-hailing industry

Xuecheng Wang[1,*]

[1]North China University of Technology, School of Economics and Management, No.5 Jinyuanzhuang Road, ShiJingShan District, 100144 Beijing, P.R.China

Abstract. E-hailing cars have become an important innovation in transportation operations. The E-hailing originated in the US and has developed rapidly around the world. As a laggard, China has experienced remarkable progress in exploring development period, rapid expansion period and adjustment period. The main reason is that China has a large user base, a sufficient labor force, a good network innovation environment, and a relatively relaxed regulatory environment. The new policy formulated by the Chinese government has caused a relatively large constraint on the E-hailing industry. Whether the E-hailing platform can maintain neutrality business and whether it can safeguarding consumer rights while developing the economy is an important challenge.
Key words: E-hailing cars, platform business model, E-hailing industry, development, business

1 Introduction

With digital and knowledge-based economy becoming the trend, innovation happens fast due to the interaction between mobile Internet technology and business platforms, in terms of which, e-hailing becomes a good example of the "Internet Plus Transportation" model, reshaping the traffic service market. What's the role of e-hailing platforms in this new industry, or is it possible to create a specific trip market by relocating some resources to match the diversified personal needs with distributed supplies? It's inevitable that economic theories have limitations explaining new phenomenon and sometimes they are divorced from real world, neglecting specific space-time factors.

There's a lot of related research about E-hailing cars. Some researchers found that drivers' collaborative models and evaluation mechanisms at companies such as Uber increase the value of flexible employment [3]. Cramer and Krueger [1] think that Uber has realized the subversive innovation of the taxi industry. A study of Brazil showed that online taxis have replaced a large number of traditional taxis and public transport trips [2]. Vivoda et al. [5] found that the elderly use Internet car very rare, that it is necessary to provide training services for the elderly in order to facilitate their travel. But few people have made research on Chinese E-hailing industries compared with other country.

* Corresponding author: wxc@outlook.com

2 E-hailing has become a new trend of business mode

Uber is usually considered as the originator and typical example of E-hailing service. UberCab was founded in Los Angeles in 2009 and launched the Uber APP in San Francisco in 2010. Passengers can get travel services through this APP, and drivers can meet the passengers' needs and earn income through their own private cars. The generation of online car has its inevitability, because most countries have stricter regulations in the traditional taxi industry [8]. Uber has developed rapidly because it broke the prevailing regulatory difficulties in the taxi industry. Uber offers a variety of services with different prices to meet different travel scenarios. In the three years after the official operation, the average monthly growth rate by over 10% in an explosive expansion period. Earnest Research estimates that 43% of people in the past year have use the E-hailing car service from US credit card data.

The Internet travel service represented by Uber has quickly become the most popular innovative business model on a global scale. It has been a model of success in the sharing economy, even named as "Uberification" in other industries. Judging from the development of the global E-hailing platform, Uber has rapidly spread its business to major developed countries with its first-mover advantage, mature information processing capability and operational organization capability, and has achieved market leading position in the United States, Canada, Australia and the United Kingdom. On the other hand, more local companies have joined the model imitation and launched a localized E-hailing car platform. For example, in the Indian market, the Olacabs penetration rate reached 14.4%, much higher than Uber's 4.49%. By the beginning of 2017, online car or similar services appeared in more than 170 countries and regions. According to Bain Consulting's estimates, the global E-hailing car market will reach approximately US$45 billion in 2016 and is expected to reach US$72 billion by 2020 [7].

At the end of 2015, the total number of Uber orders reached 1 billion. In 2017, the company achieved a total transaction volume of 37 billion US dollars, with a total order volume of about 4 billion. The number of monthly active users reached 75 million, and the number of active drivers was about 3 million. Uber currently offers a wide range of travel services in 545 large and medium cities in more than 60 countries around the world. In January 2018, Uber received a $9 billion investment from Softbank Capital. Uber has accumulated nearly $30 billion in financing and has a valuation of $72 billion, making it one of the most valuable start-ups in the United States. Uber's biggest competitor, Lyft, also had a total of 375.5 million orders in the US and Canada throughout 2017.

The foreign E-hailing car represented by Uber also faces many development problems. For example, there are insufficient measures for consumer safety protection, and the relationship with the network driver is not clear. The management of problems such as swiping orders and driver picking orders is weak, and there are many conflicts with local policies and regulations. It is difficult for the national E-hailing car platform with a small population to form a network economy effect, and overseas expansion cannot be well integrated with local culture [6].

The E-hailing car model originated in Uber in the United States, and it has evolved from concept to mature product in the United States. However, from the development status quo, China's E-hailing car market has become a global model. China's E-hailing car market capacity, major platform enterprise orders and penetration rate among the people are far better than foreign countries. By the end of 2017, the total number of Chinese netizens reached 772 million, and the total population penetration rate reached 55.8%, higher than the global average (51.7%) and Asian level (46.7%), including 753 million mobile phone users, accounting for 97.5 % of the total number of Internet users. The number of listed Internet companies in China and abroad has reached 102, with a total market value of 8.97 trillion yuan; and a total of 77 unicorn companies [7]. From the perspective of global Internet development, the number of Internet users has exceeded 3.4 billion, and has maintained a relatively fast growth rate for many years [4]. With the proliferation of the Internet, especially the mobile Internet, the information age has developed rapidly at a rate that no one in the past can match.

Uber entered China as early as February 2014, invested more than 20 billion US dollars in more than 60 cities, and launched the first UberX in China to compete with the traditional taxi industry. In its heyday, China's top five cities with the highest turnover in Uber, China occupied three seats, and Chengdu once replaced New York with the top spot. In August 2016, Uber reached a strategic cooperation agreement with Didi, and Uber officially withdrew from the Chinese market. Some scholars attribute Uber's failure in China to cultural and environmental adaptability [6]. For example, Uber discount codes are far less acceptable to consumers than WeChat red envelopes. Uber insists on using email customer service and cannot respond quickly to consumer complaints. However, from the internal reasons, the E-hailing car platform represented by Didi can fully tap the unique elements of China's economic development, accurately grasp user needs, flexibly organize supply, and play a more important role in achieving bilateral matching.

The aim (importance) of this study can be concluded on two levels, the first being the practical meaning that it can increase the understanding of E-hailing industry and provide references for the operation of Information platforms as well as the management efficiency for governments; and the second being the theoretical meaning that it can enrich the scope of economic interpretation in terms of supply-demand relations and provide support for innovative study on the way of market organizing under the background of digital and knowledge-based economy.

3 Results

3.1 The development of China's E-hailing car

China's E-hailing car has experienced three relatively obvious development stages: before 2013, it was the exploration and development period of the E-hailing car, and the offline car rental company represented the development of the Internet; from 2013 to 2015, it was the E-hailing car. During the rapid expansion period, the network of car travel services continued to improve, the number of users increased sharply, and the market response was fierce. In the second half of 2015, the E-hailing car ushered in the adjustment period, the central and local policies were introduced, and the E-hailing car industry entered. As the threshold increases, the platform assumes more regulatory responsibilities.

1) Exploring period

As early as around 2006, car rental companies took the lead in trying to integrate with the Internet. Car rental companies such as car rentals and car rental companies in China are digitally transforming business processes such as vehicle information release, user reservation, and process auditing while developing offline service outlets. At this stage, there has not been a real platform enterprise, but it has laid the foundation for platform development. In 2010, the company was established in Beijing, and Dongfang Cheyun Information Technology Co., Ltd. was established. Easy to get is the first company to organize travel service resources as an Internet information company, focusing on business travel, and vehicles mainly from offline car rental companies. It is easy to position itself as a platform to provide comprehensive services for users, vehicle leasers and drivers. By August 2012, it was easy to integrate nearly 5,000 vehicles.

In August 2012, Didi's taxi software was launched. This type of mobile APP first focuses on the car business of traditional taxis. Passengers can call the taxi quickly and conveniently through the APP, and the driver can also conveniently take orders through the APP and increase revenue by reducing the no-load rate. The emergence of taxi software optimized the traditional taxi travel service, which is helpful for solving the problem of difficult taxis, but there is no fundamental improvement. After a difficult push stage, the taxi software won the first batch of stable users, and was recognized by the market. The number of users has increased dramatically. It has become a new thing for public opinion to report, and it has become an excellent target for venture capital and capital markets. The taxi software is a typical intermediary platform. It only provides information services, which combines the supply and demand of taxis, and has obvious characteristics of light asset operation.

2) Rapid expansion period

In 2013, it was called the first year of mobile Internet. The mobile communication base station was integrated into the network. The wlan network technology was mature, the commercialization of LBS technology was mature, and the cost of smart phones continued to move down, which created a good foundation for the development of mobile Internet technology. At the commercial practice level, mobile phone shopping and mobile payment have matured. The O2O platform has reached the Internet outlet, and a large number of online and offline platforms have emerged.

Driven by platform companies, innovative products in various segments of the travel market continue to emerge. The appearance of the special car marks that the Internet App is not only an auxiliary tool for the taxi industry, but also a competitor that cannot be ignored. In July 2014, the fast taxi announced the launch of the new business brand "No. 1 Special Vehicle" targeting mid- to high-end users. Didi also launched a special car in August 2014, targeting the mid- to high-end business car group. Special vehicles have higher requirements on the value of vehicles, and drivers also need to undergo standardized business etiquette training to get on the job. While the car has gained widespread praise in the market, Didi continues to improve its product line. In 2015 alone, direct shuttle buses, express trains, shuttles, buses, drivers and chartered vehicles were launched. It provides a wide range of alternative products for different travel preferences, covering the entire field of daily travel of residents.

At this stage, the competition between the network and the car platform has become more intense. Didi and fast have achieved market leading position, launched a subsidy war, and the market has also spread rapidly. The participation of capital provided support for the subsidy war, and both companies received hundreds of millions of financing. After the subsidy war, the rapid development stage of the E-hailing car market was basically completed, and the capital force contributed to the merger of Didi and the fast. The two occupied more than 80% of the market share of the E-hailing car industry and achieved a strong market monopoly.

At the policy level, most local governments have adopted a form of informal interviews and seminars to convey the attitude of strict management of online vehicles; a few local governments such as Shanghai and Hangzhou have begun to carry out standard management work. The Ministry of Transport also tends to manage the E-hailing car strictly, but due to concerns about policy uncertainty, it has not rushed to introduce a new deal for the car, and has experienced a long wait-and-see period.

In 2015, the total number of users in the online car market was approximately 294 million. The total number of orders per year reached 1.43 billion, which is nearly twice the number of all taxi orders in the United States in 2015, surpassing the cumulative number of orders Uber has established since its establishment six years ago.

3) Specification period

With the rapid development of the travel service platform, the network of car products has been continuously enriched, especially the launch of express trains and express trains. Compared with traditional taxis, the service quality is high while the price advantage is obvious. It has fundamentally touched the interests of the traditional taxi industry, and the conflict between the old and new formats has intensified. Local taxi drivers' strikes and protests have occurred from time to time and have become an important hidden danger affecting social stability.

The Ministry of Transport has accelerated the development of the taxi industry reform and the development of E-hailing car management methods. On July 27, 2016, the Ministry of Transport, the Ministry of Industry and Information Technology, the Ministry of Public Security, the Ministry of Commerce, the State Administration for Industry and Commerce, the General Administration of Quality Supervision, Inspection and Quarantine and the National Network Information Office jointly promulgated the Interim Measures for the Administration of Online Appointment of Taxi Management Services (hereinafter referred to as "Interim measures"). China has thus become the first country in the world to recognize the legal status of the E-hailing car. Since then, the Ministry of Transport has successively established a comprehensive management system from the platform to the driver and the vehicle.

The New Deal of the Internet has given local governments the right to formulate detailed rules for local conditions. In October 2016, first-tier cities such as Beijing, Shanghai, Guangzhou, and Shenzhen took the lead in publishing the draft guidelines for the implementation of the rules for the implementation of the network. Compared with the relatively loose conditions of the Ministry of Transport, the standards for vehicles and drivers for the network were greatly improved. This move has caused widespread dissatisfaction among the public. The main E-hailing car platforms have indicated that the local rules are too strict. Driven by the first-tier cities, the major cities in the country have formulated strict implementation rules, and the development of the E-hailing car industry has been hit hard. Many legal experts have questioned the legitimacy of the establishment of administrative licenses in local areas. Some economists have also questioned whether local implementation rules can promote the healthy development of the industry. At the platform level, the E-hailing car platform gained legal status after the introduction of the national policy and ushered in the golden age of standardized development. However, following the introduction of the local rules, the vehicles and drivers available on the E-hailing car platform fell off the cliff. The heavy asset platform based on the B2C model is relatively less affected by local rules, while the C2C-based light asset platform faces a huge test. In order to gain long-term competitive advantage, the platform has made a fuss about internal management, strengthened the management of E-hailing car drivers, and improved consumer safety protection measures.

As of December 2017, the number of domestic and express passengers has reached 236 million, an increase of 40.6% over the previous year. Take Didi as an example. In 2017, more than 20 million drivers and various vehicles access the platform. The annual order volume was 7.43 billion, and the daily order volume was 25 million. The daily order volume of taxis reached 15 million. The average daily car is 6 million, and the daily average is 500,000, and the daily processing data reaches 4,500 TB. Didi Travel has become the dominant player in the global travel service platform, and has expanded to foreign markets through mergers and acquisitions. Didi's vision is to use DDT to become the world's largest one-stop travel platform for ten years, serving 2 billion users worldwide and meeting 50% of travel needs of consumers.

3.2 Why China E-hailing car development well compared with other countries

China's Internet economy innovation and development has a unique economic and social environment. The E-hailing car platform can adapt to the trend of the times, seize the crux of the Chinese travel industry, and make full use of the convenience of the digital economy to achieve relatively significant results. In the development process of China's E-hailing car platform, there are the following outstanding external advantages:

1) China has a large-scale and huge potential online consumer market. As of 2017, the total number of Internet users in China has reached 770 million, more than one-fifth of the global total. China's Internet consumption reached US$96.7 billion in 2016, only slightly lower than the US. China's Internet development has a strong stamina. Compared with developed countries, there is still about 30% of the Internet penetration rate. In addition, from the perspective of the proportion of the Internet economy in terms of overall GDP, China is as high as 6.9%, ranking second in the world [9].

2) China's abundant labor supply during the transition period is an important driving force for the rapid development of the network. The development of the E-hailing car is basically consistent with the period when the Chinese economy entered the "new normal". The traditional manufacturing capacity has released a large amount of labor, and the E-hailing car platform has absorbed a large number of low-cost labor, which has overcome the bottleneck resource constraints of the service industry development. China's Internet marketing promotion is good at combining with the tactical ground promotion of the sea, and the expansion speed is very fast. The labor-intensive matching supporting industries such as logistics, terminal distribution, and manual customer service fully enjoy the demographic dividend and ensure a stable and low-cost operating environment for the Internet platform.

3) China has a sound Internet application innovation system. From the perspective of innovation, China's innovation in Internet application is significantly ahead of the United States and other countries. Crunbase global unicorn company list shows that 90% of China's listed companies belong to application-driven innovation; from the large Internet enterprise platform Look, the Internet giants represented by Tencent, Alibaba and Baidu provide information flow, capital and other resources for the online car. From the perspective of Internet ecology, China has the most complete network payment and personal credit service system with LBS. A large number of free underlying service support, with intensive offline service sites.

4) China's open Internet innovation supervision provides an inclusive development environment for network vehicles. China regards the "Internet Plus" action plan as a national strategy and attaches great importance to the development of the digital economy. Compared with major developed countries such as Europe and the United States, the Chinese government has provided a relatively long observation period for the development of new Internet things. The government mainly manages through informal channels such as interviews, and can collect various sectors of society before formulating laws or industry management policies. Comments and opinions. The imperfections of the legal and policy systems provide a late-comer advantage for innovation [10].

After study the city transportation service system, this paper divides city transportation into four categories: public transport, sub-public transport, sub-private transport and private transport. It's obvious that current policies that encourage public transport while restrain private transport have caused many problems in sub-public transport systems, driving taxi industry into the dilemma of quantity and price restrictions, bringing about a great shortage of sub-public transport supply. As a supplement for public transport, E-hailing service has become a significant supply, replacing private cars while competing with taxies to solve the problems with market-oriented ways.

E-hailing platforms combine business platforms with mobile Internet technology, responding to specific demands with efficiency. After the space-time redundant analysis, it can be concluded that E-hailing service can reduce space-time redundancy without increasing distinct cost. It provides a virtual operation platform for drivers and cars, with a virtual occupation and virtual space-time arrangement, which is more flattening with simplified participants and business model. This platform maintains its order by eliminating the bad-behaved drivers and provides comment section for distribution reference to meet customers' demand to the maximum extent, which can also affect drivers' income due to specific algorithm. This real-time distribution mechanism is more flexible and it can encourage drivers to provide better services. This mechanism matches platform with capital, promoting innovation gradually and it has no motivation for monopoly from inside. However, there may be pressure from outside for potential monopoly.

China has issued comparatively easing policies on E-hailing services, but many problems appeared during the implementing of these policies by local governments. After the empirical analysis of local enforcement regulations, we find that randomness and restrictions are very common, without thorough consideration of residents' demands and traffic supplies. This causes obstacles for the development of Information platforms and platforms focusing on capitals appear on the contrary, which goes against China's reform on the increase of transportation models and the activation of assets. Bureaucracy and profit-oriented organizations stage a comeback, restricting the E-hailing platforms and reducing diversified virtual organizations, failing to meet residents' demand in the end. Thus, this paper suggested that Chinese government shall enhance unified management overall profit-oriented cars and implement new E-hailing policies nationwide.

3.3 Challenges that E-hailing faced in the future

However, we should also see the many challenges faced in the development of the E-hailing car platform, including external policy pressures, including how to adapt to the platform role, balance the difference between supply and demand, and fulfill the social governance function beyond the economic matching function. The main challenges and bottlenecks include:

1) The E-hailing car policy have limited the economics of platform. The foundation of the platform economy is the market spirit of "free choice". There is no absolute freedom. Any economic action must be established under certain management and control. The original intention of the E-hailing car policy was to safeguard consumer rights and regulate the industry's competitive order. However, in terms of actual development, in different regions, the competent authorities of the industry have different goals and different ways, and the effects of implementation are different. On the whole, most cities have implemented strict management, which forms a constraint on the economic nature of the E-hailing car platform.

2) As a matching organization, the E-hailing car platform should keep neutral. However, as the market scale expands, the industry standard requirements increase, and the power of the E-hailing car platform expands. Almost all E-hailing car platforms are involved in pricing. The function of bargaining through the platform is compressed. Most platforms charge commissions, management fees and other methods are more concealed, and the proportion in the price formation structure has not decreased significantly. Part of the platform that concurrently operates B2C and C2C services also has the problem of opaque and unfair distribution mechanism.

3) The platform not only undertakes the function of economic matching, but also undertakes social functions such as safeguarding consumers' lives and property, maintaining information security and protecting privacy. Due to the ineffective review of the E-hailing car platform and the imperfect security system, many vicious incidents of passengers' life and property safety have been violated, which has had a serious negative impact on the credibility of the E-hailing car. In terms of information security, in May 2015, China's Internet vulnerability exposure platform, Wuyun.com released 59 security vulnerabilities

about taxi software, involving as many as nine manufacturers. In 2016, a serious data breach occurred in Uber, with the name, email and phone number of 57 million passengers, and the names and driver's license numbers of about 600,000 American drivers leaked. Whether it is possible to establish a comprehensive and reliable safety management system has become a sword hanging on the head of the E-hailing car platform.

4 Conclusion

The E-hailing platform has shaped a new travel service market. The E-hailing platform under the participation of capital can cultivate the economic habits of both the supply and demand sides through subsidies, quickly reach the basic market capacity required for the platform economic development, and change the life cycle of the Internet platform enterprises. From the perspective of matching expansion, the E-hailing platform is based on its own diversified elements, and conducts rapid trial and error to achieve continuous and incremental innovation in business development. From the perspective of matching monopoly, the E-hailing platform has the characteristics of bilateral market, and it is easy to form a monopoly position of winners, but the implementation of mandatory monopolistic behavior will pose a threat to the diversity elements on which the platform depends, so the E-hailing platform itself not have incentive to implement monopoly. Instead, capital is induced and pressurized by the monopolistic behavior of the platform under the stimulation of profit.

As an economic organization, the E-hailing car platform undertakes the social functions of urban quasi-public transportation supply and coordination. For travelers, more travel options are available. For urban traffic managers, platform-making rules organize loose and diverse market elements and reduce the cost of managing individuals. They do not rely on government subsidies and enrich the supply market through marketization.

Innovative points of this paper include: 1) Emphasis on the individuality and peculiarity of supply-demand relations from the perspective of specific space and time, which differs from traditional price-quantity analysis. 2) Establishing a special supply-demand analysis mechanism on the basis of platform-based distribution system. 3) Combining Information technology with business models with real-time respond to specific demand and a virtual organization with selective rules that diversifies the transport market. 4) Researches on the problems after the implementation of E-hailing policies by local governments and provide an overall management system consisting of ministerial-municipal forces and E-hailing platforms.

References

1. J. Cramer, A. Krueger, Disruptive Change in the Taxi Business: The Case of Uber. *The American Economic Review*, **106**, 177-182, (2016)
2. L. A. de Souza Silva, M. O. de Andrade, M. L. Alves Maia, How does the ride-hailing systems demand affect individual transport regulation? *Research in Transportation Economics*, **69**, 600-606, (2018)
3. J. V. Hall, A. B. Krueger, An Analysis of the Labor Market for Uber's Driver-Partners in the United States. *ILR Review*, **71**(3), 705-732, (2015)
4. M. Meeker, *Internet Trends 2017 – Code Conference*, Report: Los Angeles, 355 p., (2017)
5. J. M. Vivoda, A. C. Harmon, G. M. Babulal, B. J. Zikmund-Fisher, E-hail (rideshare) knowledge, use, reliance, and future expectations among older adults. *Transportation Research Part F: Traffic Psychology and Behaviour,* **55**, 426-434, (2018)

6. J. Wirtz, Uber: Competing as Market Leader in the US versus Being a Distant Second in China. *Services Marketing*, 626-632, (2016)
7. W. M. Ceng, P. Boutot, Q. Cai, *China Travel Industry engine acceleration*. Bain & Company China, Shanghai, (2018)
8. M. Y. Chen, Review and Enlightenment of foreign taxi market regulation. *Foreign economy and management*, 41-48, (2006)
9. S. Li, Z. Z. Wu, X. L. Wu, Y. Hu, G. He, X. Cheng, F. Song, Z. P. Xie, L. L. Huang, Q. Wang, W. H. Xian, *Interpret the Chinese Internet Features: Beijing, Boston Consulting Company, Ali Research Institute, Baidu Development Research Center, Didi Policy Research Institute*, (2017)
10. China Internet Network Information Center, *41st China Internet Development Statistical Report*: Beijing, 134 p., (2018)

DID model based research on the policy effect of national independent innovative demonstration zones

Xinyu Wang[1,*] and *Yu* Ma[2]

[1]North China University of Technology, School of Economics and Management, No.5 Jinyuanzhuang Road, ShiJingShan District, 100144 Beijing, P. R. China
[2]Beihang University, School of Economics and Management, 37 Xueyuan Road, 100191 Beijing, P. R. China

Abstract. National Independent Innovation Demonstration Zone set up as combination of regional technology innovation policy in nine areas of Beijing Zhongguancun, East Lake in Wuhan, Shanghai Zhangjiang as a national independent innovation demonstration area as a sample of policy mix to when the impact of technological innovation ability by using Difference-In-Difference model. Studies have shown that compared to after the implementation of the policy, local regional technological innovation capability has increased, and fail to carry out the policy of Chongqing, Shenyang, Beijing and Wuhan, to carry out the policy of regional technological innovation capability enhancement is more obvious;The whole society investment in fixed, population density and area of technical level of regional technology innovation in inhibition, human capital, research funding, foreign real investment and government funding support to promote the regional technology innovation, After we re-estimate the same models by the stability test and conclude the similar results.
Key words: regional technology innovation, DID model, national independent innovation demonstration zone, policy effect

1 Introduction

Since the 21st century, enterprise competition intensifies, the development environment showing a high degree of uncertainty, especially in the era of economic marketization, information technology, consumer demand diversification, adjusting the industrial structure, optimizing the economy has become the main form of economy, science and technology plays a more and more important role in the modern economic transformation, role of technological innovation in transition acted as a catalyst. It can improve the technological innovation ability of enterprises, enhance the core competitiveness of enterprises, has become the determinants for enterprises to enhance their market, the comprehensive national strength competition determinants.

* Corresponding author: wangxy@ncut.cn

Since the first National Independent Innovation Demonstration Zone – the establishment of Zhongguancun in Beijing in March 2009, by the end of April 2016, there have been in Lake Donghu, Wuhan, Shanghai Zhangjiang hi-tech park, Tianjin Binhai, etc. 14 national independent innovation demonstration zone established. The establishment of demonstration zone is to play a role in the demonstration of independent innovation and the development of high and new technology, lead, radiation, driven the development of regional economy, how to measure the effect in the demonstration area, which the most important standard is The most important criterion is whether the demonstration zone policy can promote the regional technological innovation ability.

For the impact of regional technology innovation capability, the effect of National Independent Innovation Demonstration Zone of policy on regional innovation capability is immeasurable. Therefore, evaluate Net policy effect, adjust the policy and implementation direction, has a certain theoretical and practical significance, Research on the national independent innovation demonstration zone policy has reference value on Implementation and improvement. This article attempts to use of policy effect evaluation of the mainstream analysis method – double difference analysis method from the quantitative perspectives, analysis and compare the National Independent Innovation Demonstration Zone policy combination of the net effect analysis.

2 Literature review

Research on the technological innovation policy. The main part of the research on the efficiency of technology innovation policy; In the article, Tijssen proposed the of innovation effect principle (including the proposed the design principle of the system), including a specific, measurable, accept, and with time and change. Forster et al. proposed the use of scientific and technological achievements of the industrial income tax revenue as a measure of the efficiency of scientific research tools and supply the existing research benefits of the evaluation tool range [1].

Research on policy performance and policy effect. Wang Baoshun used the data envelopment analysis to evaluate the efficiency of China's local urban environmental governance, and empirically tested the environmental variables that affect the efficiency of fiscal expenditure [2]. Deng Li calculated and analyzed the total factor productivity of Guangdong province by using non-parametric DEA Malmquist index method, estimated from 1980 to 2004 in Guangdong Province 21 prefecture level cities total factor productivity growth, efficiency change and technological progress rate. The analysis showed that the economic growth of Guangdong was mainly driven by the increase of factor inputs [3]. Liu Qilin analyzed the sources of growth in China's energy industry productivity, technical progress, technical efficiency, difference and the change trend of empirical analysis by Malmquist DEA method, the selection of data in 1999-2010 includes 29 provinces in China, energy industry panel data [4]. The trend of total factor productivity and the key factors and convergence conditions are analyzed. Chenqiuying used the analysis method of DEA CCR model, BCC model and Malmquist index analysis of the Xiamen, Zhangzhou and Quanzhou City of science and technology policy performance to pure technical efficiency change, scale changes, total factor productivity, technical efficiency change, technical change index [5].

Research on the application of DID method. Cao Honghua et al. analyzed of the control effect of ecological agricultural policy on the point source pollution of industrial and service industries, selected the Erhai River Basin as the study area, and revealed the effect of ecological agriculture policy [6]. Wang Rongcheng et al. constructed the "mountain town" policy effect evaluation system, the policy effect of prediction and test of plateau towns [7]; Yang Shali et al. surveyed 2009 the transformation of value-added tax after the enterprise business environment from the micro level enterprises, economic benefit and the effect of policy [8]; Zhao Luan et al. assessed the reform of rural credit cooperatives to improve the effect of financial support for agriculture policy [9]; Lin Chen et al. studied on higher education expansion policy on technology innovation efficiency effect [10].

On the long-term development of the level of regional technology, the influence of National Independent Innovation Demonstration Zone of policy of regional technological innovation capacity is far-reaching, for that reason, this article will use the DID model, nine regions of 2010-2017 Zhangjiang hi-tech park, Zhongguancun, Beijing, Shanghai, Wuhan East Lake policy area, zone and non policy area, zone technology innovation capacity index of double difference analysis, comparing the variations between the regional policy and non policy area, the impact of measure policy of regional technological innovation capacity, in-depth analysis of policy in the implementation process of influence factors and the extent of the impact.

3 Data and methods

3.1 Model setting

The analysis of Policy effect mainly include causal model and treatment effect model, causal model although is widely used, but whether causality is true causal relationship and to much extent reflect causality was still in doubt [11].

DID model (difference - in - difference model) in recent years was used in the implementation of the policy effect evaluation, the principle is according to the divided into the experimental group and the control group before and after the implementation of the policy of different performance or the implementation of the policy and without the implementation of policy effects, according to the number of changes before and after the specified index calculated the fold difference policy on the implementation of the policy in the experimental group, the net effect can be obtained. By its basic principles, the establishment of a national autonomous innovation demonstration area for the experimental group, the country did not get the national independent innovation demonstration zone as the control group.

The general model of the model is as follows:

$$Y_{i,t} = \beta_0 + \beta_1 D_{i,t} + \beta_2 T_{i,t} + \beta_3 D_{i,t} * T_{i,t} + \beta_4 X_{i,t} + \epsilon_{i,t} \tag{1}$$

The result of the individual i is $Y_{i,t}$ in the period, the group dummy variable is $D_{i,t}$, $D_{i,t}=1$, indicate that the individual belongs to the policy value, $D_{i,t}=0$, indicate that it belongs to the non-policy group, the period of virtual variables $T_{i,t}$, the experimental period $T_{i,t}=1$, non-experimental group $Ti,t = 0$. $D_{i,t}T_{i,t}$ is effect of interastion. For the parameter β to be estimated, for random $\epsilon_{i,t}$ disturbance term. According to the purpose of use can be divided into four groups: control group before the implementation of the control group after the implementation of the policy, the policy, the implementation of the policy of the experimental group and after the implementation of the policy of the experimental group, the net impact of the policy implementation of the results β_2, as shown in Table 1.

Table 1. Policy effect and coefficient of National Independent Innovation Demonstration Zone

	Experience group	Reference group	Difference
After the policy implementation	β 0 + β1 + β2 + β3	β 0 + β1	β 2 + β3
Before the policy implementation	β 0 + β2	β 0	β 2
Difference	β 1 + β3	β 1	β 3

Source: Authors.

3.2 Variable selection and description

According to the double differential measurement model of thought, we must first select the experimental group and the control group, because currently has 14 national independent innovation demonstration zone of our country, excluding newly established policy no obvious effect of the demonstration zone, finally, select Zhongguancun, Beijing, Shanghai Zhangjiang and Wuhan East Lake as the experimental group, Xian, Shenzhen, Chengdu, Chongqing, Guangzhou, Shenyang as the control group.

The selected variables are: invention patents, high-tech enterprise scientific research personnel number, high-tech enterprise R & D expenditures, whole society fixed assets investment, torch plan government funding support, the actual foreign investment, population density and GDP.

Specific models are as follows:

$$IPA_{i,t} = \alpha_0 + \alpha_1 \ln L_{i,t} + \alpha_2 \ln R\&D_{i,t} + \beta_1 T_{i,t} + \beta_1 D_{i,t} + \beta_3 T_{i,t} * D_{i,t} + \alpha_3 \ln K_{i,t} + \alpha_4 \ln GF_{i,t} + \alpha_5 \ln PD_{i,t} + \alpha_6 \ln FDI_{i,t} + \alpha_7 \ln GDP + \eta_{i,t}$$

In this paper, the main sources of data from the China Torch Statistical Yearbook (2010-2017), the science and technology of China Statistical Yearbook (2010-2017), China Statistical Yearbook (2010-2017), as well as parts of The national economy and social development bulletin (2010-2017).

The statistical characteristics of the above variables are shown in Table 2, the sample size is the 72 National Independent Innovation Demonstration Zone 2010-2017 years in 9 cities.

Table 2. Statistical characteristics of regression variables

	Conceptual data			Experience group			Control group			Symbol
	Max	Min	Std	Max	Min	Std	Max	Min	Std	
Invention patent authorization amount	78129	354	14749.29605	78129.00	766.00	22245.70	13068	354	3242.706439	IPA
Scientific research personnel of high and new technology enterprises	431259	27121	114502.985	431259.00	45432.00	128626.35	336823.5333	27121	82722.81869	L
High tech enterprise R& D funds internal expenditure	51943354	441142	12848140.1	51943354.00	3663326.00	15174457.56	32962323	441142	7602693.432	R&D
Fixed assets investment	69625338	2353.15	22310749.9	69625338.00	3814.70	22210806.45	66203700	2353.15	21746843.46	K
The torch plan to implement the project government funding support	106983	300	22710.524	106983.00	7779.00	24680.87	102020	300	19721.7526	GF
Population density	3826	546.82	924.432	3826.00	965.47	1140.25	3328	546.82	753.6092199	PD
Actual foreign investment	904000	828.88	261920	904000	2244.16	312343.33	876000	828.88	235866.0523	FDI
Gross Regional Product	167068719	4676.1	49216638.6	100694800	9846.81	33735446.03	167068719	4676.1	51585649.5	GDP

Source: Authors.

Overall, in addition to the regional GDP, overall data in all the maximum of all data sources in the experimental group, and general data about minimum values are derived from the control group, show and control groups, the experimental group in funds and personnel input, innovation output is more significant.

4 Empirical results and analysis

This paper for panel data are variable intercept model with fixed effects regression variable intercept model and random effects regression, after using the Hausman test, test results reject the fixed effect model assumptions. Therefore, the random effects model is more appropriate.

By using Stata 12.0 to analyze the model, regression results as shown in Table 3. Model 1 is the basic model, which contains the influence of the internal expenses, the control group, the time and the cross terms of the R&D funds of the high and new technology enterprise's scientific research personnel and the high and new technology enterprises. Model 2 to 6 model is continue to join the other explanatory variables in the regression results of model 6 includes a scientific research personnel of high-tech enterprises, high-tech enterprise R&D expenditures, whole society fixed assets investment, the torch plan implement project government funding support, population density, the actual foreign investment and regional GDP, explanatory variables.

Table 3. The results of DID model of policy effects of National Independent Innovation Demonstration Zone

	Model 1	*Model* 2	*Model* 3	*Model* 4	Model 5	Model 6
	IPA	*IPA*	*IPA*	*IPA*	*IPA*	*IPA*
L	0.0557*	0.0547*	0.0466	-0.0358	0.0506	0.0211
	(1.82)	(1.78)	(1.51)	(-0.15)	(0.24)	(0.91)
R & D	0.0146	0.0172	0.0178	0.0802***	0.0430*	0.0398*
	(0.49)	(0.56)	(0.59)	(3.35)	(1.91)	(1.78)
T	-22.45	0.2304	0.23834	0.3173	-0.1555	-187.6
	(-0.01)	(0.08)	(0.74)	(1.33)	(-0.67)	(-0.08)
D	0.4675	0.48782	0.58643	-0.3249	-1976.9	-0.3279
	(0.99)	(1.02)	(1.23)	(-0.87)	(-0.60)	(-0.98)
T*D	0.6150	0.5907	0.55474	-0.0659**	0.1135***	0.1321**
	(1.06)	(1.01)	(0.96)	(2.47)	(2.95)	(2.74)
GF		-0.0255	-0.0658	0.0574	0.0883*	0.0585
		(-0.43)	(-1.02)	(1.14)	(1.98)	(1.22)
K			-0.0107	-0.039***	-0.035***	-0.027***
			(-1.59)	(-6.29)	(-6.26)	(-3.73)
PD				-0.0012***	-0.0089*	-0.0078**
				(-7.39)	(-1.69)	(-2.11)
FDI					0.0287***	0.0250**
					(4.53)	(3.71)
GDP						-0.0505
						(-1.55)
_ *CONS*	-0.4246*	-0.4084	-0.1441	0.0393***	0.1500	0.3973
	(-1.73)	(-1.63)	(-0.48)	(5.54)	(0.28)	(0.73)
N	72	72	72	72	72	72
R^2	0.5416	0.5358	0.5466	0.7532	0.8116	0.8157
F	17.78***	14.66***	13.23***	28.09***	34.98***	32.42***

Source: Authors.
Note: *t* statistics in parentheses; * $p < 0.1$, **$p < 0.05$, ***$p < 0.01$

From dummy variable cross coefficient,it can be seen that the regression coefficient is positive, indicate that output in the establishment of national independent innovation demonstration zone can indeed significantly improve the park where the local authorized invention patent, help to improve the regional technology innovation ability, and in fact also proved this point.

High technology enterprise's scientific research personnel proportion coefficient is positive. It is proved that it has a positive effect on regional technological innovation capability, which is owing to an increase in the number of researchers and need a lot of money support, human capital transform for the output, which has greatly improved on the level of R&D technology.

The coefficient of High-tech enterprise R&D expenditures is positive, which is consistent with reality, reflecting the enterprises pay more attention to technology innovation. Generally speaking, corporate R&D investment intensity bigger, the more innovation output, technology innovation ability is stronger.

The coefficient of total social investment in fixed assets is negative, although side effects had little effect on that in the more developed regions. After the development of innovative environmental infrastructure to a certain extent, it has not like at the beginning of the reform and opening up of technological innovation ability has improved significantly with the increase.

The coefficient of population density is negative, theoretically, the greater population density areas more conducive to innovation, but in many cases, for example the official term effect of pure pursuit of GDP growth and ignore the long-term is more favourable to the output of innovation education, together with the population density of the region,the more foreign population, the lower the quality of human capital accordingly, which makes the population density and innovation output negative correlation.

The coefficient of actual foreign investment is positive, showing the positive effect, foreign investment has been a conscious to the high and new technology industries and strategic emerging industry, the government has improved the market for technology strategy, while the introduction of foreign and domestic competition will promote the enterprise technology innovation desire. Under the dual stimulation, the foreign investment introducing has an influence on technology innovation output slightly outweigh the disadvantages, but only have little effect.

The coefficient of regional GDP is negative, although this result is not significant. In general, in regional GDP higher area, the more developed the economy, but also pay attention to R & D investment accounted for the ratio did not increase with the increase of GDP. At the same time, government funding support variable coefficient is positive, indicating that capital investment is the benefit for the regional technological innovation ability.

In conclusion, in view of the relationship between the cross term coefficient, we can get that the establishment of National Independent Innovation Demonstration Zone can improve the ability of regional technology innovation, namely, the policy measures are significantly effective.

4.1 Robustness test

In order To make the conclusion correct, this paper use the fixed effect model benchmark regression model estimation, Table 4 is the results of robustness test, from this table, we found coefficient of cross multiplication (T*D) are different, but little difference and bith has a positive effect, at the 1% level significantly, after robustness tests the result still holds, that is robust in the National Independent Innovation Demonstration Zone Policy effect of DIDS model, namely, the National Independent Innovation Demonstration Zone Policy has a positive role in promoting regional technology innovation output.

Table 4. The results of the robustness of DID model in the policy effects of the National Independent Innovation Demonstration Zone

	Model 7	Model 8	Model 9	Model 10	Model 11	Model 12
	IPA	*IPA*	*IPA*	*IPA*	*IPA*	*IPA*
L	0.0163	0.0162	0.0200	0.0164	0.0483	0.0880*
	(0.35)	(0.35)	(0.44)	(0.36)	(1.07)	(1.78)
R & D	0.0428*	0.0428*	0.0431*	0.0395	0.0164	0.0139
	(1.69)	(1.67)	(1.74)	(1.57)	(0.64)	(0.56)
T	-0.7697	-0.7856	0.12860	0.0476	-0.0819	-0.0917
	(-0.42)	(-0.40)	(0.61)	(0.21)	(-0.37)	(-0.04)
D	-0.1929***	-0.1931***	-0.2142***	-0.2249***	-0.21418***	-0.2546***
	(-3.14)	(-3.09)	(-3.50)	(-3.60)	(-3.60)	(-4.07)
*T *D*	0.1316***	0.1317***	0.1359***	0.1366***	0.1399***	0.12338***
	(4.01)	(3.93)	(4.18)	(4.20)	(4.52)	(3.89)
GF		0.0010	-0.0099	-0.0142	0.0817	-0.0270
		(0.02)	(-0.24)	(-0.34)	(0.20)	(-0.61)
K			-0.0181**	-0.0155*	-0.0027***	-0.0598
			(-2.21)	(-1.78)	(-2.91)	(-0.39)
PD				5.369	3.621	4.128
				(0.88)	(0.62)	(0.72)
FDI					0.0187***	0.0184***
					(2.66)	(2.67)
GDP						-0.0153*
						(-1.79)
_ *CONS*	0.2901	0.2897	0.5496*	-0.1694	-0.4770	-0.6609
	(0.93)	(0.92)	(1.69)	(-0.19)	(-0.57)	(-0.79)
N	72	72	72	72	72	72
R^2	0.4669	0.4669	0.5098	0.5165	0.5726	0.5969
F	10.16***	8.32***	8.32***	7.35***	8.04***	7.85***

Source: Authors.
Note: *t* statistics in parentheses; * $p < 0.1$, ** $p < 0.05$, *** $p < 0.01$

5 Conclusions and prospects

This paper adpot the DID model to the evaluate policy effect of the National Independent Innovation Demonstration Zone, the invention of the patent can be used as the effect of the policy variables, scientific research personnel of high-tech enterprises, high-tech enterprise R & D expenditures, whole society fixed assets investment, the torch plan implement project government funding support, population density, the actual foreign investment and regional GDP can be used as explanatory variables. Through the analysis, we draw the following conclusions:

The establishment of National Independent Innovation Demonstration Zone has a significant role in promoting the regional technological innovation ability.

This shows that our country according to all levels from the overall planning, establish demonstration area, and then extended to the country, produce technology diffusion effect, to take the role, through policy to increase output of technological innovation activities, enhance the ability of regional technology innovation. From the perspective of previous measurement, it shows that the establishment of national independent innovation demonstration zone areas brings technological innovation ability promotion, demonstration zone was established to make government funding to support more targeted and aggregation, stimulate enterprise science and technology personnel and funding of research and development investment. The establishment of the National Independent Innovation Demonstration Zone is more confident in the future of the enterprise. The demonstration zone demonstration should be in aspect of the investment policy, talent introduction, tax policy and financial policy aspects, in order to promote the technology innovation ability as the core goal to achieve, to achieve this, we must solve the current policy in the implementation process in the performance, outstanding technological innovation capability and technology innovation efficiency.

Innovative environment has a certain role in promoting the ability of regional technology innovation.

In a certain extent, the actual foreign investment will stimulate business activity, it bring enterprises to have a sense of crisis and "catfish effect". In order to survive, develop and grow, enterprises should improve their ability of technology innovation, but the same time, we also want to pay attention to wary of the purpose of occupying the Chinese market of foreign capital control the core technology, at this stage we should introduce the core technology, the initiative in their own hands, and human capital, technology digestion, knowledge spillover effect.

Scientific and technical personnel and scientific research funds has played an important role in improving the ability of regional technology innovation.

The increase of human capital on regional technological innovation capability is more important, this viewpoint in previous studies has been verified, but the promotion of population density on the quality of human capital without an obvious role in promoting, therefore, the effect of regional technological innovation capacity is not very obvious. This is mainly reason on the one hand, the quality of human capital increases are not overnight and mobility of the population also limits the government's support to education, on the other hand, the implementation of regional policy can attract more foreign population, making the quality of human capital at a lower level, it plays a negative role for regional technology innovation. The support of government funds, R&D expenditure is a kind of emphasis on R&D, which can stimulate enterprise development, give the confidence of enterprise technology innovation ability. In the future, it should be the purpose of the implementation of the special introduction of talent mechanism to accelerate the accumulation of human capital. Further aggregation of government innovation fund support, advanced industry into the focus of the object support, advocacy and corporate R&D investment at the same time, reduce the risk of enterprise development, in order to maximize the mobilization of demonstration zone enterprises technological innovation of enthusiasm. The combination of high and new technology industries together, which is conducive to the spread of technology and also easy to produce aggregation effect.

Due to the imperfect policy and short implementing period, the implementation of small area, policy and the possible delay, effect has not been fully demonstrated, the future will be tracking survey analysis from multiple perspectives of the national independent innovation demonstration zone policy effect; in addition, due to the availability of data, we did not refine the policy. In the near future will analysis the policy impact on the ability of technology innovation.

References

1. S. P. Forster, S. Seeger, Tax revenue accruing from the commercialization of research findings as an indicator for economic benefits of government financed research. *Research Evaluation*, **23**(3), 233-248, (2014)

2. Y. Guan, J. Liu, B. Wang, An Empirical Study on the spatial dependence of fiscal policy of local government in China -- Based on the perspective of government efficiency. *Contemporary Finance and Economics*, **03**, 38-47, (2014)
3. L. Deng, F. Yu, The measurement and analysis of total factor productivity in Guangdong: 1980~2004 – Based on the panel data of Malmquist DEA. *Social Sciences in Guangdong*, **05**, 39-44, (2006)
4. Q. Liu, F. Li, Research on the technology progress and technical efficiency of China's energy industry based on. *Malmquist-DEA China Science and Technology Forum*, **10**, 67-73, (2013)
5. C. Q. L. Chen, L. F. Yang, Xiamen, Zhangzhou and Quanzhou, most of the city of science and technology policy efficiency enhance. *Science and Technology Management Research*, **11**, 30-35, (2014)
6. H. Cao, R. Wang, G. Zhang, Study on the driving mechanism and policy effect evaluation system of urban mountain. *Journal of Yunnan Normal University (Philosophy and Social Science Edition)*, **04**, 54-60, (2014)
7. R. Wang, H. Cao, L. Li, Study on the policy effect of ecological agriculture in Erhai River Basin Based on DID model. *China Population, Resources and Environment*, **10**, 157-162, (2014)
8. S. Yang, P. Zhang. Micro enterprises from the perspective of value added tax transformation policy effect analysis – Based on the double difference divided DIDS model. *The Chinese Economy Problem*, **04**, 3-12, (2014)
9. D. S. Wenkai, The reform of rural credit cooperatives to improve the financial supporting agriculture policy effect evaluation – Based on the national household survey panel data of times deviation analysis. *Journal of Financial Research*, **03**, 194-206, (2010)
10. L. Chen, J. Xia, The policy effect on innovation efficiency of college enrollment expansion – Based on the measurement and test of quasi experiment and double difference model. *China's Population Science*, **05**, 45-57+127, (2015)
11. X. Wang, The spatial and temporal characteristics of the effect of housing restriction policy in Nanchang city and the effect of land spillover. *China Land Science*, **06**, 41-48, (2015)

Economic inequality according Atkinson

Barbora Wernerová[1,*]

[1]Institute of Technology and Business in České Budějovice, School of Expertness and Valuation, Okružní 517/10, 37001 České Budějovice, Czech Republic

Abstract. Tony Atkinson is one of the most prominent academics. He devoted his life to studying and researching inequality. His work has led to the recognition of the economy of equality as a new economic sector. Atkinson tried to show that economics is primarily a social and moral science, which he promotes in his book The Economy of inequality. At present, inequality occupies an important place in public debate. Atkinson comes up with concrete proposals aimed at the existence of alternatives that will contribute to the struggle for social progress and equality. The aim of this paper is to introduce Atkinson's theory of economics of inequality, with emphasis on our global responsibility.
Key words: inequality, justice, equality, economy issues

1 Introduction

Inequality is a global problem that is related to a number of other dangerous factors. In this article I will attempt a theoretical, philosophical insight into the issue of inequality. One of Anthony Atkinson's main equality experts talks about the need for an overall view of inequality. We can not perceive inequality, for example, only from an economic point of view; on the contrary, we have to look for connections between different scientific disciplines. Inequality is one of today's most pressing problems in the social sphere. After World War II, it managed to keep it in check, but recently it has returned with great vehemence. Everyone knows about the extent of this problem, but little has been said so far about what can be done if we do not want to despair. Atkinson shows that the problem is not only that the rich are getting richer. We can not deal with poverty and the whole economy is changing in such a way that most people lose it. If we want to reduce inequality, we can not settle for simply imposing new taxes on the rich people who will fund existing programs. [1].

These factors include globalization, technology development, financial services growth, changes in payment standards, increasingly smaller trade union rights, and, last but not least, the breakdown of tax reallocation and transfer payments policy. From determining these factors, it may seem that inequality grows for reasons beyond which we have complete control, which would be a misconception. That is because globalization is the result of political, economic and social processes that are influenceable, as well as the direction of technology development. [10]. The aim of the paper is to analyze Atkinson's theory of inequality economics with an emphasis on global responsibility[5].

* Corresponding author: wenerova@mail.vstecb.cz

2 Theoretical approach to economic inequality

2.1 Justice

This part of the paper focuses on the theoretical introduction of the concept of justice. Its philosophical basis is unforgettable. So what does the concept of justice mean? People have been asking this question since ancient times, and it is one of the most important virtues that relates to the other person. Plato has already included justice among the four cardinal virtues and emphasized its indispensable importance for the good functioning of social life. Aristotle saw justice as "good for others". The need for justice in society can still be perceived today. The role of justice has changed over the years in civil life, and its breakdown had to be categorized. Peter Fischer, in his article on Justice, states three meanings of justice. The oldest meaning of justice relates to being as a relation to one's own self. In this sense, justice is perceived as an individual property of physical equipment. The second meaning of justice is the relationship between me and society, which is manifested especially in relation to laws and treaties. The third meaning is the structure of human relationships where justice plays an important role in the relationship between the individual and society, with an emphasis on mutual obligations [6].

Liberal American political philosopher John Rawls sees justice as the fundamental virtue of social institutions. In his book Theory of Justice, Rawls puts a fundamental question on what principles of justice can people agree across all differences. Rawls sets the principles of justice as some kind of fair dealing, and he reveals these principles through an experiment in which he examines how things look before laws exist. This status is called primary, fairness is characterized by ignorance, because in the natural state nobody knows who they are, people do not know their social status, their prospects. At this point, people can act fairly - all to choose the options that will be fair for all and even for the most disadvantaged - people will be thinking about whether they are not the handicapped, the poor. And so logically they want laws that will be beneficial to them in all these conditions. This can be described as a hypothesis that, in the case of a person's ignorance of what he is and what life he will have, he will vote with caution and thought [13].

Principles of Rawls' Theory of Justice

1. Equal freedoms - the greatest possible freedom that does not endanger the freedom of others

2. Difference - the authorities are open to all and no one can be prevented from holding office (social and economic inequalities are organized to favor the most disadvantaged)

Rawls formulated what a fair society means in terms of institutions:

a) Social justice: serves to maintain society, it is layered, it is primarily redistribution

b) Legal justice: laws must apply equally to all

c) Exchange justice: it is at the individual level - it is much easier to act upon it than in an individual vs. state

d) Distributive justice: the relationship between the individual and society, the general good is superior to the good of an individual, the army is to fairly protect every citizen of the state, but it is very difficult to determine the degree of distributive justice (either equality: everyone equally or according to Aristotle: based on merit)[13].

Economists, like Atkinson, work with economic inequality that we can divide into vertical and horizontal. Economic justice mainly concerns tax issues. Vertical justice is defined as a fair approach to individuals of different social status, with the tax burden being derived from its income, which leads to progressive taxation. Against the principle of horizontal justice, the same tax burden is applied regardless of the particular economic environment. Individuals with the same income should therefore be equally tax-burdened [7].

2.2 Equality

If we talk about justice, we can not omit its connection to equality. Equality is a certain degree of identity between individuals. The essence of equality can be experienced in every sector of human life. All people are equal before the law, political equality allows us to vote and be elected, and equality of opportunity, which enables all citizens to gain social benefits [15]. Roger Scruton states the definition of the difference between equality and justice: "While it would be tautology to say that treating people fairly is just, it is definitely not tautology if we say it is fair to treat them impartially - although some people think so is. "[14].

Equality and justice are entities that manifest themselves in all aspects of life, and more than ever before, these concepts are fading away from civil life. Atkinson tries to speak about inequality not only in a narrow perspective, but presents a coherent theory that besides economics, also includes political philosophy, sociology, and the moral aspect of human decision-making. A hundred years ago economists solved justice from a wider perspective in a purely utilitarian spirit. Amartya Sen talks about utilitarianism as a theory focusing only on the sum of individual benefits as the least suitable for assessing inequality, since he lacks the interest in the interpersonal distribution of this amount [8].

2.3 Inequality

The term inequality can be explained from a mathematical point of view as a distinction between two or more particular characters, provided that these characters can be quantified. From another perspective, the notion of inequality by many people is perceived as a failure to achieve equal opportunities, which is determined by different circumstances that people can not influence [2].

The so-called fair average income (y_e) is determined to calculate its value. Its definition is "evenly distributed income in a group that creates the same level of well-being as the current income distribution". This can be calculated as follows:

$$\left(\frac{1}{n}\cdot\sum_{i=1}^{n} y_i^{1-s}\right) \quad (1)$$

- yi corresponds to the balanced income of the i-th group
- e is the parameter of aversion of people to inequality
- n corresponds to the number of income groups

The aversion to inequality parameter tells about the degree of society's tolerance to egalitarianism. In theory, it may have the value of <0; ∞). At zero, people are indifferent to the inequality, at the value ∞ society would only be interested in the one with the lowest income. You actually get values in the range of <0.4; 3>. The Atkinson's inequality point index is then calculated as follows:

$$I = 1 - \frac{y_e}{\mu} \quad (2)$$

- μ corresponds to the current average income per person

Atkinson's index I may have the value of <0; 1>. The more you approach the average income in the group of fair average income in society, the closer I will be to zero, the more evenly the income distribution will be, even theoretically, at point 0, totally uniform. On the other hand, in the direction to the value of 1, income will be more and more inequal [9].

2.4 Equal opportunities

Equality of opportunity is a highly discussed phenomenon that has been addressed by leading economists in recent years.

Richard Tawney in his essay, *Equality*, presents a sample model of opportunity inequality that results from inequal initial opportunities. As an example, we can introduce a student who endeavors to be admitted to a medical school by studying, while another student is admitted to a medical school because of the influence of his parents. This is an opportunity inequality, but does it mean that we could consider the inequality of the results to be entirely irrelevant? [16].

Those who do not consider the inequality of the results as an important aspect hold the fact that we all should have set equal chances and the same starting conditions at the start of our lives. Atkinson refuses this thinking for three reasons. Atkinson says that ignoring the results would not only be illogical but especially unethical. Society should not ignore the individual, whose striving results in hardship, even though the preconditions for equality of opportunity are given. The importance of the results can be demonstrated in a sport analogy, when the unequal equality of opportunity states that we each have the same chances to fulfill our life plans independently of others. Congenial equality of opportunity shows us that we all have a chance to participate in the swimming race, but only one can win. Equality of results will play the role of future generations as the impact of family background on results depends on the strength of the relationship between background and results and on the level of inequality between the conditions of individual families. Atkinson states that if we are concerned about equal opportunities in the future, we need to be concerned about the inequality of the results right now[2].

Reducing inequality in results can serve as a means of reducing inequality in opportunities. The inequality of the results has a negative impact on the whole of today's society. The increase in inequality increases the lack of social cohesion, the increase of crime and generally the deterioration of the quality of health.

3 Development of Atkinson's theory of inequality from the perspective of descriptive methods

This part of the article focuses on the specifics of global inequality, which is closely related to the proposals for its reduction. Atkinson, as a major way of fighting inequality, identifies development aid and accepting global responsibility, which is reflected in the ethical behavior of countries. To address the whole issue of global inequality is not possible in this scope of the member, but to emphasize the importance of the moral aspects of national decision-making processes.

3.1 Global Inequality

Current research shows that 1.2 billion people still live on less than $ 1.25 a day [12]. At present, 3 billion people live below the poverty line. Global inequality is greatly influenced by the existence of global poverty geographically concentrated in Sub-Saharan Africa (46.4%). Global inequality fully reflects the issues of today's globalized world. Global poverty is unconscious for the consumer's way of life, causing 25,000 victims to daily suffer from hunger and poverty. The United Nations adopted the Millennium Development Goals (MDGs) in 2001 to fight for the economic, political and social stability of the world [4].

In 2015, the proportion of the poorest people in these countries is 14 percent. Worldwide, the number of people living in extreme poverty declined by more than half, from 1.9 billion in 1990 to 836 million in 2015. The situation has changed significantly since 2000 [11].

Atkinson emphasizes the role of historical experience from which we should draw. Thanks to today's modern methods, scientists have created clear and accurate statistics that need to be scrutinized so that we can fully trust them. It is necessary to rely on statistical data and to derive from it the challenge for the future development of global inequality, which poses other possible problems. One of the most effective ways of combating inequality is development aid. Development aid is also linked to our global responsibility, which is reflected not only in the ethical behavior of specific countries but also in numbers. When we talk about the ethical aspect of development aid, it is obvious to point out the paradoxical attitude of the Western world to developing countries. It is absolutely necessary for the Western world to be confronted with global inequality with regard to future economic, social and political developments. [3].

In 2005, Europe contributed 45 billion euro to development cooperation, accounting for more than half of global aid, supporting a total of 144 countries. The fundamental problem of the implementation of the development aid by the Western world is the fact that they were not entirely willing to liberalize their market with the developing countries' market. To reduce the level of global inequality, it is necessary to remove the tariff barriers that make developing countries pay up to twice as much as countries of the European Union.

In the following we can see the development of inequality according to the Gini coefficient. Inequality is measured by Gini's coefficient based on the recalculated disposable income of households (income after taxes and transfer payments). The coefficient in Sweden is 23%, which can be compared with 59.4% in South Africa. [2].

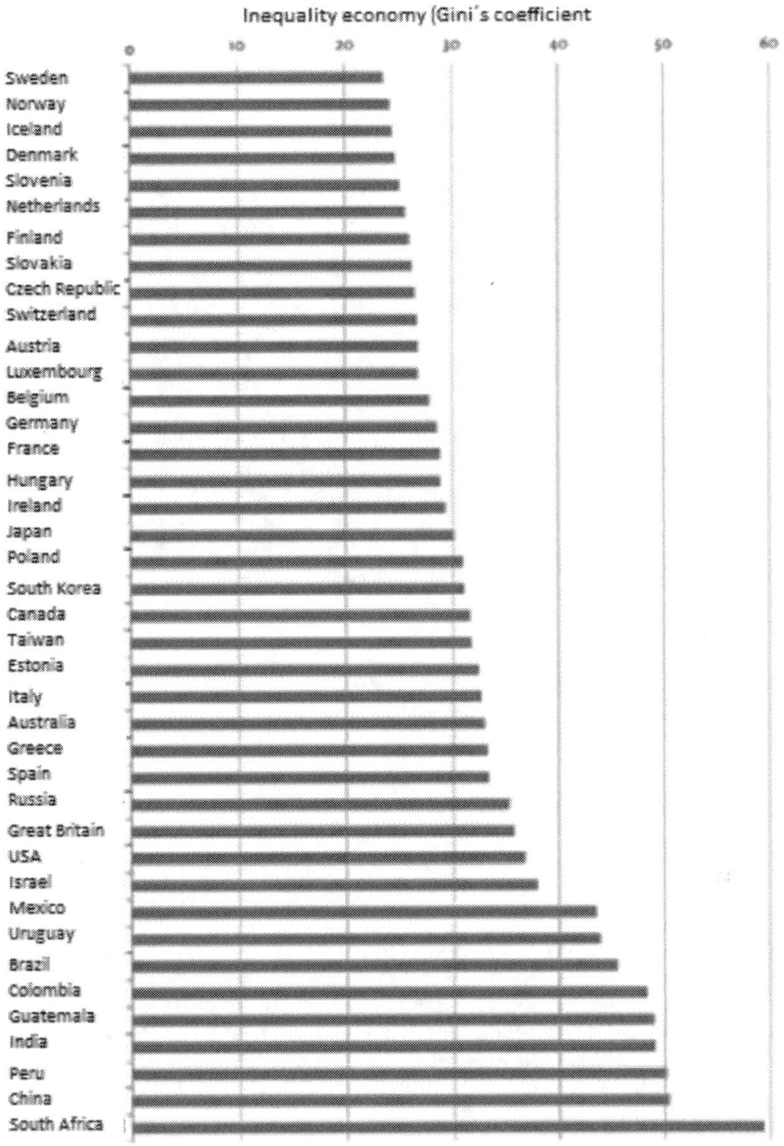

Fig. 1. LIS Key Figures

Source [17].

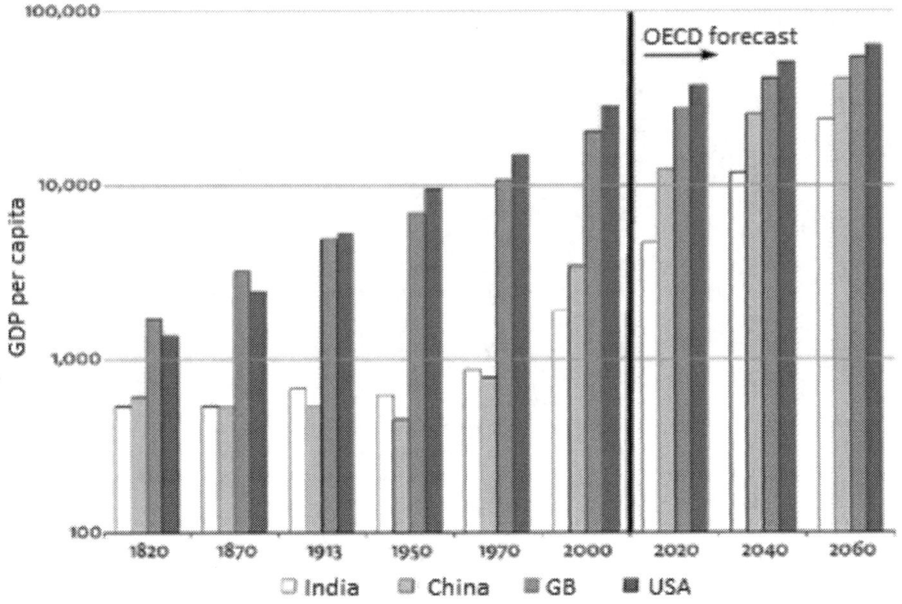

Fig. 2. Historical Statistics of the World Economy

Source: [18].

Global inequality reflects inequalities within and between countries.

4 Proposal issues

Atkinson comes up with specific proposals to significantly reduce inequality. It is not possible to include the whole text of Atkinson's proposals in this paper, but I will focus on the crucial role of capital income and the balance of ownership. Atkinson believes that most of the economic activities are carried out by a private sector that employs citizens and sells services and goods in the open market. This opinion relates to the importance of ownership as well as its control. In addition, Atkinson presents a revolutionary proposal on the wage code. The expanding wage gap is a major cause of deepening of the economic abyss. Atkinson supports the introduction of wage limits, which, according to him, reflects the company's ethical attitude. An example may be Mondragon Cooperative in Spain, which has cut managerial salaries to a maximum of 6.5 times the wage of the company's least paid employee. [2]

The introduction of wage limits British government has been discouraged by the British government following several objections. One of the main problems of wage limits is the differences between individual public authorities, depending on the nature of their workforce. Other objections included any motivation that could eventually lead the company management to remove employees with lower wages and replace them for example by outsourcing. The solution would be to raise the minimum wage threshold, which would have greater transparency, but would be redeemed by raising multiple minimum wages and increasing the maximum permitted maximum wages [2].

The introduction of the wage limit in the public sector would therefore lead to a lesser financial reward for senior officials in public administration. For the private sector, the introduction of wage limits would pose considerable problems, but on the other hand, wage councils would not have to address the fears of a dizzying rise in management salary. The principles of the wage code should regulate the difference between the top and the bottom of the ladders, and at the same time check that people are paid the same value for work corresponding to the same value. This topic is becoming more and more discussed and has a major impact on the future development of employee remuneration. In the course of years, employee salaries have become an individual matter that has led to a demographic fragmentation of companies and workplaces. It should be remembered that an alarming aspect is the company's offering of equal opportunities in the recruitment of employees, but the question remains whether or not it will actually be reflected in remuneration? Now, the notion of justice interferes with wage policy in terms of gender, ethnicity, or age. Atkinson also comes up with a warning about the declining tendency of gender equality and calls for a fair wage reward [3].

Atkinson's vision of the wage code includes great emphasis on the aspect of free will, which is reflected in the preference of a voluntary agreement that would be very difficult to achieve, but its changes would be more effective than the statutory control mechanisms. Atkinson also admits that change in the form of an agreement requires not only public support but, above all, the national debate on the widening of income from rising economic growth. The national discussion should include, in particular, an ethical approach to income policy and its transparency.

One of the other proposals to reduce inequality concerns guaranteed employment. According to Atkinson, the government should seek to reduce unemployment by creating jobs in the public sector for anyone interested in the job. Creating public employment for minimum wages addresses two issues at once. Today it is very popular way to employ citizens in so-called protected workshops where socially weaker people can realize their opportunities.

Atkinson's attempt to reduce inequality with his suggestions is undoubtedly a great achievement, which, as well as praise, does not avoid criticism either. To make Aktinson's proposals feasible, they should primarily include Rawls' theory of justice, which is almost impossible to apply to today's globalized world. Of course, specific steps in both social and political terms are feasible, but we should not forget the basic principle of good that every citizen must accept. Atkinson presents equality of opportunity as one of the proposals to reduce inequality, in which we can also see the absence of a fundamental lack of both theories. In both Rawls and Atkinson's theory, it is a utopian idea of the society's organization [3].

5 Conclusion

Atkinson's drive to reduce inequality in economic outputs leads in particular to the possibility of realizing equal opportunities that we can consider to be a major player in a democratic society. The functioning of democratic principles is unsustainable in the event of extreme inequality. For Atkinson, the notion of good society is incompatible with the current level of economic inequality. His book The Economics of inequality is progressive and tries to show ways that can effectively lead to economic equality. To help reduce inequality, we need to focus on analyzing distribution problems. Atkinson appeals to government officials to take action that would lead to the regulation of economically negative influences such as unemployment, poverty, globalization. Its action plan in the form of wage limits and an increase in gross national income for development aid from 0.7% to 1%.

The relationship between inequality and politics is very critical and will lead to fears of concentration of assets and income that lead to political power and influence. To reduce inequality, we need to work globally with the help of a new dialogue between nations. The issue of global inequality is a very topical issue that affects not only the economic but, above all, political situations. If we are talking about global inequality, we should be aware that it is also the global responsibility that every citizen of our country should feel. Global inequality coalesces as a result of individuals' choices, and so much awareness of the balance of freedom and moral responsibility, which is undoubtedly a civic right, is the foundation for creating a well-functioning society.

References

1. R. Aaberge, F. Bourguignon, F. Brandolini, *Tony Atkinson and his Legacy*, **63**(3), (2017)
2. T. Atkinson, *Ekonomika nerovnosti.* Bizbooks – Albatros Media, (2016)
3. T. Atkinson, *Inequality: What Can Be Done About It?* [online], Available at: https://www.socialeurope.eu/inequality-can-done (2016)
4. M. Balabán, Perspektivy, tendence a hlavní aktéři vývoje bezpečnostní situace vesvětě do roku 2020. *Sborník Bezpečnostní budoucnost České republiky. Otázky, výzvy, problémy,* (2000)
5. C. J. Bennett, Ignorance, lotteries, and measure of economic inequality. *Journal of economic inequality*, **13**(2), (2015)
6. P. Fischer, *O spravedlnosti.* Spor o spravedlnost, Filosofia, Praha, (1997)
7. M. Hrubec, John Rawls (1921-2002). *Filosofický časopis*, **50**(6), 1047-1051, (2002)
8. M. Hrubec, Za Robertem Nožíčkem. *Filosofický časopis*, **50**(2), 333-334, (2002)
9. J. Jílek, J. Moravcová, *Ekonomické a sociální indikátory: od statistik k poznatkům.* Prague: Futura, (2007)
10. V. Labudová, Income Inequality Measures. *Conference on Inequality and Proverty in the European Union and Slovakia*, 107-112, (2012)
11. OSN, *Informační centrum OSN v Praze* [online], Available at: http://www.osn.cz/mdgs-2015-fakta/ (2015)
12. OSN, *Millenium Development Goals Report,* (2014)
13. J. Rawls, *Teorie spravedlnosti.* Prague: Victoria Publishing, (1995)
14. R. Scruton, *Slovník politického myšlení.* Atlantis, (1999)
15. B. H. Sutor, *Politická etika.* Oikoymenh, (1996)
16. R. Tawney, *Equality,* (1998)
17. *Inequality & Poverty, Employment by Gender*, Cross-National Data Center in Luxembourg [online]. Available at: http:/www.lisdatacenter.org/data-access/key-figures/download-key-figures/ (2018)
18. Angus Maddison, University of Groningen [online]. Available at: http://www.ggdc.net/maddison/oriindex.htm (2018)

The analysis for Beijing´s service industry and its effects on regional development

Deng Xin[1,*]

[1]North China University of Technology, School of Economics and Management, No.5 Jinyuanzhuang Road, ShiJingShan District, 100144 Beijing, P. R. China

Abstract. This paper calculates the influence and induction coefficient based on the input-output theory, in order to evaluate the current situation of Beijing's service industry and its effects on regional development. With online comparison, the author concludes that the service sector is getting more and more important on economic development, while the second industry is still the pillar of growth. The author also finds that the impact of service sector's influence coefficient on the growth rate of regional economy and the scope of industry is weaker than the induction coefficient. At last, this paper shows that the fundamental compelling power for Beijing's service development is from the demand side of market, and this gives the benchmark point for policy making in service sector.
Key words: total input coefficients, influence coefficient, induction coefficient, service industry, regional development

1 Introduction

Beijing which is a key economic area in China adjusts the economics structure actively, and promotes the transformation of economic development patterns positively. Review studies have shown that Beijing has the comparative advantage to undertake the international service transfer, and makes it fully use is the best way for Beijing to development the service sectors [1]. But compared with the Beijing performance in FDI using, the FDI using of Beijing service is just the beginning, so it has a considerable to grow [2]; the study based on the space structure shows that there are multicenter on the development of Beijing service sector [3]; although there exists difference on the spatial distribution of services and manufacturing, the positive relationship between the input of Beijing producer service and manufacturing efficiency was verified [4].The statistics based the time series data show services sector become to be one of the three major factors that affect economic growth [5],but structural flaws is pointed and the competitive advantage is gradually weakened [6].

The existing studies pay more attention on the situations and the characteristics of Beijing service sectors; make full positive analysis on the roles of services to economic development and the competitiveness. But the research on the relationship of economy and technology in different -specific services industry are not enough, this paper is guided by the theory of input and output, analyze the relationship of agricultural, industry and service industry, the

* Corresponding author: yuanyuandx@sina.com

relationship between the sub-industries in service sector, in order to make an contribution to the transformation of Beijing economic development patterns.

2 An analysis on the situation of Beijing service sector

The statistics practices and existing research show that the service industry development is mainly manifested by the tertiary industry in Beijing. So the quantitative analysis and research on Beijing service sector are based on the third industry datum listed on the Beijing Statistical Yearbook and Beijing input-output tables. This paper follows previous practice.

2.1 The overall observation on Beijing service sector

The proportion of service industry in Beijing's GDP is close to the level of the developed countries, increased from 67.84% in 2004 to 76.07% in 2011. The added value of Beijing services sector reaches 12363.18 in 2011, increased 200.7 percent compare to 2004. The average annual growth rate reaches to the 28.7%, which is higher than the annual GDP growth rate of 4.6% (See Chart 1).

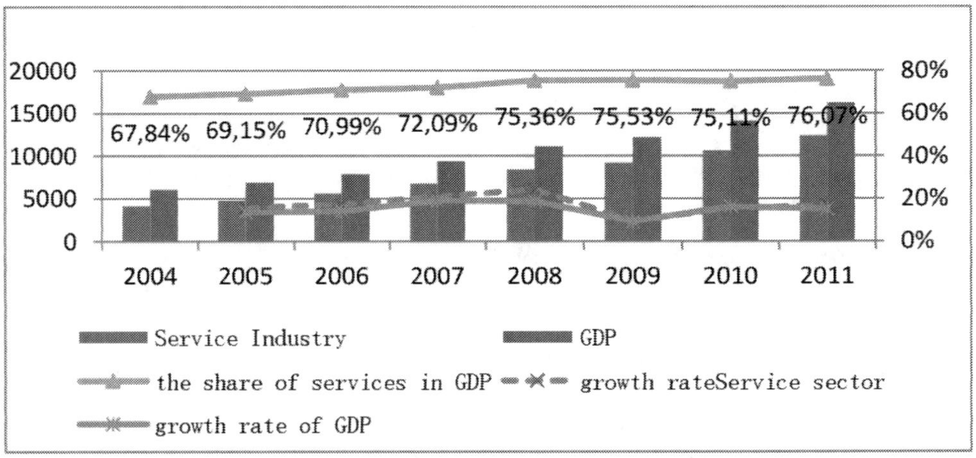

Chart 1. The relate indexes changes from 2004 to 2011 (unit: billion, %)
Source: Author.

2.2 The analysis of internal structure in Beijing service industry

From 2005 to 2011, every sub-service industry in Beijing has shown good performances except for the two sub-industries, such as Residents service and other services. The rest 12 industries' average annual growth rate are more than two digits. Especially, the average annual growth rate of such industries named Leasing and business services, R&D service and geological prospecting exceed 30%, the other's average annual growth rate exceed 20 % (See Chart 2). The contribution of each industry to added value of the whole service is different; the financial sector has a higher contribution in every year, but the growth rate of contribution is only 1.99% during the period; three sectors named Wholesale and Retail Trades, Leasing and business services, R&D service and geological prospecting get 20% increase in the contribution, the sector of Health and social welfare and security makes an 3.24% increase in the contribution to the added value of whole service. Apart from anything else, the remain nine industries' contribution to the added value of wholes service reduces differently. This shows that the advantage of Beijing service gradually reveal, the resources flow to the

 https://doi.org/10.1051/shsconf/20196101035

Comparative Advantage Industry. It was proved by statistical analysis: The variance of industries accounted for the proportion of whole service industry rose from 0.049 in 2005 to 0.056 in 2011.

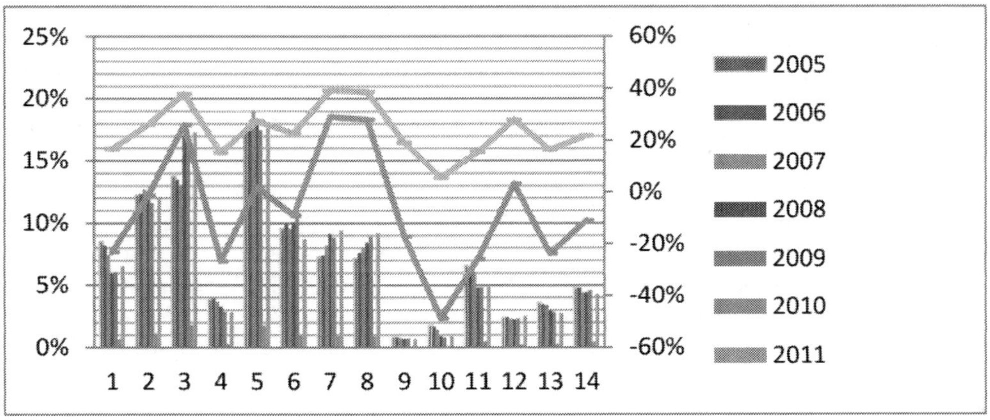

Chart 2. The share of each sector within service industry and the relates indexes from 2005 to 2011
Source: Author.
Note:1: Warehousing postal transportation; 2: Information transmission, computer services;3: Wholesale and Retail Trade; 4: Hotels and Catering Service; 5: Financial; 6: Real Estate; 7: Leasing and business services; 8: Scientific research and technology service of geological exploration; 9: Water conservancy environment public facility management; 10: Residents service and other services; 11: Education; 12: Social welfare and health protection; 13: Culture, sports and entertainment; 14: Public Management and Social Organization.

3 The analysis of industrial interrelation in Beijing service

3.1 The analysis of correlation between industry and service industry

According to the Beijing input-output tables, the complete consumption coefficient we counted show that the products from second industry are leading consumer goods for each sectors. The complete consumption coefficient decreased compared with 2007, but still higher than 2002's. At the same time, the three major industries' consumption from agricultures experienced a lowered after the first process, and the complete consumption coefficient has been a dramatic fall comparing with 2002 (see Table 1). While, the three major industries' consumption from service industry has increased significantly, especially in the agricultural recorded an impressive sales growth of 72%.

The complete consumption coefficients and its change show that: with the development of Beijing's economy and its science and technology, the supportive role of agriculture in economic development is becoming smaller and smaller, but the dependence of agricultural to service industry is getting stronger. The main reason for this is that Beijing's agricultural policy concentration on sightseeing and leisure agriculture, the evidence we can get from Table 2. Although the second industry's contribution to the GDP of Beijing in reducing, but its supporting role in economic development is still very significant; the role of service in Beijing Economic development is growing, all departments' complete consumption coefficient to service sector grew by a large margin.

Table 1. The complete consumption coefficient Beijing's industry in 2002, 2007 and 2010

	Agricultural			Growth rate	Manufacture			Growth rate	Service			Growth rate
	2002	2007	2010		2002	2007	2010		2002	2007	2010	
Agricultural	0.44	0.55	0.18	-59	0.04	0.08	0.03	-25	0.02	0.04	0.02	0
Manufacture	0.86	0.98	0.84	-2	1.65	2.14	1.97	19	0.71	1.07	1.06	49
Service	0.40	0.44	0.69	72	0.66	0.58	1.08	64	0.67	0.62	1.06	58

Source: Author.

The relationships of different services within industry with the first and second industry are different, the complete consumption coefficients agricultural to transportation. Warehousing, wholesale and retail, accommodation catering, scientific research are higher than 0.01; and there are great growth on the complete consumption coefficients of wholesale and retail, tourist industry. Scientific research as well as the public administration and social organization, all of the coefficients rise more than 500% (see table 2). The same situation happened in the second industry, some sectors' complete consumption coefficients are higher than 0.1, such as transportation, finance, leasing, business service and comprehensive technical. Some sectors' complete consumption coefficients get a larger growth rate, which is higher than 300%, especially in transportation, wholesale and retail, tourist industry as well as the public administration and social organization. But other sectors within service industry decreased in complete consumption coefficients comparing with 2002. The changes of service industry's consumption coefficients to the first and second industry show that the service industry had far less correlation to agricultural and the second industry. From 2002 to 2010, the complete consumption coefficients to agricultural increased only appears in the leasing and commercial sector, tourism and comprehensive technical services, all other sector's coefficient reduced on different degrees; to the industry, the coefficient grew only appears in leasing and commercial sector and tourism.

Overall, the correlation between second industry and others is more stable, and the same index increases gradually in Beijing service industry to other industry. But, the correlation of service with two other industries is low and fluctuating, and concentrates on a few industries within service.

Table 2. The change of each sectors complete consumption coefficient within services in 2002 and 2010

	Agricultural		Growth rate (%)	Manufacture		Growth rate (%)
	2002	2010		2002	2010	
Transportation and warehousing	0.057	0.110	93	0.082	0.104	28
Post industry	0.005	0.001	-89	0.008	0.000	-95
Information transmission computer services and software	0.034	0.005	-86	0.056	0.004	-93
Wholesale and retail establishments	0.016	0.122	661	0.026	0.125	371
Hotels and Catering Service	0.017	0.013	-25	0.027	0.011	-60
Finance and Insurance	0.073	0.016	-79	0.104	0.023	-78
Real estate	0.021	0.003	-86	0.035	0.003	-90
Leasing and Business Service	0.074	0.046	-38	0.142	0.054	-62
Tourism	0.001	0.006	503	0.001	0.007	509
Scientific study	0.011	0.075	583	0.029	0.030	4
Polytechnical Services	0.064	0.026	-59	0.104	0.002	-98
Other Social Services	0.004	0.002	-44	0.010	0.003	-71
Education	0.003	0.003	-2	0.004	0.002	-53
Health, social security and social welfare	0.004	0.000	-98	0.005	0.000	-99
Culture, sports and entertainment	0.017	0.005	-72	0.031	0.004	-86
Public Management and Social Organization	0.000	0.040	9795	0.001	0.008	923

Source: Author.

3.2 The analysis on driving force of service industry development of Beijing

The indexes which reflect the inter-affects between different industries are the influence coefficient and response coefficient. The influence coefficient means that how many demands will be deduced in the whole national economy when one department produces an additional product; and the response coefficient represents that how much the department will be affected when each department produce an additional product. According to the Beijing input-output tables 2002 and 2010, we calculate the two indexes of three major industries in Beijing and listed in Table 3. The table show that both indexes of agricultural declined during the inspection period. This means that not only the function of pull on the economic development in the agricultural sector has been weakened, but the demand deduced from the development of Beijing economic for agriculture descent. The influence coefficient of industrial and response coefficient is the highest in the three industries. From 2002 to 2010, industrial influences coefficient gets a small growth, which increases 3.53 percent; but the response coefficient declined 1.67 percent. In 2002, the influence coefficient of service industry is the lowest in the three major industries, but there is a substantial increase; by 2010, the index is second only to industry, and higher than 15.6% of agricultural. The absolute value of the response coefficient of service is the highest in the three industries, and gets a largest growth equal to 19.42%.

There are sixteen sectors in the service industry, the influence coefficient of nine sectors has significantly increased, and 11 sectors' response coefficient has increased. There into, some sectors influence coefficient grow more than 15%, such as wholesale and retail, leasing and business services. The sectors which increasing significantly on the response coefficient is including transportation shipping industry, wholesale and retail, leasing and business services, tourism, scientific research, education, and public administration and social organizations. Especially, some sectors response coefficient rose more than 49%, such as scientific research, tourism, wholesale and retail and transportation shipping industry. So in general, Beijing's service industry gets an increasingly significant role in the economic development, especially in the pulling and driving force of the economic progress. In addition, no matter on the number of sectors or the grow rate, the response coefficients are higher than the influence coefficients. It suggests that the driving force to economy coming from service is stronger than the pulling force in Beijing, and the development of Beijing's service industry mainly depends on the expansion of demand.

Table 3. The influence coefficient and response coefficient and the changes of the all industries

	Influence coefficient		Growth rate%	Response coefficient		Growth rate%
	2002	2010		2002	2010	
Agricultural	0.959	0.820	-14.49	0.532	0.370	-30.45
Manufacture	1.190	1.232	3.53	1.499	1.474	-1.67
Service	0.851	0.948	11.40	0.968	1.156	19.42
Transportation and warehousing	1.078	1.152	6.93	0.964	1.843	91.24
Post industry	0.951	0.969	1.85	0.445	0.508	14.05
Information transmission computer services and software	0.958	0.950	-0.89	0.904	0.614	-32.05
Wholesale and retail establishments	0.818	0.996	21.84	0.541	1.345	148.64
Hotels and Catering Service	1.060	1.004	-5.30	0.656	0.738	12.41
Finance and Insurance	0.876	0.915	4.42	1.294	0.808	-37.53
Real estate	0.987	0.887	-10.09	0.838	0.584	-30.33
Leasing and Business Service	0.860	1.165	35.45	1.167	1.394	19.52
Tourism	0.888	1.038	16.86	0.403	0.604	49.92
Scientific study	1.000	1.083	8.34	0.576	0.859	49.19
Polytechnical Services	1.048	0.974	-7.08	0.750	0.527	-29.70
Other Social Services	0.978	0.934	-4.44	0.468	0.544	16.26
Education	0.955	0.871	-8.82	0.441	0.580	31.68
Health, social security and social welfare	1.160	0.942	-18.82	0.447	0.494	10.44
Culture, sports and entertainment	0.962	0.996	3.55	0.697	0.628	-9.82
Public Management and Social Organization	1.004	1.095	9.08	0.404	0.710	75.68

Source: Author.

4 Conclusion

The foregoing research shows that the contribution of Beijing service industry to economic progress increased significantly, but the difference between the internal services industry has become increasingly apparent. At the same time, the correlation between the service industry and others is reinforced obviously, but it is limited to the a handful of sectors within service industry, for instance, transportation shipping industry, leasing and business services, tourism, polytechnic services. The other characteristic is that the development of Beijing's service industry mainly depends on the expansion of demand. In order to meet the goal of Beijing social and economic development, the service industry in Beijing should witness further development of the more sectors.

Improve the performance of producer service based on the industry development

Industry is the foundation of economic development; it has been confirmed by the world economic development practice. It will cause the regional economy of industrial hollowing out if the development of industry is ignored. So, strengthen the basic role of industrial development will not only meet the inherent demands of economic development, but also the premise and foundation of service industry development. The sectors including transportation, warehouse and postal service, information transmission, computer services and software, wholesale and retail, finance, leasing and business services, technical services, scientific research and geological exploration are producer services, which provide intermediary services for the production process directly or indirectly. But if the second industry's development is not enough and lack of science and technology, the development of producer service industry is lack of necessary demand. Also, the development of producer services' efficient is a requirement of the optimization of industrial developing. But the calculation of the influence coefficient based on the input-output shows that the influence coefficient of producer service in Beijing is generally low. The index of sectors which includes the financial insurance, wholesale and retail get an weak growth, but is still less than 1. And worse still, there is an negative grow on the sectors of information transformation and polytechnical services during the period of 2002 to 2010.the situation of transportation and the scientific study which influence coefficient is higher than 1 slightly. This result may have negative impact to upgrade service industries in Beijing.

Accelerate the construction of environment, create a good social environment for the development of service industry

The key factor of the service industry's developing is the construction of talent team on service, because the talents have higher requirements on the environment for life. Although Beijing is the administrative capital of our country and its competitiveness are higher than other provinces on economic, science and technology and culture, but compares with the requirements of economic development, there is a big gap. Some sectors which provide service to the are lag behind in Beijing's services development, they are low on the correlation with the sectors within the service industry, agricultural and manufacture. Although the education's contribution to the development of Beijing Service is in the medium level in the 16 sectors (see chart 2), but it is in decline, and weak of correlation between agriculture and industry, and gradually decline. In addition, the influence coefficients of consumer service are also low, and reduced year by year.

The goal of Beijing services development is not the scale but to be a winner on the international competition, the accelerating of environment' construction is not the need to attract high talent team on services, but also the necessary to build the competitiveness system. For all these efforts, the factors of Porter's diamond model are perfect. Because the sectors above are not only the auxiliary and supporting factor to the industry development, but to cultivate the reasonable consumptive conceptions.

References

1. K. Wang, J. Wang, The strategy of undertaking the International service transfer in Beijing. *Journal of Beijing Polytechnic University*, **6**, 15-21, (2006)
2. Q. Wu, Modern Servicing Industry in Beijing. *Commercial Research*, **19**, 163-166, (2006)
3. J. Han, The empirical of Beijing services using FDI. *International Economic Cooperation*, **4**, 57-59, (2008)

4. L. Qiu, Industrial Relevancy and spatial distribution between producer services and manufacturing in Beijing city. *Acta Geographical Sinica*, **12**, 96-106, (2008)
5. Ch. Hao, The statistical analysis economic level of Beijing. *Journal of Anhui Agricultural Sciences*, **17**, 92-95, (2010)
6. X.-B. Wu, J. Chenjin, An analysis on structure and competitiveness of services industry in Beijing Based on dynamic shift – share method. *Journal of International Trade*, **3**, 96-106, (2012)

Impact of foreign economic sanctions on the Russian market of bonds denominated in foreign currencies

Alla Yakunina[1,*], *Yulia* Semernina[1], *Sergey* Yakunin[1] and *Elena* Ermakova[1]

[1]Saratov Socio-Economic Institute of Plekhanov Russian University of Economics, Saratov, Russia

Abstract. The market of Russian bonds denominated in foreign currencies was considered as the primary target for the initial package of the USA and EU economic sanctions in 2014. The analyses showed that, by the end of 2017, there was a significant reduction in the number of outstanding bond issues denominated in foreign currencies and a sharp increase in the bonds nominal volume evaluated in Russian rubles. The sanctions did not lead to a significant change in the structure of the market either from the standpoint of the number of bond issues denominated in foreign currencies or in terms of their nominal value. However, the number of foreign currencies in which bond issues were denominated was reduced from 11 to 6.

Key words: Russian bond market, bonds denominated in foreign currencies, sanctions, bond market structure

1 Introduction

Bonds are popular instruments in the Russian financial market [1]. The restrictive measures (economic sanctions) imposed by the USA and the EU against Russia in 2014 as a result of the Crimea crisis have had non-equal impacts on different segments of the Russian bond market.

It is worth to note that, when the first package of the restrictive measures was introduced, the market of Russian bonds denominated in foreign currencies (DFC bonds) was considered as the main "sanction goal" as it was the most closely related to foreign capital markets in terms of both the structure of its participants and its capital sources. The potential great impact of the USA and EU economic sanctions on the Russian domestic DFC bonds market was expected due to the influence of the following factors. First, the beginning of the "war of sanctions" meant an almost complete "closure" of foreign capital markets for Russian borrowers. Secondly, expecting further escalation of the "war of sanctions", foreign investors began to reduce the shares of Russian assets, including DFC bonds, in their investment portfolios. The logical outcome of the simultaneous impact of these factors was the more complicated process of attracting capital for Russian borrowers through placing the above financial instruments and the increasing competition between the latter.

* Corresponding author: alla.yackunina@yandex.ru

At the same time, domestic borrowers increased their demand for financial resources, as they simultaneously faced the need to refinance existing debt denominated in foreign currencies and the need to get a kind of "replacement", albeit to a limited extent, via alternative foreign sources of financing (for example, loans from foreign banks received in foreign currency).

However, the sanctions' impact on the Russian ruble bond market is much less obvious. The main reason is the historical orientation of this market to local investors.

The main purpose of the study is to identify, substantively describe and systematize the changes that have occurred on the Russian bond market during the period of foreign economic sanctions (years of 2014-2017).

2 Literature review

A bond market is getting quite a lot attention in the literature. In his famous book, F. Fabozzi presented "an applied approach to understanding bond markets" [2.]. There are many interesting studies of a bond market development in different countries. P. G. Szilagyi and J. A. Batten [3; 4] investigate bond market development in Japan. I. Mathur and S. De study the dim sum bond market in Hong Kong [5]. The development of the corporate bond market in Croatia has been investigated by D. M. Sprčić and I. Wilson [6]. The paper of I. Matei is devoted to the relationship between the government bond market and real GDP growth for 16 euro zone countries [7]. Sh. Thukral, Sh. Sridhar, and M. Sh. try to understand the factors that have limited the development of Indian corporate bond markets [8]. H. Blommestein studies the impact of new regulatory factors on liquidity in government bond markets since the onset of the global financial crisis [9]. S. Babatunde and S. Perera identify and critically assess the barriers to bond financing for public–private partnership infrastructure projects in Nigeria [10].

The Russian bond market development has received attention primarily in the studies of the Russian authors. T.I. Starodubtseva and S.M. Nekrasov [11] consider certain aspects of the Russian Federation public debt management in the conditions of the geopolitical instability and foreign economic sanctions. M. V. Petrov [12] compares the attractiveness of various financing sources for corporate borrowers in the context of foreign economic sanctions, drawing attention to the attractiveness of bond financing, including DFC bonds. I.A. Alekseeva and E.V. Makarov [13] distinguish trends and evaluate the prospects for the national corporate bond market development. The authors consider foreign economic sanctions as one of the factors that provoked instability in the bond market. The mentioned above authors touch on certain aspects of the impact of the foreign economic sanctions on the domestic debt market (or its certain segments), but the authors don't get into the detailed analysis of the sanctions' impact on the Russian market of DFC bonds although it is a very important issue.

3 Methods and Data

We used a number of the simplest statistical methods for our analyses, among which there are methods of groupings, comparison, and dynamics analysis. In particular, the grouping method was used to make two simple groupings of bonds:

1) depending on the bonds' issuer, the following groups were indicated: government bonds, subfederal and municipal bonds, corporate bonds;

2) depending on the currency in which the bond issue is nominated, we separated bonds denominated in foreign currencies and bonds denominated in national currencies.

Sub-federal and municipal bonds were merged into one group to comply with the rule of the absence of "empty groups" (during the entire analyzed period there were no issues of municipal bonds denominated in foreign currency on the Russian market).

On the basis of simple groupings, a two-dimensional grouping has been built, implying the allocation of six groups of bonds on the Russian bond market as a whole. The object of the research was the market for bonds denominated in foreign currency, within which the three indicated above groups of bonds were studied.

The data for the analyses was taken out of the following sources:

- data of the Internet project RusBonds managed by the information agency "Finmarket" [14];
- data reported by the analytical department of the National Association of Stock Market Participants (NAUFOR) [15];
- the Moscow Exchange data [16].

The data reported by the Finmarket's Internet project RusBonds was used to compare the structure of the Russian market of DFC bonds at the end of 2013 (before the foreign economic sanctions were imposed) and at the end of 2017 (after several years of foreign economic sanctions, which, although without regularity, were gradually tightened). The comparison of the structure of the national bond market was carried using data of the bonds nominal value, ignoring depreciation.

4 Results and Discussion

The Russian DFC bonds market has historically developed very disproportionately, and there were serious imbalances in the development of its separate segments yet before the introduction of foreign economic sanctions. The disproportionality of the development of the Russian DFC bonds market is obvious when one is looking at the distribution of bond issues (by their number) among individual groups, formed by type of issuers (Table 1).

Table 1. The structure of the Russian DFC bond market by the number of bond issues, as of the end of the year

Segments of the Russian DFC bonds market	Number of bond issues		The share of bond issues in their total number, %	
	2013	2017	2013	2017
Federal Bonds	12	13	2.66	5.60
Sub-federal and Municipal Bonds	1	0	0.22	0.00
Corporate Bonds	438	219	97.12	94.40
Total	451	232	100.00	100.00

Source: Authors.

At the beginning of the period under consideration, the undoubtedly dominant segment (by the number of bond issues) was a segment of corporate bonds, which accounted for 97.12% of their total, while the total share of the remaining market segments was only 2.88%. The introduction of the USA and EU economic sanctions impacted primarily the number of circulating bond issues, but did not lead to a radical change in its structure in the context of separate segments (Table 1).

Nevertheless, a comparison of the available data on the national DFC bonds market allows one to draw a number of meaningful conclusions:

1) during the analyzed period, there was a sharp decrease in the total number of outstanding DFC bond issues: 232 issues by the end of 2017 as compared with 451 bond issues in 2013 (decrease by 219 issues or 48,56%);

2) the marked reduction was quite disproportional with respect to particular market segments: it affected the corporate DFC bonds segment to the greatest extent: the number of outstanding bond issues decreased by 219 issues (or 50.00%) over the 4 years, while the sub-federal and municipal bonds segment decreased by only one bond issue (but it actually disappeared), and, at the same time, the number of outstanding federal bond issues increased by one issue (or by 8.33%);

3) during 2013-2017, there was a significant reduction in the "scale gap" between the corporate DFC bonds segment and the other segments of this market (by number of bond issues): at the beginning of this period, the corporate bonds segment was bigger than the two remaining segments by 33.7 times, while at the end of the period the corporate bonds segment still remained undoubtedly dominant but exceeded the other two segments by "only" 16.8 times.

It is also worthwhile to look at the DFC bonds market structure using face (nominal) value of bonds.

Since there were bonds denominated in different currencies at the Russian bond market in 2013, we used the Russian currency to estimate the DFC bonds market structure with respect to face value of bonds. The face value of bonds denominated in foreign currencies was recalculated using the Bank of Russia official exchange rate as of December 31, 2013 and December 31, 2017, respectively (Tables 2).

Table 2. Structure of the Russian market of bonds denominated in foreign currencies, by face value of bond issues at the end of the year

Segments of the Russian DFC bonds market	**Face value of bond issues, billion rubles**		**The share of bond issues in their total number, %**	
	2013	2017	2013	2017
Federal Bonds	1528,947	3167,197	19,88	14,49
Sub-federal and Municipal Bonds	18,303	0,00	0,24	0,00
Corporate Bonds	6143,550	18 689,472	79,88	85,51
Total	7690,800	21856,669	100.00	100.00

Source: Authors.

At the beginning of the analyzed period, the structure of the Russian DFC bond market at face value was similar (but not identical) to its structure in terms of the number of bond issues. As the most significant difference of these structures, it should be noted that the share of the federal DFC bonds segment at the face value (19.88%) is 7.5 times its share by the number of bond issues (2.66%), while the share of the corporate bond segment at face value, on the contrary, is lower than the share of the number of bond issues (79.88% versus 97.12%). The discrepancy is due to the fact that the average face volume of a single bond issue in the segment of corporate DFC bonds estimated in Russian rubles (it was 14.026 billion rubles at the end of 2013) was significantly lower than the average face volume of a single bond issue in the government bond segment (127.412 billion rubles). As for the segment of sub-federal and municipal bonds, its share is 0.24% at face value and 0.22% by the number of bond issues, which are quite close.

At the same time, a number of important trends developing in this market should be highlighted:

- after the four years, there was a sharp increase in the nominal volume of the DFC bonds market estimated in Russian rubles: it amounted to 7,690.800 billion rubles at the end of 2013, and it was already 21,856 669 billion rubles at the end of 2017 (an increase of 14165.869 billion rubles or 184.19%);

- the sharp increase in the nominal volume of the analyzed market was caused not only by an increase in the level of activity of issuers interested in attracting bond financing in foreign currency, but also by a change in the exchange rate of the Russian ruble against foreign currencies. During this period, the Russian ruble significantly weakened against all currencies in which bonds traded on the market were nominated (for example, the US dollar against the Russian ruble rose during this period by 75.99%, the Swiss franc - by 60.71%, and the euro - by 53.14%). In parallel, there was a "depletion" of the national DFC bond market in terms of the number of currencies represented on it: at the end of 2013, bonds were denominated in 11 currencies (USD , euro, Swiss franc, British pound sterling, Kazakh tenge, Norwegian krone, Singaporean dollar, Chinese yuan, Australian dollar, Turkish lira, Czech koruna) and, at the end of 2017, there were bond denominated in only 6 currencies (US dollar, euro, Swiss franc, British pound sterling, Turkish Lira, Romanian Leu) at the Russian bond market;

- while particular segments of the Russian DFC bond market were experiencing significant growth in absolute values, the change in their weights was following different directions; in particular, the share of the federal bond segment decreased by 5.39% (to 14.49%) although its absolute volume "jumped up" by 1638.250 billion rubles (or by 107.15%), the share of the sub-federal bond segment decreased to zero, and the corporate bond segment grew by 5.63% (to 85.51%) while its absolute value grew by 12,545.922 billion rubles (or 204.21%).

5 Conclusions and Recommendations

To sum up we can state that the USA and EU economic sanctions against Russia had the following effects of the national DFC bond market:

1) the sanctions did not lead to a significant change in the structure of the DFC bond market either from the standpoint of the number of circulating bond issues or in terms of their nominal value (in fact, the structural hierarchy of the Russian DFC bond market segments remained unchanged: the segment of corporate liabilities retained a dominant position in absolute and relative terms, followed by a segment of federal bonds, and the last position was occupied by the segment of sub-federal and corporate bonds);

2) during the analyzed period, the structure of the national market for bonds denominated in foreign currencies was simplified, and it was because the only issue presented in the sub-federal and municipal bond segment in 2013 had been redeemed by 2017;

3) from 2013 to 2017, there was a significant reduction in the number of outstanding bond issues denominated in foreign currencies;

4) during the analyzed period, there was a sharp increase in the nominal volume of the DFC bond market evaluated in Russian rubles, which was due to a weakening of the national currency against foreign currencies and increasing activity of issuers in the bond market;

5) the only obvious effect of the imposed foreign economic sanctions was the "currency impoverishment" of the Russian DFC bond market, expressed in the nearly two times reduction in the number of foreign currencies in which bond issues were nominated.

Thus, the sanctions have a deterrent effect on the development of the DFC segment of the Russian bond market. At the same time, an increase in the number of foreign investors willing to buy Russian bonds would increasing the inflow of relatively cheap investment resources and, hence, would contribute to the Russian economy growth.

In our opinion, it is possible to significantly reduce the negative impact of the sanctions through applying digital technologies for Russian bonds trading. In this regard, it is worth to follow the experience of Australia. In August 2018, the World Bank, jointly with Commonwealth Bank of Australia (CBA), the largest bank in Australia, announced the release of the world's first public blockchain bonds, called Bondi, for a total of $ 73 million. According to the CBA Executive Manager James Wall, they are moving away from traditional bonds and manual labor of a registrar and a custodian to a new process that can be completely online [17].

In the post-Soviet space, there is also experience in using blockchain technologies to distribute bonds. In March 2018, the National Bank of Kazakhstan started to sell its short-term notes within the country through a mobile application using the blockchain [18].

We believe that this experience can be used to distribute bonds issued in Russia under sanctions. In order to expand the number of potential investors, non-residents should be allowed to purchase Russian bonds. This will help attract funds from individuals and companies of other jurisdictions. To make it possible to purchase and sale bonds from any device (without brokers, dealers, commissions and taxes) it is possible to follow the experience of Kazakhstan, where investors use the Invest application, available in the Play Market and the App Store.

References

1. Y. Semernina, A. Yakunina, E. Nesterenko, S. Yakunin, E. Korobov, Evaluation of reinvestment risk for bond portfolios (in Computer Modelling in Decision Making 2017). *CEUR Workshop Proceedings,* (2018)
2. F. Fabozzi, *Bond Markets, Analysis and Strategies*. Pearson, (2013)
3. P. Szilagyi, J. Batten, Disintermediation and bond market development in Japan. *International Finance Review*, **4**, (2003)
4. P. Szilagyi, A Perspective on Japan's Corporate Bond Market. *Contemporary Studies in Economic and Financial Analysis*, **86**, 417-434, (2005)
5. I. Mathur , S. De, The Dim Sum Bond Market in Hong Kong. *Contemporary Studies in Economic and Financial Analysis*, **96**, 367-388, (2014)
6. D. Sprčić, I. Wilson, The development of the corporate bond market in Croatia. *EuroMed Journal of Business*, **2**(1), 74-86, (2007)
7. I. Matei, Euro Zone Bond Market and Economic Growth: Evidence from a Time Series Analysis. *International Symposia in Economic Theory and Econometrics*, **22**, 39-51, (2012)
8. S. Thukral, S. Sridhar, M. Joshi, Review of factors constraining the development of Indian corporate bond markets. *Qualitative Research in Financial Markets*, **7**(4), 429-444, (2015)
9. H. Blommestein, Impact of regulatory changes on government bond market liquidity. *Journal of Financial Regulation and Compliance*, **25**(3), 307-317, (2017)
10. S. Babatunde, S. Perera, Barriers to bond financing for public-private partnership infrastructure projects in emerging markets: A case of Nigeria. *Journal of Financial Management of Property and Construction*, **22**(1), 2-19, (2017)
11. T. Starodubtseva, S. Nekrasova, The state and evolution of the public debt of the Russian Federation. *Politics, Economics and Innovations*, **7**(17), (2017)
12. M. Petrov, Financing of companies in the context of economic slowdown and sanctions. *Finance: Theory and Practice*, **3**, (2018)

13. I. Alekseeva., E. Makarova, Russian corporate bond market: trends and development prospects. *News of the Baikal State University*, **3** (2017)
14. *Internet project of the information agency "Finmarket" RusBonds* [online]. Available at: http://www.rusbonds.ru/srch_simple.asp (2018)
15. Russian stock market: 2017. *Events and Facts* [online]. Available at: http://www.naufor.ru/download/pdf/factbook/ru/RFR2017.pdf (2017)
16. *Annual report of PJSC "Moscow Exchange MICEX – RTS" for 2017* [online]. Available at: https://report2017.moex.com/ru (2017)
17. *World Bank and CBA to issue $ 73 million blockchain bonds* [online]. Available at: https://forklog.com/vsemirnyj-bank-i-cba-vypustyat-blokchejn-obligatsii-na-73-mln/ (2018)
18. *Central Bank of Kazakhstan launched the sale of securities on the blockchain* [online]. Available at: https://bitnovosti.com/2018/03/28/tsb-kazahstana-zapustil-prodazhu-tsennyh-bumag-na-blokchejne/ (2018)

AUTHOR INDEX